Basic Linear Algebra

Basic Linear Algebra

Gregor Olšavský

BASIC LINEAR ALGEBRA

iUniverse books may be ordered through booksellers or by contacting:

iUniverse
1663 Liberty Drive
Bloomington, IN 47403
www.iuniverse.com
1-800-Authors (1-800-288-4677)

ISBN: 978-0-5954-0644-9 (sc)
ISBN: 978-0-5958-5010-5 (e)

Print information available on the last page.

iUniverse rev. date: 02/13/2018

This book is dedicated to my parents, to whom I owe everything:

Frank Joseph Olšavský and
Betty Jane (os. Seamon) Olšavská

CONTENTS

CHAPTER 1

LINEAR SYSTEMS

CHAPTER OVERVIEW

1.1 Gaussian elimination

1.2 Geometry of solutions

1.3 System analysis

1.4 Chapter outline and review

In this chapter we learn how to solve **systems** of linear equations. We learn that every linear system can be represented by a unique matrix (a rectangular array of numbers) called its **augmented matrix**. Putting the augmented matrix in triangular form (we will call it **echelon** form) is called Gaussian elimination (after C.F. Gauss). This involves **row operations** on the matrix which represents the system. Then we simply solve the corresponding equations (starting from the bottom and working upwards) for the lowest-indexed variable in terms of higher-indexed variables. This is called "**back-substitution**".

A system is **consistent** if it has a solution and *inconsistent* if it does not have a solution. The echelon form of a matrix will let us know if we have an inconsistent system or not; we can also use the echelon form to deduce whether the system has only one solution or infinitely many solutions. Many practical problems will have only one solution.

Next we will look at the *geometry* of the solutions that we have found and we will not feel constrained to stay in 2-space or 3-space but operate in n-dimensional space; there we will define what we mean by a **hyperplane** (a generalization of a plane in 3-space).

Finally we will come up with a systematic way of determining whether or not we have a unique solution or infinitely many. We will distinguish systems with infinitely many solutions by the number of parameters necessary to describe the solution; at the same time we will discuss the difference between **free** and *determined* variables.

1.1 Gaussian Elimination

Linear systems is one of the most important topics of this book. They arise naturally in many areas, such as electrical networks, economics, population models, differential equations, etc. A **linear equation** in $x_1, x_2, ..., x_n$ is simply one of the form

$$a_1 x_1 + a_2 x_2 + ... + a_n x_n = b$$

where the a_i's are the coefficients, the variables are $x_1, x_2, ..., x_n$ and the constant term is b. A linear system is a set of two or more linear equations, such as

$$x_1 + 2 x_2 + 3 x_3 = 0$$

$$2 x_2 + 4 x_3 = 6$$

The geometry of systems in general will be discussed in the next section. However, you may recognize these equations as representing planes in 3-space and therefore the solution of the system represents the points of intersection (if any) of these two planes. To facilitate the solution of this system, we introduce the concept of its **augmented matrix**; a matrix being simply a rectangular array of (real) numbers. Here the augmented matrix of the system is:

$$\begin{bmatrix} 1 & 2 & 3 & : & 0 \\ 0 & 2 & 4 & : & 6 \end{bmatrix}$$

Each column represents a variable; column 1 represents x_1, column 2 represents x_2, etc.; the last column does NOT represent a variable since it is the column of *constants*. Each **row** represents an equation in the given system. The matrix is of size 3 x 4 (3 rows, 4 columns).
Systems are described by telling the number of *equations* versus the number of variables; so this system is of size 2 x 3 and we say it is a nonsquare system. A square system has the same number of equations as unknowns. When solving the system algebraically, there are essentially THREE operations that will guarantee that the solution of the system remains unchanged(ie. the systems are *equivalent*); these are the so-called **equivalent operations**:

1) interchanging any two equations
2) multiplying any equation by a non-zero constant
3) adding a multiple of one equation to another (this is probably the most important one!)

When dealing with the augmented matrix, we simply need to change the description of the maneuver:
1) interchanging any two *rows*
2) multiplying any *row* by a non-zero constant
3) adding a multiple of one *row* to another

These are our **row operations**; if one can take the matrix A and, using finitely many row operations, produce the matrix B, the resulting matrix thus obtained is not equal to the original matrix but rather *row equivalent* (we write A ~ B) to it, much like 1/2 is equivalent to 4/8 but not actually the same fraction. But we need a goal in mind when applying these row operations.

What do we want the matrix to look like when we are done? For example, if we had the augmented matrix

$$\begin{bmatrix} 1 & 2 & 3 & 4 \\ 0 & 1 & 2 & 3 \\ 0 & 0 & 1 & 7 \end{bmatrix}$$

it is in a particularly nice form; it represents the system

$$x_1 + 2 x_2 + 3 x_3 = 4$$

$$x_2 + 2 x_3 = 3$$

$$x_3 = 7$$

and we can see right away that $x_3 = 7$ from which we can use equation (2) to find x_2 and then equation (1) to find x_1. This is called **back-substitution**; why was it so easy? Well, the augmented matrix was triangular in form, and the leading nonzero entries were all one's so the algebra was a piece of cake. The matrix is said to be in **row-echelon** form; it is a type of triangular form and a nice form for a matrix to be in if we wish to solve the corresponding linear system that it represents. We have to agree on some criteria that will make a matrix have this simple "row-echelon" form.

| **DEFINITION** | A matrix is in **row-echelon form** if:
1) all rows of zeros at the bottom of the matrix
2) the leading nonzero entry in a nonzero row is a ONE and only zeros |

occur below this one (called a *leading one*)
3) leading one's *move to the right* as we look from the top to the bottom of the matrix (ie. a staircase pattern)
If in addition, only zeros occur **above** the leading ones as well, the matrix is in *reduced* row echelon form. One can prove that reduced row-echelon form is *unique* but that row-echelon form is not. That is, there can be two different row-echelon forms of the same matrix but *only one* reduced row-echelon form.

EXAMPLE 1 Determine if the given matrices are in row-echelon form or not. If in echelon form, determine if it is reduced or not.

$$\begin{bmatrix} 1 & 2 & -3 & 4 \\ 0 & 1 & 5 & 6 \\ 0 & 0 & 1 & 9 \\ 0 & 0 & 0 & 0 \end{bmatrix} \quad \begin{bmatrix} 1 & 0 & 0 & 0 \\ 0 & 1 & 0 & 0 \\ 0 & 0 & 1 & 0 \end{bmatrix} \quad \begin{bmatrix} 0 & 0 & 0 \\ 0 & 0 & 1 \\ 0 & 1 & 0 \\ 1 & 0 & 0 \end{bmatrix}$$

The first matrix is in echelon form but not reduced (the leading one's have non-zero entries above them). The second matrix is in reduced row-echelon form. The last matrix is NOT in echelon form (the leading ones do not move right as we look from top to bottom and the row of zeros is not at the bottom).

EXAMPLE 2 Put the matrix $\begin{bmatrix} 1 & 0 & 2 \\ 0 & 4 & 12 \\ 2 & 2 & 2 \\ 3 & 3 & 7 \end{bmatrix}$ in row-echelon form and then in reduced row-echelon

form. We are happy that the first entry in column one is a "one" - we can easily use this "leading one" (called a pivot element) to eliminate the entries below it. Without switching rows, we can first carry out the operations -2 row1 + row3 and then -3row1 + row4:

$$\begin{bmatrix} 1 & 0 & 2 \\ 0 & 4 & 12 \\ 2 & 2 & 2 \\ 3 & 3 & 7 \end{bmatrix} \begin{array}{c} \sim \\ -2\,row_1 + row_3 \\ -3\,row_1 + row_4 \end{array} \begin{bmatrix} 1 & 0 & 2 \\ 0 & 4 & 12 \\ 0 & 2 & -2 \\ 0 & 3 & 1 \end{bmatrix}$$

We now have column one in good shape so we move over to column 2. We want the second entry in column2 to be a one so let's multiply row2 by 1/4:

$$\begin{bmatrix} 1 & 0 & 2 \\ 0 & 4 & 12 \\ 0 & 2 & -2 \\ 0 & 3 & 1 \end{bmatrix} \begin{array}{c} \sim \\ \frac{1}{4}\,row_2 \end{array} \begin{bmatrix} 1 & 0 & 2 \\ 0 & 1 & 3 \\ 0 & 2 & -2 \\ 0 & 3 & 1 \end{bmatrix}$$

Using the leading one in column2, we carry out -2row2 + row3 and then -3row2 + row4 to get rid of the "2" and "3" below the leading one there:

$$\begin{bmatrix} 1 & 0 & 2 \\ 0 & 1 & 3 \\ 0 & 2 & -2 \\ 0 & 3 & 1 \end{bmatrix} \begin{array}{c} \sim \\ -2\,row_2 + row_3 \\ -3\,row_2 + row_4 \end{array} \begin{bmatrix} 1 & 0 & 2 \\ 0 & 1 & 3 \\ 0 & 0 & -8 \\ 0 & 0 & -8 \end{bmatrix}$$

Now we can eliminate row4 using row 3:

$$\begin{bmatrix} 1 & 0 & 2 \\ 0 & 1 & 3 \\ 0 & 0 & -8 \\ 0 & 0 & -8 \end{bmatrix} \begin{array}{c} \sim \\ -row_3 + row_4 \end{array} \begin{bmatrix} 1 & 0 & 2 \\ 0 & 1 & 3 \\ 0 & 0 & -8 \\ 0 & 0 & 0 \end{bmatrix}$$

4

and then fix up row3 by multiplying it by -1/8 (to get a leading one in row3):

$$\begin{bmatrix} 1 & 0 & 2 \\ 0 & 1 & 3 \\ 0 & 0 & -8 \\ 0 & 0 & 0 \end{bmatrix} \underset{-\frac{1}{8}\,row_3}{\sim} \begin{bmatrix} 1 & 0 & 2 \\ 0 & 1 & 3 \\ 0 & 0 & 1 \\ 0 & 0 & 0 \end{bmatrix}$$

The reader should be able to show that $\begin{bmatrix} 1 & 0 & 2 \\ 0 & 1 & -1 \\ 0 & 0 & 1 \\ 0 & 0 & 0 \end{bmatrix}$ is another (correct) echelon form of the

original matrix, gotten by first switching row2 and row3. To get reduced row-echelon form, we need to get rid of the two nonzero entries above the leading one in column 3; we carry out -2row3 + row1 and -3row3 + row1 and finally have:

$$\begin{bmatrix} 1 & 0 & 2 \\ 0 & 1 & 3 \\ 0 & 0 & 1 \\ 0 & 0 & 0 \end{bmatrix} \underset{\substack{-3\,row_3+row_2 \\ -2\,row_3+row_1}}{\sim} \begin{bmatrix} 1 & 0 & 0 \\ 0 & 1 & 0 \\ 0 & 0 & 1 \\ 0 & 0 & 0 \end{bmatrix}$$

All the matrices we produced here are row-equivalent to the original one. But this last matrix is unique, since it is the *reduced* row-echelon form of the original matrix. In any case, the process of putting the augmented matrix in row-echelon form and then using back-substitution to find $x_1, x_2, \ldots, x_n$ is called **Gaussian elimination** after the famous mathematician Carl Friedrich Gauss[1]. The advantages of using the matrix form for the system is that even larger systems can be done in a systematic fashion and the procedure is amenable to computer algorithms so that even a large system can be solved on a computer using Gaussian elimination. Putting the augmented matrix in reduced row-echelon form is time-consuming and unnecessary so that computer algorithms usually only use the row-echelon form of the matrix. Using the reduced row-echelon form followed by back-substitution to solve a system is called *Gauss-Jordan elimination*.

---

[1]Using observations of the asteroid Pallas taken between 1803 and 1809, Gauss obtained a system of 6 equations in 6 unknowns and gave a systematic method for solving such equations which is precisely Gaussian elimination.

EXAMPLE 3 Solve the system

$$2 x_1 + x_2 \qquad - x_4 = 3$$

$$x_1 + 2 x_2 + 3 x_3 + 4 x_4 = 0$$

$$3 x_1 + 3 x_2 + 3 x_3 + 4 x_4 = 5$$

using Gaussian elimination.

Here we don't have any intuition about the geometry involved (we are now in 4-space!) but we can solve the system none the less by writing down the augmented matrix and putting it in row-echelon form using row operations. The idea is to have a "one" in the pivot spot (first row, first entry) and get zeros under that pivot element. Then proceed to the next column and get a one in the second row, second entry, and zeros under that pivot element, and so forth. So we work column-wise from left to right using row operations. First let's switch rows 1 and 2 to get a "one" in the pivot position:

$$
\begin{bmatrix}
2 & 1 & 0 & -1 & 3 \\
1 & 2 & 3 & 4 & 0 \\
3 & 3 & 3 & 4 & 5
\end{bmatrix}
\underset{row_1 - row_2}{\sim}
\begin{bmatrix}
1 & 2 & 3 & 4 & 0 \\
2 & 1 & 0 & -1 & 3 \\
3 & 3 & 3 & 4 & 5
\end{bmatrix}
$$

We can then eliminate the entries below the leading one in column1:

$$
\begin{bmatrix}
1 & 2 & 3 & 4 & 0 \\
2 & 1 & 0 & -1 & 3 \\
3 & 3 & 3 & 4 & 5
\end{bmatrix}
\underset{-2\,row_1 + row_2\ \ -3\,row_1 + row_3}{\sim}
\begin{bmatrix}
1 & 2 & 3 & 4 & 0 \\
0 & -3 & -6 & -9 & 3 \\
0 & -3 & -6 & -8 & 5
\end{bmatrix}
$$

$$
\begin{bmatrix}
1 & 2 & 3 & 4 & 0 \\
0 & -3 & -6 & -9 & 3 \\
0 & -3 & -6 & -8 & 5
\end{bmatrix}
\underset{-\,row_2 + row_3}{\sim}
\begin{bmatrix}
1 & 2 & 3 & 4 & 0 \\
0 & -3 & -6 & -9 & 3 \\
0 & 0 & 0 & 1 & 2
\end{bmatrix}
$$

$$
\begin{bmatrix}
1 & 2 & 3 & 4 & 0 \\
0 & -3 & -6 & -9 & 3 \\
0 & 0 & 0 & 1 & 2
\end{bmatrix}
\underset{-\frac{1}{3}\,row_2}{\sim}
\begin{bmatrix}
1 & 2 & 3 & 4 & 0 \\
0 & 1 & 2 & 3 & -1 \\
0 & 0 & 0 & 1 & 2
\end{bmatrix}
$$

The last matrix is in echelon[2] form so we can use back-substitution to find the solution, always solving for the lower-subscripted variable in terms of higher-subscripted ones. Clearly $x_4 = 2$. Moving up to the second row, we can solve for x_2;

$$x_2 + 2 x_3 + 3 x_4 = -1 \Rightarrow x_2 = -1 - 2 x_3 - 3 x_4 = -1 - 2t - 3(2) = -7 - 2t$$

Finally from the first row we have:

$$x_1 = -2 x_2 - 3 x_3 - 4 x_4 = -2(-7-2t) - 3t - 4(2) = 14 - 8 + t = 6 + t$$

So the solution can be written as a 4-tuple and we enclose it in braces to indicate that we actually have a *set* of solutions- the **solution set** of the system:

$$\{ (6+t, -7-2t, t, 2) : t \in \Re \}$$

We call this the **general solution**, since it contains ALL solutions. If we replace t by a particular number, then we get a *particular solution*. For example, if t = 1 we get
(6+t, -7-2t, t, 2) = (7,-9,1,2).
It is a good practice to pick a simple value for t and then check that this solution works in at least one of the original equations in the system; better to check it in *all* of them, if you have the time.
From the general solution, we observe that the system has infinitely many solutions in 4-space. This is to be expected since the system had fewer equations than unknowns; it could NOT have had a unique solution.

Possible outcomes when solving a linear system

It is easy to convince yourself that, for two lines in 2-space, there are only three possibilities: 1) two non-parallel lines, 2) two coincident lines and 3) two parallel lines. These same outcomes can occur for any system, no matter what size it is.
Thus for any linear system there are only 3 possible outcomes:
1) one solution - consistent
2) infinitely many solutions - consistent and dependent
3) no solution - inconsistent
If a system has one solution or infinitely many it is called a consistent system. If it has no solution, it is called *inconsistent*. We may further describe a system with infinitely many solutions as consistent and dependent to emphasize the fact that it does not have just one solution.

EXAMPLE 4 Describe the solution of the system whose augmented matrix is

$$\begin{bmatrix} 1 & 1 & 0 & 3 \\ 0 & 1 & -1 & 4 \\ 0 & 0 & 0 & 1 \end{bmatrix}$$

The system represented is

---

[2] From now on, echelon form means row-echelon form; we have no use for column operations.

$$x_1 + x_2 + \quad = 3$$

$$x_2 - x_3 = 4$$

$$0\, x_1 + 0\, x_2 + 0\, x_3 = 1$$

which is a 3 by 3 system. The matrix is already in echelon form, but the last row represents the equation 0 = 1 which is clearly absurd; thus the corresponding system is *inconsistent*.

EXAMPLE 5 Discuss the solutions of the system whose augmented matrix is

$$\begin{bmatrix} 1 & 0 & 2 & a \\ 0 & 1 & 3 & b \\ 1 & 1 & 5 & c \end{bmatrix}$$

This system is square of size 3 by 3. Let's put the matrix in echelon form:

$$\begin{bmatrix} 1 & 0 & 2 & a \\ 0 & 1 & 3 & b \\ 1 & 1 & 5 & c \end{bmatrix} \underset{-\,row_1 + row_3}{\sim} \begin{bmatrix} 1 & 0 & 2 & a \\ 0 & 1 & 3 & b \\ 0 & 1 & 3 & c-a \end{bmatrix} \underset{-\,row_2 + row_3}{\sim} \begin{bmatrix} 1 & 0 & 2 & a \\ 0 & 1 & 3 & b \\ 0 & 0 & 0 & c-a-b \end{bmatrix}$$

From the last row, we must have c - a - b = 0 or c = a + b; if not the system is inconsistent. If c = a + b, we cannot get a unique solution (fewer equations than unknowns) so the system will be consistent and dependent.

EXERCISES 1.1
Part A. Determine if A is in row-echelon form; if it is not, put it in row-echelon form.

1. $A = \begin{bmatrix} 1 & 0 & 2 & 3 \\ 0 & 1 & 2 & 0 \\ 0 & 0 & 0 & 1 \end{bmatrix}$ ans. in echelon form

2. $A = \begin{bmatrix} 2 & 0 & 4 & 6 \\ 0 & -1 & -2 & 0 \\ 2 & -1 & 2 & 7 \end{bmatrix}$

3. $A = \begin{bmatrix} 0 & 1 & 4 & -3 \\ 1 & 1 & 2 & 4 \\ 0 & 2 & 6 & 1 \end{bmatrix}$ ans. $A = \begin{bmatrix} 1 & 1 & 2 & 4 \\ 0 & 1 & 4 & -3 \\ 0 & 0 & 1 & -\dfrac{7}{2} \end{bmatrix}$

4. $A = \begin{bmatrix} 1 & 0 & 2 & 3 \\ 1 & 1 & 2 & 3 \\ 2 & 3 & 4 & 5 \end{bmatrix}$

5. $A = \begin{bmatrix} 0 & 1 & 2 \\ -2 & 0 & 4 \\ 2 & 3 & 4 \\ 3 & 3 & 3 \end{bmatrix}$ ans. $A = \begin{bmatrix} 1 & 0 & -2 \\ 0 & 1 & 2 \\ 0 & 0 & 1 \\ 0 & 0 & 0 \end{bmatrix}$

6. $A = \begin{bmatrix} -2 & 0 & -4 \\ 0 & 1 & 2 \\ 3 & 3 & 3 \\ 2 & 3 & 4 \end{bmatrix}$

7. $A = \begin{bmatrix} 1 & 0 & 2 \\ 0 & 2 & -4 \\ 0 & 0 & 0 \\ -2 & 2 & -2 \end{bmatrix}$ ans. $A = \begin{bmatrix} 1 & 0 & 2 \\ 0 & 1 & -2 \\ 0 & 0 & 1 \\ 0 & 0 & 0 \end{bmatrix}$

8. $A = \begin{bmatrix} -3 & 6 & -3 \\ 4 & 5 & 6 \\ 7 & 7 & 14 \end{bmatrix}$

9. $A = \begin{bmatrix} 1 & 0 & 2 & 3 \\ 0 & 1 & 2 & 0 \\ 0 & 1 & 4 & 4 \end{bmatrix}$ ans. $A = \begin{bmatrix} 1 & 0 & 2 & 3 \\ 0 & 1 & 2 & 0 \\ 0 & 0 & 1 & 2 \end{bmatrix}$

Part B. Solve each linear system by putting its augmented matrix in row-echelon form and then using back-substitution.

10.
$$x_1 + 2x_2 + \quad = 0$$
$$-x_1 + 3x_2 + 4x_3 = 6$$
$$-x_1 + 2x_2 - x_3 = 7$$

$$x_1 + 2\,x_2 + \quad = 0$$

11. $- x_1 + 3\,x_2 + 4\,x_3 = 0$ ans. $\{(0,0,0)\}$

$$- x_1 + 2\,x_2 - x_3 = 0$$

$$x_1 + 2\,x_2 + \quad = 0$$

12. $- x_1 + 2\,x_2 + 4\,x_3 = 5$

$$- 4\,x_2 - 4\,x_3 = 5$$

$$x_1 + 2\,x_2 + \quad + x_4 = 4$$

13. $- x_1 + 2\,x_2 + 4\,x_3 - 3\,x_4 = 2$ ans. $\{(-t, 2, -1/2 + 1/2t, t)\}$

$$- x_1 + \quad - 4\,x_3 + x_4 = 2$$

$$x_1 + 2\,x_2 + \quad + x_4 = 0$$

14. $- x_1 + 2\,x_2 + 4\,x_3 - 3\,x_4 = 6$

$$- x_1 + 2\,x_2 - x_3 + x_4 = 7$$

$$x_1 + 2\,x_2 + \quad = 0$$

15. $- x_1 + 2\,x_2 + 4\,x_3 = 6$ ans. $\{\,\}$

$$- x_1 + 2\,x_2 - x_3 = 7$$

$$- x_1 - x_2 + x_3 = 0$$

Part C. Solve each system by putting its matrix in reduced row-echelon form (this is called Gauss-Jordan elimination). Compare the amount of work you did with that done in part B.

16. system of #12

17. system of #13 ans. $\begin{bmatrix} 1 & 0 & 0 & 1 & 0 \\ 0 & 1 & 0 & 0 & 2 \\ 0 & 0 & 1 & \frac{-1}{2} & \frac{-1}{2} \end{bmatrix}$

18. system of #14

Part D. Each matrix is the augmented matrix of a linear system. Give the size of each system; then put the matrix in echelon form. Determine whether the system is consistent or inconsistent. If it is consistent, further classify as having only one solution or infinitely many (consistent and dependent). If possible, give a relationship involving a,b, and c which guarantees the system has a solution.

19. $\begin{bmatrix} 1 & 0 & 2 & a \\ 2 & 3 & 4 & b \end{bmatrix}$ ans. 2 x 3, consistent and dependent for all a, b.

20. $\begin{bmatrix} 1 & 0 & 2 & a \\ 0 & 1 & 3 & b \\ 1 & 1 & 5 & c \end{bmatrix}$

21. $\begin{bmatrix} 1 & 0 & -2 & a \\ 1 & 1 & 4 & b \\ 2 & 1 & 7 & c \end{bmatrix}$ ans. 3 x 3, consistent(one solution) for any a,b, & c.

22. $\begin{bmatrix} 1 & 0 & -2 & a \\ 1 & 1 & 4 & b \\ 2 & 1 & 2 & c \end{bmatrix}$

23. $\begin{bmatrix} 1 & 1 & a \\ -1 & 1 & b \\ 0 & 2 & c \\ 2 & 2 & d \end{bmatrix}$ ans. 4 x 2, consistent(one solution) if c=a+b and d=2a, othewise inconsistent

24. $\begin{bmatrix} 1 & 1 & 0 & a \\ -1 & 1 & 1 & b \\ 0 & 2 & -1 & c \\ 2 & 2 & 0 & d \end{bmatrix}$

Part E. Application: The steady state temperature at each point in metal plate is the average of the temperatures at the 4 points surrounding it (West, North, East, South).

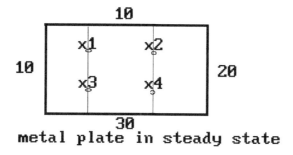
metal plate in steady state

25. Set up the system of equations for each point, starting with x1. Hint:
equ1: $x1 = 1/4(10 + 10 + x2 + x3)$, etc.
26. Write down the augmented matrix.
27. Solve the corresponding system for x1, x2, x3 and x4 using Gaussian elimination.
ans. (205/48,25/6,65/12,235/48)

Part F. Curve fitting:
28. A cubic function is of the form $y = f(x) = a x^3 + b x^2 + c x + d$; find a,b, c and d if the graph passes through the points (-2,-5),(0,1),(1,4) and (2,15).
29. A quadratic function is of the form $y = f(x) = a x^2 + b x + c$; find a,b, and c if the parabola passes through the points (-1,0),(1,8) and (2,6). ans. $y = -2 x^2 + 4 x + 6$

Part G. Given the matrix $\begin{bmatrix} 1 & 0 & 2 & -3 \\ 2 & 1 & 5 & -7 \\ 3 & 0 & 7 & -17 \end{bmatrix}$;

30. Put this matrix in echelon form, carefully keeping track of all your row operations.
31. Now take your answer from #30 and "undo" the row operations (in the reverse order!) in order to recover the original matrix given. This shows that if A~B then B~A.
Hint: If you do the operation -k row1 + row2, undoing it means doing the operation k row1 + row2

Part H. Phone numbers in Mayberry: Suppose in Mayberry the phone numbers have only 4 digits. It is known that the sum of the digits is 20, the sum of the first 3 digits is 16 and the sum of the last 2 digits is 10.

32. Find the form of any phone number in Mayberry.

33. How many phones are there in Mayberry? ans. 9 (it's a small town!)

1.2 Geometry of Solutions

In this sections we wish to explore the geometry of the solutions of linear systems. If a system is consistent and has a unique solution, it is simply a point in n-space and is not that interesting. However, if there are infinitely many solutions, the set of solutions has a definite structure in n-space and we want to learn a little about this structure.

In 2-space, we can think about the solution of a system as the ordered pair (x,y) or the vector[3] (x,y). The difference is illustrated by the diagram below for the simple linear equation $y = 1/2x + 4$. This one equation has infinitely-many solutions, thought of as ordered pairs. However, we can also view the solutions as *vectors* in a parametric form. Since the solutions of consistent and dependent systems are expressed in parametric form, it is nice to be able to consider the solutions of an equation like $y = 1/2x + 4$ as vectors in 2-space (instead of just points). There is no problem with this if all vectors have their "tails" at the origin.

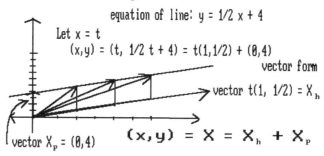

Homogenous systems

EXAMPLE 1 Let's solve the following 2 by 3 system:

$$x_1 - x_2 + 3\,x_3 = 0$$

$$x_2 + 2\,x_3 = 0$$

This system is called a **homogeneous** system. Since all the constants are zero, one solution is (0,0,0) and it therefore cannot be inconsistent. And in fact, since there are fewer equations than unknowns, there must be infinitely many solutions.

THEOREM For a homogeneous m by n system, if m < n the system is *consistent and dependent* (ie. infinitely many solutions).

---

[3] This same point of view is taken in *Topology and Modern Analysis* by G. Simmons

14

Back to **EX.1**; we solve the system using Gaussian elimination. The augmented matrix is

$$\begin{bmatrix} 1 & -1 & 3 & 0 \\ 0 & 1 & 2 & 0 \end{bmatrix}$$

which is already in row-echelon form. So we can proceed directly to back-substitution:

$$x_2 + 2 x_3 = 0 \Rightarrow x_2 = -2 x_3 = -2t$$

$$x_1 - x_2 + 3 x_3 = 0 \Rightarrow x_1 = x_2 - 3 x_3 = (-2t) - 3t = -5t$$

So the solution is all ordered triples of the form (-5t,-2t,t). If we factor out the parameter t, we have a set of solutions (vectors) which are all multiples of the vector (-5,-2,1):

$$X = t(-5,-2,1)$$

This set of points or vectors forms a *line* through the origin; this line is in the direction of the vector (-5,-2,1). And in fact something else interesting is true about the set of solutions. Suppose we pick two different solutions of the system, say X_1 & X_2:

$$X_1 = (-5,-2,1) \quad X_2 = (10,4,-2)$$

If we add these two solutions (addition being defined by just adding corresponding coordinates) we would get:

$$X_1 + X_2 = (-5,-2,1) + (10,4,-2) = (5,2,-1) = X_3$$

then the sum X_3 is actually a solution of the system too. This is the mark of a homogeneous system; the *sum* of any two solutions is a solution and a *scalar multiple* of any solution is a solution. This is not true for an inhomogeneous system which is consistent and dependent. For example, let's just change the constants in the original system to make it inhomogeneous:

EXAMPLE 2

$$x_1 - x_2 + 3 x_3 = 4$$

$$x_2 + 2 x_3 = 5$$

The solution is easily found to be:

$$X = (9 - 5t, 5 - 2t, t)$$

This can be rewritten in a suggestive fashion:

$$X = (\ 9 - 5t \ , 5 - 2t \ , t \ ) = (-5t , -2t , t \ ) + (\ 9,5,0 \ ) = X_h + X_p$$

Here we see the solution from the corresponding homogeneous system is part of the solution of this new inhomogeneous system. It consists of the part with the parameter t (belonging to X_h) and a single (particular) vector X_p. To get X_p, just set all parameters equal to *zero* in the general solution.

THEOREM Any solution of an inhomogeneous linear system is of the form $X = X_h + X_p$ where X_h is the solution of the corresponding homogeneous system and X_p satisfies the given system.

It is easy to show that X_p alone satisfies the inhomogeneous system; but that is not the whole story! The complete solution $X = X_h + X_p$ satisfies the inhomogeneous system. However, the *sum of two such solutions* is **not** a solution! Looking back at EX2, let

$$X_1 = (-5,-2,1 \ ) + (\ 9,5,0 \ ) \quad X_2 = (-10,-4,2 \ ) + (\ 9,5,0 \ )$$

$$X_1 + X_2 = [\ (-5,-2,1 \ ) + (9,5,0) \ ] + [\ (-10,-4,2 \ ) + (9,5,0) \ ]$$

$$X_1 + X_2 = (\ -15,-6,3 \ ) + (18,10,0 \ ) = (3,4,3 \ ) = X_3$$

The vector X_3 does **not** work in the system:

$$x_1 - x_2 + 3 \, x_3 = 4 \Rightarrow 3 - 4 + 3(\ 3 \ ) = 8 \neq 4$$

Moral of the story- for an inhomogeneous system, adding two solutions (or scalar multiplying a solution) does NOT produce a solution of the system.

EXAMPLE 3 Solve and discuss the geometry of the solution of the system whose augmented matrix is

$$\begin{bmatrix} 1 & -1 & 3 & 0 \\ 2 & -2 & 6 & 0 \\ 3 & -3 & 9 & 0 \end{bmatrix}$$

The system represented is a 3 by 3 homogeneous square system. Putting the augmented matrix in echelon form:

$$\begin{bmatrix} 1 & -1 & 3 & 0 \\ 2 & -2 & 6 & 0 \\ 3 & -3 & 9 & 0 \end{bmatrix} \begin{array}{c} \\ -2 \ row_1 + row_2 \end{array} -3 \ row_1 + row_3 \quad \sim \quad \begin{bmatrix} 1 & -1 & 3 & 0 \\ 0 & 0 & 0 & 0 \\ 0 & 0 & 0 & 0 \end{bmatrix}$$

Using back-substitution on the echelon form implies

$$x_1 - x_2 + 3 \, x_3 = 0 \Rightarrow x_1 = x_2 - 2 \, x_3$$

Letting $x_2 = s$ and $x_3 = t$ we have X = (s -3t,s,t) as the general solution. This represents a *plane* through the origin in 3-space, since

$$X = (s - 3t, s, t)=(s, s, 0)+(-3t, 0, t)= s(1, 1, 0)+t(-3, 0, 1)$$

In general, any solution set with a "two parameter family" as the solution represents a plane. We need not limit ourselves to three dimensions however.

EXAMPLE 4 Solve and discuss the geometry of the solutions for the system whose augmented matrix is

$$\begin{bmatrix} 1 & -1 & 2 & 2 & 0 \\ 2 & -1 & 7 & 2 & 0 \\ 3 & -3 & 4 & 4 & 0 \end{bmatrix}$$

The system represented is a 3 by 4 homogeneous system; so m = 3 and n = 4 and since m < n, the system will be consistent and dependent. Putting this matrix in echelon form gives us:

$$\begin{bmatrix} 1 & -1 & 2 & 2 & 0 \\ 2 & -1 & 7 & 2 & 0 \\ 3 & -3 & 4 & 4 & 0 \end{bmatrix} \sim \begin{bmatrix} 1 & -1 & 2 & 2 & 0 \\ 0 & 1 & 3 & -2 & 0 \\ 0 & 0 & 1 & 1 & 0 \end{bmatrix}$$

The third row says

$$x_3 + x_4 = 0$$

If we let $x_4 = t$ then $x_3 = -x_4 = -t$. Now from the second row we have

$$x_2 + 3 x_2 - 2 x_4 = 0$$

thus

$$x_2 = -3 x_3 + 2 x_4 \Rightarrow x_2 = -3(-t)+2t = 5t$$

so finally from the first row

$$x_1 = x_2 - 2 x_3 - 2 x_4 = 5t - 2(-t)-2t = 5t$$

Thus the solution is of the form X = (5t,5t,-t,t) = t(5,5,-1,1). The solutions here represent a *line* (in 4-space) in the direction of (5,5,-1,1) and go through the origin (just let t = 0).

Geometry in n-dimensional space

We call the graph of any linear equation in n variables a *hyperplane*. In EX4, we have shown that 3 different hyperplanes (in 4-space) intersect in a line. The 3 equations for the hyperplanes come from the three rows of the augmented matrix:

$$x_1 - x_2 + 2\,x_3 + 2\,x_4 = 0$$

$$2\,x_1 - x_2 + 7\,x_3 + 2\,x_4 = 0$$

$$3\,x_1 - 3\,x_2 + 4\,x_3 + 4\,x_4 = 0$$

Each row represents a hyperplane whose equation is written out above.
Claim: A hyperplane in n-space is determined by (n - 1) parameters.
Consider one of the equations in the above system, say equ.(1)

$$x_1 - x_2 + 2\,x_3 + 2\,x_4 = 0$$

Solve the equation for x_1:

$$x_1 = x_2 - 2\,x_3 - 2\,x_4$$

Now let $x_2 = r, x_3 = s, x_4 = t$; then we have

$$x_1 = r - 2\,s - 2\,t$$

If we replace x_1 in a 4-tuple which represents any point in this hyperplane, we would get:

$$X = (\,x_1, x_2, x_3, x_4\,) = (\,r - 2\,s - 2\,t, r, s, t\,) = (\,r, r, 0, 0\,) + (-2\,s, 0, s, 0\,) + (\,-2\,t, 0, 0, t\,)$$

Factoring out the scalars r, s and t then finally gives us:

$$X = r(\,1, 1, 0, 0\,) + s(-2, 0, 1, 0\,) + t(\,-2, 0, 0, 1\,)$$

As a check on this simple algebra, one could replace X by (1,1,0,0), (-2,0,1,0) or (-2,0,0,1) and show it works in equ(1). Then there can be no doubt about it. Thus we can thus think of a hyperplane in 4-space as any 4-tuple created from one linear equation; it will have 3 parameters in it. We could deal similarly with any linear equation with n unknowns:

$$a_1\,x_1 + a_2\,x_2 + \ldots + a_n\,x_n = b$$

Therefore in n-space, the graph of any n-tuple which has (n - 1) parameters in it is a hyperplane.
We thus make the following generalizations for the general solution of a system:
1) any solution with no parameter is a *point* (this happens only if the system is consistent)
2) a solution with one parameter represents a line
3) a solution with two parameters represents a *plane*
4) a solution with n - 1 parameters represents a **hyperplane**
It is helpful to use these ideas so that we can retain some semblance of our geometric intuition, even though we may be in 4-space and cannot draw any pictures. We trust our algebra; after that the rest is analogy. Using the above classifications, in 2-space a line is a hyperplane. In 3-space a plane is a hyperplane. In 4-space, a hyperplane is really something different than either a line or a plane; it is like a copy of 3-space sitting inside 4-space.

The point is that if a system is consistent and dependent, then the solution set is an infinite collection of vectors, which in some sense we can regard as a geometric object, even though we may not be able to draw it. The *size* of this collection of vectors is indicated by the number of parameters necessary to describe it. For example, in 3-space we think a line as "smaller" than a plane, since a plane contains infinitely many lines.

If we have a consistent and dependent system with m different equations in n unknowns (m < n), we will be able to show that the general solution will contain (n - m) parameters. The following chart illustrates this idea for 5-space.

n = 5	m = no. of equations	n - m = no. of parameters	solution of system
	4	1	line
	3	2	plane
	2	3	NA[4]
	1	4	hyperplane

EXAMPLE 5 Create a system in 5 unknowns whose solution is a plane. We would need 3 different equations to do this:

$$x_1 - x_2 + x_3 + 2 x_4 - x_5 = 1$$

$$x_2 + 6 x_3 + 2 x_4 + x_5 = 0$$

$$x_1 + 7 x_3 + 5 x_4 + 2 x_5 = 3$$

To keep getting different equations, be sure that the second is not a multiple of the first, the third is not a sum of the first two, etc. The solution will have two parameters in it and represent a plane in 5-space.

EXERCISES 1.2

A. Describe each linear system as homogeneous or inhomogeneous. Then 1) solve it by putting its augmented matrix in row-echelon form and 2) using back-substitution. Write the solution in the form X = Xh + Xp; then describe geometrically what your solution set is: a point, a line, a plane or a hyperplane or none of these.

---

[4] There is no special name in n-space for an object defined by n - 2 parameters.

$2 x_1 + 4 x_2 = 6$

1.   ans. $\{(-1,2)\}$,point

$3 x_1 + 5 x_2 = 7$

$x_1 + 2 x_2 = 0$

2.

$- x_1 - 2 x_2 = 0$

$x_1 + 2 x_2 + \quad = 0$

3.   ans. $\{(2t,-t,t) + (-4,2,0)\}$,line

$- x_1 + 3 x_2 + 5 x_3 = 10$

$x_1 + 2 x_2 - x_3 + 2 x_4 = 1$

4.

$- x_1 + 3 x_2 + 4 x_3 - x_4 = 4$

$x_1 + 2 x_2 - x_3 + 4 x_4 = 1$

5.   ans. $\{(3s-2t,s-t,s,t)+(-3,2,0,0)\}$,plane

$- x_1 + 2 x_2 + 5 x_3 \quad = 7$

$x_1 + 2 x_2 + \quad = 0$

6. $- x_1 + 3 x_2 + 4 x_3 = 6$

$- x_1 + 2 x_2 - x_3 = 7$

$x_1 + 2 x_2 + \quad = 0$

7. $- x_1 + 3 x_2 + 4 x_3 = 0$ ans.$\{(0,0,0)\}$,point

$- x_1 + 2 x_2 - x_3 = 0$

$x_1 + 2\,x_2 - x_3 + 2\,x_4 = 1$

8. $-x_1 + 3\,x_2 + 4\,x_3 - x_4 = 4$

$-2\,x_1 + x_2 - x_3 + x_4 = 0$

$x_1 + 2\,x_2 - x_3 + \quad x_5 = 0$

9. $-x_1 + \quad\quad 4\,x_4 - x_5 = 0$ ans. $\{(4s - t, -11 + 2t, 4s - 18t, s, t)\}$, plane

$-2\,x_1 + x_2 - x_3 + x_4 \quad = 0$

$x_1 + 2\,x_2 + \quad = 0$

10. $-x_1 + 2\,x_2 + 4\,x_3 = 5$

$-4\,x_2 - 4\,x_3 = 5$

$x_1 + 2\,x_2 + \quad + x_4 = 4$

11. $-x_1 + 2\,x_2 + 4\,x_3 - 5\,x_4 = 4$ ans. $\{(-t, 0, t, t) + (-2, -3, -1, 0)\}$, line

$-x_1 \quad - x_3 \quad = 3$

$x_1 + 2\,x_2 + \quad + x_4 = 0$

12. $-x_1 + 2\,x_2 + 4\,x_3 - 3\,x_4 = 6$

$-x_1 + 2\,x_2 - x_3 + x_4 = 7$

$$x_1 + 2\,x_2 + \ \ = 0$$

$$-x_1 + 2\,x_2 + 4\,x_3 = 6$$

13.   ans. inconsistent

$$-x_1 + 2\,x_2 - x_3 = 7$$

$$-x_1 - x_2 + x_3 = 0$$

B. Solve each of the homogeneous systems below. In 4-space, 3 different hyperplanes (each represented by a linear equation in 4 variables) trying to intersect produces a line; 2 different hyperplanes trying to intersect produces a plane. Verify that this is true by solving each system, observing the number of parameters in each solution set (which should be consistent and dependent!). Describe your solution geometrically (use point, line, plane or hyperplane)

$$x_1 + x_2 + \ \ + x_4 = 0$$

14. $-x_1 + 2\,x_2 + 4\,x_3 - 3\,x_4 = 0$

$$3\,x_2 - x_3 + x_4 = 0$$

$$x_1 + x_2 + \ \ + x_4 = 0$$

15. $-x_1 + 2\,x_2 + 3\,x_3 - 4\,x_4 = 0$ ans. $\{(-8t, 7t, 6t, t)\}$, line

$$3\,x_2 + 4\,x_3 + 3\,x_4 = 0$$

$$x_1 + x_2 + \ \ + x_4 = 0$$

16. $-2\,x_1 - 2\,x_2 \ - \ 2\,x_4 = 0$

$$3\,x_1 + 3\,x_2 + 3\,x_4 = 0$$

22

17.
$$x_1 + x_2 + \quad + x_4 = 0$$
ans. $\{(s-2t, -s+t, s, t)\}$, plane
$$-x_1 + 3x_2 + 4x_3 - 5x_4 = 0$$

C. Solve each of these systems below. In 5-space, 4 different hyperplanes (each represented by a linear equation in 5 variables) trying to intersect produces a line; 3 different hyperplanes trying to intersect produces a plane. Verify that this is true by solving each system, observing the number of parameters in each solution set (which should be consistent and dependent!). Describe your solution geometrically (use point, line, plane or hyperplane)

18.
$$x_1 + x_2 + \quad + x_5 = 0$$

$$-x_1 - 2x_2 + x_3 - \quad x_5 = 0$$

$$x_1 + x_2 + x_3 - x_4 \quad = 0$$

$$x_1 + \quad x_3 - x_4 + x_5 = 0$$

19.
$$x_1 + x_2 + \quad + x_5 = 0$$

$$-x_1 - 2x_2 + x_3 - \quad x_5 = 0 \quad \text{ans. } \{(-s - 2t, s + t, s + t, s, t)\} \text{ plane}$$

$$x_1 + x_2 + x_3 - x_4 \quad = 0$$

20.
$$x_1 + x_2 + \quad + x_5 = 2$$

$$-x_1 - 2x_2 + x_3 - \quad x_5 = 6$$

$$x_1 + x_2 \quad + \quad x_5 = 1$$

$$-x_1 - 2x_2 + x_3 - x_4 = 2$$

21. $\quad x_1 + x_2 + x_3 - x_4 \quad = 3$ ans.$\{(2,-1,4,2,0)\}$, point

$$x_1 + \quad x_3 - x_4 + x_5 = 4$$

$$x_3 - 2x_4 + 4x_5 = 0$$

D. Given the one linear equation in x,y,z and w of the form x + 4y - 3z + 5w = 10;

22. Solve the equation for x.

23. If X = (x,y,z,w), find the form of any vector X which satisfies the given linear equation in terms of x, y and z. ans. X = (10 - 4y + 3z - 5w, y,z, w)

24. Now replace y by r, z by s, and w by t and rewrite your answer to #23 in parametric form.

25. Finally write the parametric version of the original equation in the form $X = X_h + X_p$; show that X_h represents a hyperplane in 4-space which goes through (10,0,0,0).

ans. (x,y,z,w) = [r(-4,1,0,0) + s(3,0,1,0) + t(-5,0,0,1)] + (10,0,0,0); 3 parameters means a hyperplane
NB. When you do this, you are finding THREE vectors which somehow create this hyperplane- ie. the vectors (-4,1,0,0), (3,0,1,0), and (-5,0,0,1).

26. Use the ideas in part D to find the parametric version of the plane x + 2y - 3z = 6; can you give the two vectors which create this plane? Hint: solve for x, let y = s and z = t, its easy!

24

1.3 System Analysis

In this section we wish to analyze the structure of a system by looking at certain non-negative integers that can be connected with the system. These numbers will be useful in determining under what conditions the system will be consistent and whether or not the solution will be unique. First, let's agree to the following notation- if we have a 3 by 5 system like

$$x_1 - x_3 + x_5 = 1$$
$$x_2 + x_4 = 2 \qquad augmented\ matrix = \begin{bmatrix} 1 & 0 & -1 & 0 & 1 & 1 \\ 0 & 1 & 0 & 1 & 0 & 2 \\ 0 & 0 & 1 & 0 & -1 & 3 \end{bmatrix}$$
$$x_3 - x_5 = 3$$

we'll call the matrix of coefficients in the system the *coefficient matrix*

$$A = \begin{bmatrix} 1 & 0 & -1 & 0 & 1 \\ 0 & 1 & 0 & 1 & 0 \\ 0 & 0 & 1 & 0 & -1 \end{bmatrix}$$

and the matrix of constants B:

$$B = \begin{bmatrix} 1 \\ 2 \\ 3 \end{bmatrix}$$

Notice that the augmented matrix A is already in echelon form (not reduced) and we can solve it by back-substitution: we would get

$$X = \{ \ (\ 4, -s+2, t+3, s, t \ ) : s, t \in \Re \ \}$$

This general solution represents a *plane* in 5-space, since two parameters (s and t) are needed to describe the set. The system is consistent and dependent (which we sort of expected since there are fewer equations than unknowns). We could rewrite the solution in the form

$$X = (\ x_1, x_2, x_3, x_4, x_5 \ ) = (\ 0, -s, t, s, t \ ) + (\ 4, 2, 3, 0, 0 \ ) = X_h + X_p$$

Now the variables x_1, x_2, x_3 are called <u>determined</u> variables, because once s and t (the parameters here) are chosen, $x_1, x_2, \& x_3$ have specific values. However, $x_4 \& x_5$ have no constraint on them and are free to be whatever they like; these are *free* variables. So the system has three determined variables (corresponding to the <u>leading ones</u> in columns 1, 2 and 3 as seen in the row-echelon form of the augmented matrix) and two free variables (columns 4 and 5 have *no leading ones*). The first three columns of the augmented matrix have leading ones (or pivot elements) in them and are thus called **pivot columns**; columns 4 and 5 have no leading ones and are therefore *non-pivot* columns (the last column is NOT in this game, since it represents the constants, not a variable). The number

3 then is significant to the structure of the coefficient matrix A (and any system formed from A). This number is called the **rank** of the matrix A.

rank(A) = number of leading ones in the echelon form of A

The maximum value of the rank is determined by the number of *rows* of the matrix; so for an m by n matrix, the rank satisfies $rank(A) \leq m$.

From the previous section, we know that the solution of the corresponding homogeneous system

$$x_1 \ - \ x_3 \ + \ x_5 = 0$$

$$x_2 \ + \ x_4 \quad = 0$$

$$x_3 \ - \ x_5 = 0$$

is X_h, which is described by two parameters s and t. The reader should be able to show fairly easily that $X_h = s(0,1,0,1,0) + t(0,0,1,0,1)$. Why two parameters here (and not just one or maybe three or more)? The answer is another number which is important to the matrix A- the nullity of A. This number is simply n - rank(A); it counts *how many parameters there are in* X_h (and thus in the general solution of the system).

nullity(A) = number of columns of A - rank(A) = n - r

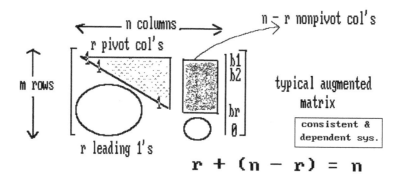

26

This means, for any coefficient matrix A;

$$\boxed{\text{rank(A) + nullity(A) = n (no. of columns of A)}}$$

RANK + NULLITY THEOREM

So once the rank is known, the nullity can be determined by just subtracting the rank from the number of columns of the coefficient matrix A. Clearly the system CANNOT have a unique solution if X_h has even one parameter in it; so if the system is consistent, a necessary and sufficient condition for a unique solution of the system is that the *nullity of A must be ZERO* (ie. rank(A) = n). This means that there are no non-pivot columns and $X_h = \theta$.

Another important advantage to knowing the rank of the matrix A is that, if the right-hand side B comes from $\mathfrak{R}^m$ (that is the vector B has m components), and r = rank(A) = m, then the system *cannot* be inconsistent. Or, put another way, the system can be *inconsistent* if rank(A) < m.

EXAMPLE 1: Given the system

$$x_1 \qquad = a$$

$$x_2 = b$$

$$2\,x_1 + 3\,x_2 = c$$

determine for what $B = \begin{bmatrix} a \\ b \\ c \end{bmatrix}$ the system has a solution. Also determine if the solution will be unique or not.

The augmented matrix of this system is $\begin{bmatrix} 1 & 0 & a \\ 0 & 1 & b \\ 2 & 3 & c \end{bmatrix}$. Putting this matrix in echelon form:

$$\begin{bmatrix} 1 & 0 & a \\ 0 & 1 & b \\ 0 & 0 & c-2a-3b \end{bmatrix}$$

This means that the system is inconsistent unless c - 2a - 3b = 0 or c = 2a + 3b. Here the coefficient matrix A is of the form:

$$A = \begin{bmatrix} 1 & 0 \\ 0 & 1 \\ 2 & 3 \end{bmatrix} \sim \begin{bmatrix} 1 & 0 \\ 0 & 1 \\ 0 & 0 \end{bmatrix}$$

so that rank(A) = 2 and nullity(A) = 2 - 2 = 0. Thus here if the entries of B satisfy c = 2a + 3b, the solution must be unique (there are no non-pivot columns so no free variables). If the entries of B don't satisfy this criterion, then the system must be inconsistent (probable since B is a 3-vector and rank(A) = 2). The reader should try to solve this system with a B like (1,2,13); you will be in for a disappointment!

EXAMPLE 2 Analyze the system

$$x_1 + 2 x_2 + 3 x_3 + x_4 = 1$$

$$x_2 + 2 x_3 + 4 x_3 = 5$$

The augmented matrix is

$$\begin{bmatrix} 1 & 2 & 3 & 1 & 1 \\ 0 & 1 & 2 & 4 & 5 \end{bmatrix}$$

and the coefficient matrix is

$$A = \begin{bmatrix} 1 & 2 & 3 & 1 \\ 0 & 1 & 2 & 4 \end{bmatrix}$$

and A is in row-echelon form so we know that rank(A) = 2 and nullity(A) = 4 - rank(A) = 2. So x_1 & x_2 are determined variables and x_3 & x_4 are free. The system will be consistent and dependent and its solution will have two parameters in it. It is a *plane* in 4-space. The solution of the corresponding homogeneous system

$$x_1 + 2 x_2 + 3 x_3 + x_4 = 0$$

$$x_2 + 2 x_3 + 4 x_4 = 0$$

would also have two parameters in it, since the augmented matrix again has no leading ones in columns 3 and 4:

$$\begin{bmatrix} 1 & 2 & 3 & 1 & 0 \\ 0 & 1 & 2 & 4 & 0 \end{bmatrix}$$

Since $B \in \Re^2$ (the augmented matrix is of size 2 by 5) and rank(A) = 2, it is impossible to construct an inconsistent system using the coefficient matrix A. Any other right-hand side B (besides the original) would also have given rise to a consistent (and dependent) system.

EXAMPLE 3 Analyze the system

$$x_1 + 2x_2 + 3x_3 = 1$$

$$-x_1 + x_2 + 2x_3 = 5$$

$$-2x_1 + 4x_2 + 5x_3 = 0$$

The augmented matrix is

$$\begin{bmatrix} 1 & 2 & 3 & 1 \\ -1 & 1 & 2 & 5 \\ -2 & 4 & 5 & 0 \end{bmatrix}$$

and the coefficient matrix is

$$A = \begin{bmatrix} 1 & 2 & 3 \\ -1 & 1 & 2 \\ -2 & 4 & 5 \end{bmatrix}$$

The row-echelon form of A is:

$$\begin{bmatrix} 1 & 2 & 3 \\ 0 & 1 & \dfrac{5}{3} \\ 0 & 0 & 1 \end{bmatrix}$$

so that rank(A) = 3 and therefore nullity(A) = 3 - 3 = 0. The system is consistent and has a unique solution since there are no free variables in the solution. The solution of the original system is then

$$X = (-1, -8, 6) = (0,0,0) + (-1, -8, 6) = X_h + X_p$$

Hence the only solution of the corresponding homogeneous system is the zero vector (0,0,0). In this case, since rank(A) = 3 and the system was a 3 by 3 (square) system, the only solution of the corresponding homogeneous system (ie. X_h) is the zero vector. And since the matrix A is a **3** by 3 (and the right-hand side B comes from **3-space**) any right-hand side B would have produced a consistent system, since rank(A) = 3 = m (no. of rows of the coefficient matrix A).

Overdetermined versus underdetermined

If a system is *underdetermined*, it has fewer equations than unknowns(m < n). Such a system

29

clearly cannot have a unique solution so it is either inconsistent or consistent and dependent. Most likely it is consistent and dependent.

On the other hand, if a system has more equations than unknowns($m > n$), it is called overdetermined and is likely to be inconsistent. For example, if we consider the system

$$x_1 - 2\,x_2 = 3$$

$$-x_1 + 3\,x_2 = 4$$

it is square($m = n$) and it would be consistent no matter what constants would be on the right hand side(since rank(A) = 2). If we add a third equation however (to make the system overdetermined) it is very likely that the system will become inconsistent.

EXAMPLE 4 Consider the overdetermined system below:

$$x_1 - 2\,x_2 = 3$$

$$-x_1 + 3\,x_2 = 4$$

$$2\,x_1 + 4\,x_2 = 6$$

Putting the augmented matrix in echelon form:

$$\begin{bmatrix} 1 & -2 & 3 \\ -1 & 3 & 4 \\ 2 & 4 & 6 \end{bmatrix} \begin{array}{c} \\ row_1 + row_2 \\ -2\,row_1 + row_3 \end{array} \begin{bmatrix} 1 & 2 & 3 \\ 0 & 1 & 7 \\ 0 & 8 & 0 \end{bmatrix} \begin{array}{c} \\ \\ row_2 - row_3 \end{array} \begin{bmatrix} 1 & 2 & 3 \\ 0 & 8 & 0 \\ 0 & 1 & 7 \end{bmatrix} \begin{array}{c} \\ \frac{1}{8}\,row_2 \\ \end{array} \begin{array}{c} - \end{array}$$

$$\begin{bmatrix} 1 & 2 & 3 \\ 0 & 1 & 0 \\ 0 & 1 & 7 \end{bmatrix} \begin{array}{c} \\ \\ -row_2 + row_3 \end{array} \begin{bmatrix} 1 & 2 & 3 \\ 0 & 1 & 0 \\ 0 & 0 & 7 \end{bmatrix} \begin{array}{c} \\ \frac{1}{7}\,row_3 \\ \end{array} \begin{bmatrix} 1 & 2 & 3 \\ 0 & 1 & 0 \\ 0 & 0 & 1 \end{bmatrix}$$

From the augmented matrix, one can see that the system is inconsistent, since the last equation says 0 = 1. As more equations would be added, this danger increases. The reason is that as m increases (more rows) the rank of A stays the same, so the inequality rank(A) < m becomes more lopsided, the overdetermined system is more and more likely to be inconsistent. But be careful, an overdetermined system can be consistent.

EXERCISES 1.3

A. Given the **coefficient** matrix A (not the augmented matrix!); find the rank and nullity of A. Determine the number of determined and free variables in the corresponding system with coefficient matrix A.

1. $A = \begin{bmatrix} 1 & 0 & 2 & 3 \\ 0 & 1 & 2 & 0 \\ 0 & 0 & 0 & 1 \end{bmatrix}$ ans. rank(A) = 3, nullity(A) = 1, 3 determined variables, one free

2. $A = \begin{bmatrix} 2 & 0 & 4 & 6 \\ 0 & -1 & -2 & 0 \\ 2 & -1 & 2 & 7 \end{bmatrix}$

3. $A = \begin{bmatrix} 0 & 1 & 4 & -3 \\ 1 & 1 & 2 & 4 \\ 0 & 2 & 6 & 1 \end{bmatrix}$ ans. rank(A) = 3, nullity(A) = 1, 3 determined variables, one free

4. $A = \begin{bmatrix} 1 & 0 & 2 & 3 \\ 1 & 1 & 2 & 3 \\ 2 & 3 & 4 & 5 \end{bmatrix}$

5. $A = \begin{bmatrix} 1 & 0 & 2 \\ 0 & 1 & 2 \\ 0 & 0 & 0 \\ 2 & 2 & 2 \end{bmatrix}$ ans. rank(A) = 3, nullity(A) = 0, 3 determined variables, no free

6. $A = \begin{bmatrix} 1 & 0 & 2 \\ 0 & 1 & 2 \\ 1 & 1 & 1 \\ 2 & 3 & 4 \end{bmatrix}$

7. $A = \begin{bmatrix} 1 & 0 & 2 \\ 0 & 2 & -4 \\ 0 & 0 & 0 \\ 1 & 2 & -2 \end{bmatrix}$ ans. rank(A) = 2, nullity(A) = 1, 2 determined variables, 1 free

8. $A = \begin{bmatrix} 1 & 2 & 3 \\ 4 & 5 & 6 \\ 7 & 8 & 9 \end{bmatrix}$

$9. A = \begin{bmatrix} 1 & 0 & 2 & 3 \\ 0 & 1 & 2 & 0 \\ 0 & 1 & 4 & 4 \end{bmatrix}$ ans. rank(A) = 3, nullity(A) = 1, 3 determined variables, 1 free

B. Solve each linear system using Gaussian elimination. If the system is inhomogeneous, write the solution X in the form X = X_h + X_p. Tell which variables are free and which are determined. State rank(A) and nullity(A), where A is the coefficient matrix A of the system. Describe *geometrically* what the solution of the original system is (ie. point, line, plane, hyperplane or none of these)

$x_1 + 2 x_2 + = 0$

10. $- x_1 + 3 x_2 + 4 x_3 = 6$

$- x_1 + 2 x_2 - x_3 = 7$

$x_1 + 2 x_2 + = 0$

11. $- x_1 + 3 x_2 + 4 x_3 = 0$ ans. {(0,0,0)}, rank(A) = 3, nullity(A) = 0, all variables determined, point

$- x_1 + 2 x_2 - x_3 = 0$

$x_1 + 2 x_2 + = 0$

12. $- x_1 + 2 x_2 + 4 x_3 = 5$

$- 4 x_2 - 4 x_3 = 5$

$x_1 + 2 x_2 + + x_4 = 4$

13. $- x_1 + 2 x_2 + 4 x_3 - x_4 = 4$

$- x_1 + 2 x_2 - x_3 + 9 x_4 = 14$

ans. {(3t,-2t,2t,t) + (-4,4,-2,0)}, rank(A) = 3, nullity(A) = 1, x_1,x_2,x_3 determined, x_4 free, line

$$x_1 + 2\,x_2 + \quad + x_4 = 0$$

14. $\ -x_1 + 2\,x_2 + 4\,x_3 - 3\,x_4 = 6$

$$-x_1 + 2\,x_2 - x_3 + x_4 = 7$$

$$x_1 + 2\,x_2 + \quad = 0$$

$$-x_1 + 2\,x_2 + 4\,x_3 = 8$$

15.
$$-x_1 + 2\,x_2 - x_3 = 13$$

$$-x_1 + 6\,x_2 + 3\,x_3 = 21$$

ans. $\{(-6,3,-1)\}$, rank(A) = 3, nullity(A) = 0, all variables determined, point

$$x_1 + x_2 + \quad + x_4 + x_5 = 0$$

$$-x_1 + 2\,x_2 + 4\,x_3 - 3\,x_4 - x_5 = 0$$

16.
$$3\,x_2 - x_3 + x_4 + 2\,x_5 = 0$$

$$-x_1 + 6\,x_2 + 3\,x_3 - x_4 \quad = 0$$

C: In 4-space, 4 different hyperplanes meet in a point, 3 different hyperplanes meet in a line, and 2 different hyperplanes meet in a plane. Show this is true by solving each system and describing the solution geometrically.

$$x_1 + 2\,x_2 + \quad + x_4 = 4$$

17. $\quad x_2 + 4\,x_3 - 3\,x_4 = 2$

$$x_1 + 3\,x_2 + 4\,x_3 + x_4 = 6$$

ans. $\{8t, 2-4t, t, 0)\}$, line

$$x_1 + 2\,x_2 + \quad + x_4 = 4$$

18.
$$-x_1 + 2\,x_2 + 4\,x_3 - 9\,x_4 = 8$$

33

$$x_1 + x_2 + \quad + x_4 = 0$$

$$-x_1 + 2\,x_2 + 4\,x_3 - 3\,x_4 = 0$$

19.

$$3\,x_2 - x_3 + x_4 = 0$$

$$-x_1 + 6\,x_2 + 3\,x_3 - x_4 = 0$$

ans. {(0,0,0,0)}, point

D: Given the linear system

$$x_1 \quad + x_3 = a$$

$$2\,x_1 + 2\,x_2 + 3\,x_3 = b$$

$$3\,x_1 + 2\,x_2 + 4\,x_3 = c$$

20. Write down the augmented matrix and put it in echelon form.
21. Write down the condition which guarantees a solution exists. Is it unique?
ans. c = a + b, NO (nullity(A) = 1)
22. What happens if $c \neq a + b$?
23. What happens if a = b = 0 ? ans. homogeneous system is consistent and dependent

E: Given the linear system

$$x_1 + 2\,x_2 = a$$

$$x_1 + x_2 = b$$

$$x_2 = c$$

24. Write down the augmented matrix and put it in echelon form.
25. What condition guarantees a solution? Is it unique? ans. c = a - b, YES
26. Can the system be inconsistent? If so, what condition guarantees this?
27. Can the system be consistent and dependent? ans. NO, nullity(A) = 0

34

Part F: Given the linear system

$$x_1 + x_2 + x_3 + x_4 = a$$

$$-2 x_1 + 2 x_2 + 4 x_3 - 3 x_4 = b$$

$$-x_1 + 3 x_2 + 5 x_3 - 2 x_4 = c$$

28. Write down the coefficient matrix A and gets its rank and nullity.
29. Determine for what B = (a,b,c) the system has a solution X. ans. B = (a,b,a+b)
30. If the system has a solution, can it be unique? Explain why/why not.
31. Find X if a = 1, b = 3, c = 4; show that Xh satisfies the corresponding homogeneous system. Describe Xh geometrically. ans. Xh = (1/2s - 5/4t,-3/2s + 1/4t,s,t)}, plane

G: Given the linear system

$$x_1 + 2 x_2 + 3 x_3 = 0$$

$$2 x_1 - x_2 + x_3 = 4$$

32. Add a third equation which will make the system consistent and dependent.
33. Add a third equation which will make the system inconsistent. ans. $3 x_1 + x_2 + 4 x_3 = 5$ (there are infinitely many other answers)
34. Add a third equation which will make the system have only one solution.

H. Given the system

$$x_1 - 2 x_2 - x_3 = 3$$

$$-x_1 + 3 x_2 + x_3 = 4$$

35. This system is *underdetermined*; show it is consistent and dependent.
36. Add a third equation to this system, which is neither a sum of the first two nor a multiple of equations one or two. Solve this system; does it have one solution or none or infinitely many???
37. Add a fourth equation chosen at random with no zero entries(so your system is now *overdetermined*); see if your system is inconsistent.

38. Show that the (overdetermined) system

$$x_1 - 2x_2 - x_3 = 3$$

$$-x_1 + 3x_2 + x_3 = 4$$

$$2x_1 + x_3 = a$$

$$x_1 + x_2 + 2x_3 = b$$

is inconsistent unless a = b + 10. Then show that if a = b + 10, the system has a unique solution. Find this solution in terms of a and b.

1.4 Chapter One Outline and Review

1.1 Gaussian elimination
A linear system is represented by its *augmented matrix;* if the system is of size m x n (m equations, n variables) then its augmented matrix is of size m x (n + 1).
Putting the augmented matrix in **row-echelon form** means having:
1) a "one" as the first nonzero entry in a nonzero row and only zeros below this (leading) one
2) row(s) of zeros at the bottom
3) leading ones move *right* as you look from top to bottom
The system is then solved by *back-substitution*- using the last nonzero row, solve for the lowest subscripted variable and then work your way up, row by row. Putting the augmented matrix in echelon form and then using back-substitution is called *Gaussian elimination.*
When done, write the answer as an n-tuple in set notation.
Reduced row-echelon form means the matrix is in echelon form and has only zeros ABOVE the leading ones (as well as below); it is unique for a particular augmented matrix.
A consistent system has at least one solution; a consistent and dependent system has infinitely many. An inconsistent system has no solution.

1.2 Geometry of solutions

Homogeneous system- has all constants ZERO. A homogeneous system cannot be inconsistent, since (0,0,...,0) always works.
Inhomogeneous- has at least one nonzero constant. The solution to such a system can always be written in the form $X = X_h + X_p$ where

1) X_h - solution to corresponding homogeneous system

2) X_p - a particular solution to the given system

Geometry (for a system with m equations in n variables)- the solution of a linear system is a vector(n-tuple) in n-space.
Geometrically it may represent a 1)point(no parameter), 2)line(1 parameter), 3)plane(2 parameters) or 4) hyperplane (n - 1 parameters).
THEOREM An m by n homogeneous system has infinitely many solutions if m < n.

1.3 System analysis

Given the *augmented* matrix in echelon form for a linear system:
1) any column (except the last) with a **leading one** is called a pivot column
it represents a **determined** variable
2) any column (except the last) with no leading one is a *non-pivot column*
it represents a *free* variable
Given the **coefficient** matrix A of size m x n (for an m by n inhomogeneous system) in *echelon* form;
1) rank(A) = number of pivot columns of A (leading ones)
2) rank(A) + nullity(A) = n (number of COLUMNS of A) so
3) nullity(A) = n - rank(A) (number of non-pivot columns) tells number of *free variables* (ie.

parameters) in the solution of the corresponding homogeneous system X_h;

if n = rank(A) then nullity(A) = 0 and the only solution of corresponding homogeneous system is zero vector

A system is either inconsistent or has a unique solution if nullity(A) = 0.

rank(A) satisfies $rank\ (\ A\ ) \le \min\ (\ m, n\ )$.

Review Problems

Part A: Given the augmented matrix $\begin{bmatrix} 1 & 0 & 0 & 2 & 4 \\ 0 & 1 & 0 & 3 & -5 \\ 0 & 0 & 1 & 0 & 6 \\ 0 & 0 & 0 & 0 & 0 \end{bmatrix}$ of a certain linear system:

1. What size is the system? ans. 4 by 4
2. What form is the matrix in? ans. reduced echelon
3. Find the general solution. ans.{(4-2t,-5-3t,6,t)}.
4. Write the solution in the form $X_h + X_p$. ans. {(-2t,-3t,0,t) + (4,-5,6,0)}
5. What geometric object does the solution represent? ans. line
5. Identify the pivot columns and determined variables. ans. 1,2,& 3; x_1, x_2, x_3
6. Identify the non-pivot columns and the free variables. ans.4; x_4
7. Find the rank and nullity of the corresponding coefficient matrix A. ans. 3, 1

Part B: Solve each system using Gaussian elimination. Describe the solution (if possible) as
a) point b) line c) plane c) hyperplane

$x_1 + 2\ x_2 + \quad + x_4 = 5$

8. $-x_1 + 2\ x_2 + 4\ x_3 - 3\ x_4 = 5$ ans. {(2t,-t,t,0)+(10,0,0,-5): t real},line

$x_1 + 3\ x_2 + x_3 + 2\ x_4 = 0$

$x_1 + 2\ x_2 + \quad + x_4 = 4$

9. $\qquad\qquad\qquad$ ans. {(-4s-2t,-2s+1/2t,s,t)+(-6,5,0,0): s,t real},plane

$-2\ x_1 - 2\ x_2 + 4\ x_3 - 3\ x_4 = 2$

$$x_1 + 2\,x_2 + \quad + x_4 = 4$$

$$-x_1 + 2\,x_2 + 4\,x_3 - 3\,x_4 = 2$$

10.   ans. {(1,1,1,1)},point

$$-x_1 + 2\,x_2 - x_3 + 2\,x_4 = 2$$

$$-x_1 + 6\,x_2 + 3\,x_3 + x_4 = 9$$

Part C: Given $\begin{bmatrix} 1 & 0 & 1 & 2 & 3 \\ 0 & 1 & 2 & -2 & 2 \\ -1 & -1 & 0 & 0 & 1 \\ 0 & 0 & 3 & 0 & 6 \end{bmatrix}$ is the **augmented** matrix of a certain system, where A is the

coefficient matrix:

11. Put the augmented matrix in echelon form. ans. $\begin{bmatrix} 1 & 0 & 1 & 2 & 3 \\ 0 & 1 & 2 & -2 & 2 \\ 0 & 0 & 1 & 0 & 2 \\ 0 & 0 & 0 & 0 & 0 \end{bmatrix}$

12. Put the augmented matrix in reduced echelon form. ans. $\begin{bmatrix} 1 & 0 & 0 & 2 & 1 \\ 0 & 1 & 0 & -2 & -2 \\ 0 & 0 & 1 & 0 & 2 \\ 0 & 0 & 0 & 0 & 0 \end{bmatrix}$

13. Identify the pivot columns, find rank(A) and list the determined variables.
ans. columns 1, 2 & 3, rank(A) = 3, x_1, x_2, & x_3
14. Identify the free variables and find nullity(A). ans. x_4, nullity(A) = 1
15. Use back-substitution to solve the system; write the answer in the form $X = X_h + X_p$.
ans. $X = (-2t, 2t, 0, t) + (1, -2, 2, 0)$
16. Check your answer by substitution in the original system represented by the given augmented matrix.
17. Geometrically what does the solution represent in 4-space? ans. a line

Part D: Given $\begin{bmatrix} 2 & 3 & -1 & 0 & 4 & 5 \\ 0 & 1 & 2 & -2 & 2 & 6 \\ 2 & 4 & 1 & -2 & 6 & 11 \end{bmatrix}$ is the **augmented** matrix of a certain system with

coefficient matrix A;

18. Put the augmented matrix in echelon form. ans.
$$\begin{bmatrix} 1 & \dfrac{3}{2} & -\dfrac{1}{2} & 0 & 2 & \dfrac{5}{2} \\ 0 & 1 & 2 & -2 & 2 & 6 \\ 0 & 0 & 0 & 0 & 0 & 0 \end{bmatrix}$$

19. Put the augmented matrix in reduced echelon form. ans.
$$\begin{bmatrix} 1 & 0 & -\dfrac{7}{2} & 3 & -1 & -\dfrac{13}{2} \\ 0 & 1 & 2 & -2 & 2 & 6 \\ 0 & 0 & 0 & 0 & 0 & 0 \end{bmatrix}$$

20. Identify the pivot columns, find rank(A) and list the determined variables. ans. columns 1, 2 , rank(A) = 2, x_1 & x_2

21. Identify the free variables and find nullity(A). ans. x_3, x_4 & x_5, nullity(A) = 3

22. Use back-substitution to solve the system; write the answer in the form X = X_h + X_p. ans.{(7/2r-3s+t,-2r+2s-2t,r,s,t)+(-13/2,6,0,0,0), r,s,t real}

23. Verify your answer by substitution in the original system represented by the augmented matrix.

Part E: Given the augmented matrix $\begin{bmatrix} 1 & 0 & a \\ 0 & -2 & b \\ 2 & 1 & c \end{bmatrix}$;

24. Put the matrix in echelon form.

25. For what a,b, c is the system consistent? Is the solution unique? ans. c = 2a-b/2,YES

26. Solve the system if a = b = c = 0. ans. {(0,0)}

27. Solve the system if a = 5, b = 6, c = 7. ans. {(5,-3)}

Part F: Solve each system; if it is inhomogeneous, write the answer in the form
X = X_h + X_p. Check your answer by using a particular solution if the system is consistent and dependent. Describe your solution set geometrically.

28.
$$x_1 + 2\,x_2 + \quad + x_4 = 0$$
$$-x_1 + 2\,x_2 + 4\,x_3 - 9\,x_4 = 0$$
ans. {(2s-5t,-s+2t,s,t): s, t real}, plane

$x_1 + 2\,x_2 + \quad + x_4 = 0$

$-2\,x_1 + 2\,x_2 + 4\,x_3 - 3\,x_4 = 0$

29.   ans. $\{(0,0,0,0)\}$, point

$-x_1 + 2\,x_2 - x_3 + x_4 = 0$

$-2\,x_1 \qquad - x_4 = 0$

$x_1 + 2\,x_2 + \quad + x_4 = 4$

30. $-x_1 + 2\,x_2 + 4\,x_3 - 9\,x_4 = 4$ ans. $\{(-2t,-t,t,0) + (0,2,0,0)\}$, line

$4\,x_2 + 4\,x_3 - 4\,x_4 = 8$

Part G: Given the augmented matrix $\begin{bmatrix} 1 & 0 & 2 & a \\ 0 & 1 & -1 & b \\ 1 & 2 & 1 & c \\ 2 & 3 & 2 & d \end{bmatrix}$;

31. What size system is represented by this augmented matrix? What is being described geometrically by this system of equations? ans. 4 by 3, 4 planes in 3-space

32. Put the matrix in row echelon form. ans. $\begin{bmatrix} 1 & 0 & 2 & a \\ 0 & 1 & -1 & b \\ 0 & 0 & 1 & c-a-2b \\ 0 & 0 & 0 & d-a-b-c \end{bmatrix}$

33. What is the rank and nullity of the corresponding coefficient matrix? ans. 3, 0
34. For what a,b,c & d does the system have a solution? Is it unique? ans. d = a+b+c, YES
35. If a = b = c = d = 0, solve the system. ans. $\{(0,0,0)\}$
36. If a = b = c = 1 & d = 3, solve the system.   ans.$\{(5,-1,-2)\}$
37. Show the system is inconsistent if a = b = c = 1 & d = 5.

Part H: Telephone numbers - Suppose my phone number satisfies the following criteria;
1) the sum of the first 2 digits is 10, 2) the sum of the second and third digits is also 10,
3) the sum of the first 3 digits is 14, 4) the sum of the fourth and fifth digits is 9, and 5) the sum of the last 2 digits is also 9
38. Write down the augmented matrix that represents the corresponding system.

41

ans. $\begin{bmatrix} 1 & 1 & 0 & 0 & 0 & 0 & 0 & 10 \\ 0 & 1 & 1 & 0 & 0 & 0 & 0 & 10 \\ 1 & 1 & 1 & 0 & 0 & 0 & 0 & 14 \\ 0 & 0 & 0 & 1 & 1 & 0 & 0 & 9 \\ 0 & 0 & 0 & 0 & 0 & 1 & 1 & 9 \end{bmatrix}$

39. Put the augmented matrix in echelon form.

40. Find the rank and nullity of the coefficient matrix. ans. 5, 2

41. Use back-substitution to solve the system; write down 2 possible phone numbers which satisfy the criteria in the problem.

ans. $\{(4,6,4,9\text{-}s,s,9\text{-}t,t)\}$, 464-9090, 464-4518

42. What does the solution represent geometrically? ans. plane in 7-space

CHAPTER 2

MATRIX ALGEBRA

CHAPTER OVERVIEW

2.1 Matrix operations

2.2 Inverses and square systems

2.3 Elementary Matrices

2.4 Determinants

2.5 Special matrices

2.6 Chapter outline and review

In this chapter we learn the basic rules of **matrix algebra**- how do matrices differ when compared with real numbers under operations like addition and multiplication? We learn basically that the difference between real algebra and matrix algebra is that we need to be concerned about the **size** of the two matrices we are dealing with; otherwise the main difference is that matrix multiplication is, in general, *not commutative*. We proceed to learn how to find the **inverse** of a square matrix and to see how this is useful if we wish to solve a square system of equations.

Then we get introduced to a very special simple type of square matrix which is called an **elementary** matrix. These types of matrices can be used to keep track of *elementary row operations*, which were used in the previous chapter. The most important result about elementary matrices is an analog of the *fundamental theorem of arithmetic*- a non-singular square matrix can be "factored" as a product of elementary matrices.

Next we look at **determinants**; mainly we wish to observe that a non-zero determinant is the hallmark of a non-singular matrix. Then we learn the rules of determinants, the most important of which is that the determinant of a product is the product of the determinants. Also the determinant of a *triangular* matrix is terribly easy to calculate- so we learn how to use *row operations* to put a matrix in triangular form, so that we can avoid the traditional tedious calculations involved in evaluating determinants.

Finally we look at the algebraic structure that different types of matrices may form. We define the concept of "**group**"- probably the most important (and simplest) of algebraic structures. We try to determine what types of matrices form groups under either addition or multiplication.

2.1 Matrix Operations

In the previous chapter, we were primarily concerned with using matrices to represent systems via the augemented matrix. In this chapter we wish to study matrices per se to see how they behave with respect to basic algebraic operations, such as addition and multiplication. Everyone is familiar with the algebra of real numbers. How much of this carries over into the world of matrices? Well, let's get started.

Matrix Basics

Recall that a matrix[1] is a rectangular array of (real) numbers. We indicate the elements of a matrix by a double subscript notation:

$$A = \begin{bmatrix} a_{11} & a_{12} & \cdots & a_{1n} \\ a_{21} & a_{22} & \cdots & a_{2n} \\ \cdots & \cdots & \cdots & \cdots \\ a_{m1} & a_{m2} & \cdots & a_{mn} \end{bmatrix}$$

The first subscript refers to the row the element is in and the second refers to the column in which it is located; for eg. a_{34} refers to the element in the third row and fourth column. A has m rows and n columns so it is an m by n matrix; we may write dim(A) = m by n or dim(A) = m x n.

Two matrices A and B are **equal** iff they have the same dimensions and corresponding entries are equal; this means dim(A) = dim(B) and

$$[\ a_{ij}\ ] = [\ b_{ij}\ ] \text{ iff } a_{ij} = b_{ij} \text{ for all } i, j$$

EXAMPLE 1 Given the two matrices

$$A = \begin{bmatrix} 1 & 2 \\ 3 & 4 \end{bmatrix} \quad B = \begin{bmatrix} a & a+b \\ c-d & d \end{bmatrix}$$

determine what must be true for them to be equal.

Without writing down the augmented matrix of the implied system of equations, it is rather easy to observe that $a = 1$, $b = 1$, $d = 4$ and $c = d + 3 = 7$ in order that the matrix B is equal to the matrix A.

Matrix Operations

Addition of matrices is very simple- if the two matrices are of the same size (equal dimensions) then we can add them by simply adding corresponding entries:

$$A = [\ a_{ij}\ ],\ B = [\ b_{ij}\ ] \Rightarrow A + B = [\ a_{ij} + b_{ij}\ ]$$

J. Sylvester is credited with coining the term "matrix" in 1850.

EXAMPLE 2 Find the sum:

$$\begin{bmatrix} 1 & 3 \\ -1 & 4 \end{bmatrix} + \begin{bmatrix} -1 & 5 \\ 6 & 7 \end{bmatrix} = \begin{bmatrix} 0 & 8 \\ 5 & 11 \end{bmatrix}$$

As you can see, matrix addition is terribly trivial and since we are just adding corresponding entries, matrix addition has all the properties of real addition.

Properties of matrix addition:
1) A + B = B + A (commutative property)
2) A + (B + C) = (A + B) + C (associative property)

Matrix notation is especially convenient when we think of *column matrices* as **vectors** and wish to do vector addition; for example, we can use 3 by 1 column matrices to represent X = (1,2,3) and Y = (5,-6,7) and then add the corresponding matrices:

$$X + Y = \begin{bmatrix} 1 \\ 2 \\ 3 \end{bmatrix} + \begin{bmatrix} 5 \\ -6 \\ 7 \end{bmatrix} = \begin{bmatrix} 6 \\ -4 \\ 10 \end{bmatrix}$$

We are doing the same thing with the entries in X and Y, whether we write them horizontally (vector notation) or vertically (as column matrices).

The analog of the number zero for real addition is the **zero matrix**; it is simply a matrix of all zeros. It will be denoted in this text by "theta":

$$\theta = \begin{bmatrix} 0 & 0 & ... & 0 \\ 0 & 0 & ... & 0 \\ ... & ... & ... & ... \\ 0 & 0 & ... & 0 \end{bmatrix}$$

For all matrices A, $A + \theta = A$ where the size of θ matches the size of A. Once we have the zero matrix, we also can also find the *additive inverse* of any matrix A:

$$A = [\, a_{ij}\,] \Rightarrow -A = [\, -a_{ij}\,]$$

Of course we then have that

$$A + (-A) = \theta$$

Matrix multiplication is probably the most useful of all the matrix operations. Here, we have to be careful about the dimensions of the two matrices A & B which we are multiplying:

$$A \quad B \ = \ C$$
$$[\, m \, x \, p\,] \ [\, p \, x \, r\,] = [\, m \, x \, r\,]$$

The golden rule of matrix multiplication is:
multiply row by column

45

That is why the number of columns of A must match the number of rows of B; if not, the matrices are *incompatible* for multiplication (and no matrix C exists).

EXAMPLE 3 Find the product:

$$AB = \begin{bmatrix} 1 & 2 & -4 \\ -1 & 2 & 8 \end{bmatrix} \begin{bmatrix} 1 & 2 \\ -3 & 4 \\ 0 & 5 \end{bmatrix} = \begin{bmatrix} 1\cdot1 + 2\cdot(-3) + (-4)\cdot0 & 1\cdot2 + 2\cdot4 + (-4)\cdot5 \\ (-1)1 + 2\cdot(-3) + 8\cdot0 & (-1)\cdot2 + 2\cdot4 + 8\cdot5 \end{bmatrix} = \begin{bmatrix} -5 & -10 \\ -7 & 46 \end{bmatrix} = C$$

Note that A is a **2** x *3* and B is a *3* x **2** so the product is a **2** x **2**. To get the elements in the first row of the product, we take row one of A and multiply its entries by the corresponding entries in each of the columns of B; for example

$$c_{11} = a_{11}b_{11} + a_{12}b_{21} + a_{13}b_{31}$$

and

$$c_{12} = a_{11}b_{12} + a_{12}b_{22} + a_{13}b_{32}$$

We may observe that reversing the order of the product may give a product which is of a *different size* or even *undefined*. Here we would have

$$\begin{bmatrix} 1 & 2 \\ -3 & 4 \\ 0 & 5 \end{bmatrix} \begin{bmatrix} 1 & 2 & -4 \\ -1 & 2 & 8 \end{bmatrix} = \begin{bmatrix} -1 & 6 & 12 \\ -7 & 2 & 44 \\ -5 & 10 & 40 \end{bmatrix}$$

$$(3 \times 2)(2 \times 3) = 3 \times 3$$

so the reverse order in **EX3** gives a product which is not even the *same size* as the original product. It is always important to remember that matrix multiplication in general is <u>NOT commutative</u>- in fact this is why matrix multiplication can be used to represent operations which in general do not commute. Square matrices of the same size are easy to deal with, since if A and B are of size n by n, then the product is of size n by n.

EXAMPLE 4 Given $A = \begin{bmatrix} 1 & 1 \\ 2 & 2 \end{bmatrix}$; show there are infinitely many 2 x 2 matrices B such that $AB = \theta$. Let $B = \begin{bmatrix} a & b \\ c & d \end{bmatrix}$; then we wish to solve the matrix equation

$$\begin{bmatrix} 1 & 1 \\ 2 & 2 \end{bmatrix} \begin{bmatrix} a & b \\ c & d \end{bmatrix} = \begin{bmatrix} 0 & 0 \\ 0 & 0 \end{bmatrix}$$

If we multiply out on the left, we have:

$$\begin{bmatrix} a+c & b+d \\ 2a+2c & 2b+2d \end{bmatrix} = \begin{bmatrix} 0 & 0 \\ 0 & 0 \end{bmatrix}$$

We could write out the augmented matrix (a 4 by 5) but it is easy to see that we need to solve

$$a + \ c \ = 0$$

$$b + \ d = 0$$

which means that a = - c and b = -d so the desired matrix B is of the form $B = \begin{bmatrix} -c & -d \\ c & d \end{bmatrix}$. Note that

even though neither A nor B is the zero matrix, the product AB equals the zero matrix!

The **identity matrix**: We need the matrix analog of the number "1" under ordinary multiplication, that is, we wish to find a matrix (call it I) such that AI = IA = A. This matrix has one's on the main diagonal and zeros elsewhere; it is the identity element under matrix multiplication. Here we need to restrict A to be an n x n matrix – otherwise, if A is not square, we could not reverse the order of the product. For example, if A is a 2 x 2 matrix, the identity matrix is:

$$I_2 = \begin{bmatrix} 1 & 0 \\ 0 & 1 \end{bmatrix}$$

We can use the subscript on I to indicate its size, if we wish; above we used I_2 to stand for the identity matrix of size 2 x 2.

Properties of matrix multiplication:

1) A(BC) = (AB)C associative
2) A(B + C) = AB + AC left distributive
3) (A + B)C = AC + BC right distributive
4)

$$A I_n = I_n A = A$$

for any square matrix A (property of the multiplicative identity)
It would be instructive for the student to give the dimensions of the factors involved in order for the multiplication to make sense; for example, for A(B + C) = AB + AC, B and C must be of the same size (say p x r) and A must then be an m x p matrix.

Powers of a matrix can be defined for a positive integer n if A is square:

$$A^n = A \ A \dots A(\ n \ factors \ ) \quad n \in Z^+$$

For example, we have

$$A = \begin{bmatrix} 1 & 2 \\ 3 & 4 \end{bmatrix} \Rightarrow A^3 = \begin{bmatrix} 37 & 54 \\ 81 & 118 \end{bmatrix}$$

Since a matrix commutes with itself, it doesn't matter if we calculate A^3 as $A^2 A$ or $A A^2$.
One good use for matrix multiplication is to represent a *linear system*, especially a square system.

EXAMPLE 5 Write the following linear system in a matrix multiplication format:

$$x_1 + 2 x_2 + 3 x_3 = 4$$

$$x_2 + 2 x_3 = 3$$

$$\begin{bmatrix} 1 & 2 & 3 \\ 0 & 1 & 2 \end{bmatrix} \begin{bmatrix} x_1 \\ x_2 \\ x_3 \end{bmatrix} = \begin{bmatrix} 4 \\ 3 \end{bmatrix}$$

Here we can use the letter A for the *coefficient* matrix, X for the column matrix of unknowns and B for the (column) matrix of constants and get the simplistic form **AX = B**. Any linear system can thus written in the form AX = B.
It is always a good idea to check out the *dimensions* of any matrix equation, especially ones involving multiplication. Here A is 2 x 3, X is 3 x 1 and the product must have dimensions 2 x 1. Using Gaussian elimination, one quickly obtains the solution:

$$X = \begin{bmatrix} t-2 \\ -2t+3 \\ t \end{bmatrix} = \begin{bmatrix} t \\ -2t \\ t \end{bmatrix} + \begin{bmatrix} -2 \\ 3 \\ 0 \end{bmatrix} = X_h + X_p$$

We can check the solution using matrix multiplication:

$$A X = \begin{bmatrix} 1 & 2 & 3 \\ 0 & 1 & 2 \end{bmatrix} \begin{bmatrix} t-2 \\ -2t+3 \\ t \end{bmatrix} = \begin{bmatrix} (t-2)+2(-2t+3)+3t \\ (-2t+3)+2t \end{bmatrix} = \begin{bmatrix} 4+t-4t+3t \\ 3-2t+2t \end{bmatrix} = \begin{bmatrix} 4 \\ 3 \end{bmatrix} = B$$

We could also have checked the solution using

$$A X = A(X_h + X_p) = A X_h + A X_p = \theta + B = B$$

since X_h is the solution of the corresponding homogeneous system so $A X_h = \theta$.

Other Matrix Maneuvers

There are a few other operations on matrices that we need to define for completeness sake. Both are very simple; the first is called **scalar multiplication**.

By a scalar in this text we always mean a real number. The very name suggests that we are thinking of a matrix as a vector- then we can conjecture that scalar multiplication simply means multiplying all entries in the matrix by the scalar.

$$k A = [\ k \, a_{ij} \ ]$$

To multiply an m x n matrix A by the scalar k, just *multiply each entry* by k. The *size* of the matrix does not change.

EXAMPLE 6 Find kA if k = 3 and $A = \begin{bmatrix} 1 & 2 \\ 4 & 6 \end{bmatrix}$.

$$k A = 3 \begin{bmatrix} 1 & 2 \\ 4 & 6 \end{bmatrix} = \begin{bmatrix} 3 & 6 \\ 12 & 18 \end{bmatrix}$$

This is really too easy! We should mention a few properties of this operation. These are easily proved.

Properties of scalar multiplication

1) k(A + B) = kA + kB
2) (k + j)A = kA + jA
3) k(AB) = (kA)B = A(kB)

The second matrix operation is called **transposition**. It is a way of taking an m x n matrix and forming an n x m matrix, using the same entries. All you need do is change every row to a column. The notation A^t is read "A transpose".

$$A = [\ a_{ij} \ ] \Rightarrow A^t = [\ a_{ji} \ ]$$

EXAMPLE 7 Get the transpose of A if $A = \begin{bmatrix} 1 & -1 & 0 \\ 2 & 0 & -3 \end{bmatrix}$.

$$A^t = \begin{bmatrix} 1 & 2 \\ -1 & 0 \\ 0 & -3 \end{bmatrix}$$

Note that row 1 becomes *column 1* and row 2 becomes column 2. The original **2** x 3 matrix becomes a *3* x **2** matrix. Below are a few properties of transposition.

Properties of transposition

1) $(\ A^t \ )^t = A$
2) $(\ k A \ )^t = k \ A^t$
3) $(\ A + B \ )^t = A^t + B^t$
4) $(\ A B \ )^t = B^t A^t$

The third operation is called the **trace** of a matrix, denoted tr(A). The definition is very simple; if A is a square matrix, then the trace of A is the *sum of the entries on the main diagonal.*
EXAMPLE 8 Find tr(A) if

$$A = \begin{bmatrix} 1 & 2 \\ 3 & 4 \end{bmatrix}$$

Very simply tr(A) = 1 + 4 = 5.

The following properties of calculating the trace are easily verified:
1) tr(A + B) = tr(A) + tr(B)
2) tr(kA) = ktr(A) for any scalar k
3) tr(AB) = tr(BA) for any two square matrices A and B of the same size

Application: Incidence matrices
Consider the flight schedule of Allegheny Airlines.

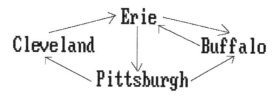

Allegheny Airlines

Let's try to find a matrix that will represent this flight schedule; if a flight from city i to city j is possible, fill in the entry in the i-th row and j-th column with a "one"- if this flight is NOT possible, fill in the entry with a "zero". Use the order Erie-Pitt-Cleveland-Buffalo. This will create an incidence matrix of size 4 by 4:

$$A = \begin{bmatrix} 0 & 1 & 0 & 1 \\ 0 & 0 & 1 & 1 \\ 1 & 0 & 0 & 0 \\ 1 & 0 & 0 & 0 \end{bmatrix}$$

What does A^2 represent?

$$A^2 = \begin{bmatrix} 0 & 1 & 0 & 1 \\ 0 & 0 & 1 & 1 \\ 1 & 0 & 0 & 0 \\ 1 & 0 & 0 & 0 \end{bmatrix} \begin{bmatrix} 0 & 1 & 0 & 1 \\ 0 & 0 & 1 & 1 \\ 1 & 0 & 0 & 0 \\ 1 & 0 & 0 & 0 \end{bmatrix} = \begin{bmatrix} 1 & 0 & 1 & 1 \\ 2 & 0 & 0 & 0 \\ 0 & 1 & 0 & 1 \\ 0 & 1 & 0 & 1 \end{bmatrix}$$

It should tell us from what city to what city we can fly *in exactly two* flights.

For example, the first row tells us that we can fly from Erie to Erie, Cleveland or Buffalo (city 4) in only one way. The second row tells us that we can fly from Pittsburgh only to Erie in two different ways: either Pittsburgh-Cleveland-Erie or from Pittsburgh-Buffalo-Erie. We cannot fly to any other city in exactly two flights from Pittsburgh. The third row says that we can fly from Cleveland to Pittsburgh or from Cleveland to Buffalo in exactly two flights. The fourth row says that we can fly from Buffalo to Pittsburgh or from Buffalo to Buffalo in exactly two flights.

One can show that $A + A^2$ would allow you to interpret the entries as being able to fly from city i to city j in two (or less) flights.

Application: Systems of differential equations (or how to get matrices to do calculus)

Suppose we have a linear system of differential equations like

$$\frac{dx}{dt} = x + 2y$$

$$\frac{dy}{dt} = 2x + y$$

We can write this system using matrix multiplication by defining the derivative of a matrix:

$$X = \begin{bmatrix} x \\ y \end{bmatrix} \quad \Rightarrow \quad \frac{d}{dt}\begin{bmatrix} x \\ y \end{bmatrix} = \begin{bmatrix} \dfrac{dx}{dt} \\ \dfrac{dy}{dt} \end{bmatrix}$$

Then we can write the system of differential equations as

$$\frac{dX}{dt} = AX \ where \quad \frac{dX}{dt} = \begin{bmatrix} \dfrac{dx}{dt} \\ \dfrac{dy}{dt} \end{bmatrix} = \begin{bmatrix} 1 & 2 \\ 2 & 1 \end{bmatrix}\begin{bmatrix} x \\ y \end{bmatrix} = AX \quad if \quad A = \begin{bmatrix} 1 & 2 \\ 2 & 1 \end{bmatrix}$$

Here A is the **coefficient matrix** and X is a **column matrix** (here we let X represent a vector in 2-space). Check out the compatibility of the multiplication on the right; X is the sought after solution to the system. It must be a 2 x 1 column matrix such that when substituted back into the original system, an identity is produced (of course the components must be differentiable functions of t).

Actually solving these systems in the general case can only be done after you learn the material in Chapter 6. The reader should be able to verify that the solution of the system of differential equations is

$$X = \begin{bmatrix} x \\ y \end{bmatrix} = \begin{bmatrix} e^{-t} & e^{3t} \\ -e^{-t} & e^{3t} \end{bmatrix} \begin{bmatrix} c_1 \\ c_2 \end{bmatrix} = \begin{bmatrix} c_1 e^{-t} + c_2 e^{3t} \\ -c_1 e^{-t} + c_2 e^{3t} \end{bmatrix}$$

by showing that dx/dt = x + 2y and dy/dt = 2x + y; for example,

$$\frac{d x}{d t} = -c_1 e^{-t} + 3 c_2 e^t = 1(c_1 e^{-t} + c_2 e^{3t}) + 2(-c_1 e^{-t} + c_2 e^{3t})$$

Application: Matrix polynomials.

One can turn a polynomial function into a matrix function by thinking of the coefficients as scalars and the powers of x as powers of the square matrix A by replacing x by A. For example, the polynomial

$$p(x)= x^2 - 4 x + 3$$

can be thought of as

$$p(A)= A^2 - 4 A + 3 I$$

where A is a square matrix and I is the identity matrix of the same size as A.

For example, if $A = \begin{bmatrix} 1 & 2 \\ 3 & 4 \end{bmatrix}$, then p(A) is found thus:

$$p(A)= A^2 - 4 A + 3 I = \begin{bmatrix} 1 & 2 \\ 0 & 3 \end{bmatrix}^2 - 4 \begin{bmatrix} 1 & 2 \\ 0 & 3 \end{bmatrix} + 3 \begin{bmatrix} 1 & 0 \\ 0 & 1 \end{bmatrix} = \begin{bmatrix} 1 & 8 \\ 0 & 9 \end{bmatrix} + \begin{bmatrix} -4 & -8 \\ 0 & -12 \end{bmatrix} + \begin{bmatrix} 3 & 0 \\ 0 & 3 \end{bmatrix} = \begin{bmatrix} 0 & 0 \\ 0 & 0 \end{bmatrix}$$

Amazingly we have found a solution of the matrix equation $p(A)= \theta$.

EXERCISES 2.1

Part A. Given $A = \begin{bmatrix} 1 & 0 & 2 \\ 0 & 1 & 2 \end{bmatrix}$ $B = \begin{bmatrix} -2 & 4 & 12 \\ -10 & 7 & -8 \end{bmatrix}$ and $C = \begin{bmatrix} 1 & -1 \\ 0 & 1 \\ 4 & 6 \end{bmatrix}$; find (if possible)

1. A + B ans. $\begin{bmatrix} -1 & 4 & 14 \\ -10 & 8 & -6 \end{bmatrix}$

2. 2A + 3B

3. C(A + B) ans. $\begin{bmatrix} 9 & -4 & 20 \\ -10 & 8 & -6 \\ -64 & 64 & 20 \end{bmatrix}$

4. (A + B)C

5. $A B^t$ ans. $\begin{bmatrix} 22 & -26 \\ 28 & -9 \end{bmatrix}$

6. $B^t A$

7. $A^t + B^t$ ans. $\begin{bmatrix} -1 & -10 \\ 4 & 8 \\ 14 & -6 \end{bmatrix}$

8. $(A + B)^t$

9. $(A C)^t$ ans. $\begin{bmatrix} 9 & 8 \\ 11 & 13 \end{bmatrix}$

10. $C^t C$

11. $(A A^t)^t$ ans. $\begin{bmatrix} 5 & 4 \\ 4 & 5 \end{bmatrix}$

Part B. Given the system $\quad x_1 + 2 x_2 = 5$;
$\qquad\qquad\qquad\qquad\qquad 3 x_2 + 4 x_2 = 6$

12. Write the system in the form $AX = B$

13. Solve the system using Gaussian elimination and show using matrix multiplication that your solution X is correct. ans.(-4,9/2)

14. Find the inverse of A by brute force(ie. write down a 2 by 2 matrix C with entries $x_1, ..., x_4$, then set AC = I (2 x 2) and solve the corresponding 4 by 4 system for $x_1, ..., x_4$.

15. Use the inverse of A from # 14 to solve the system again, this time by a simple matrix multiplication.

Part C. Matrix algebra:

16. Show that 2 x 2 scalar matrices commute with any 2 x 2 matrix. A scalar matrix is of the form $S = \begin{bmatrix} a & 0 \\ 0 & a \end{bmatrix}$.

17. Show that 2 x 2 upper triangular matrices *do not commute* under multiplication by exhibiting a pair of such matrices for which commutivity fails. An upper triangular 2 x 2 is of the form $T = \begin{bmatrix} a & b \\ 0 & c \end{bmatrix}$.

18. Expand (A + 3I)(A + 4I) using valid rules of matrix arithmetic (assume all A and I are 2 x 2 matrices). Conclude that such a matrix product behaves like polynomial multiplication (ie. the "FOIL" rule of high school).

19. Expand (A + 4I)(B - 6I) using valid matrix algebra (A,B, I square of same size).

53

ans. AB + 4B - 6A - 24I

20. Given the matrix $A = \begin{bmatrix} 1 & 2 \\ 2 & 4 \end{bmatrix}$; try to find an arbitrary 2 x 2 matrix B such that AB = I. Show that no such matrix exists. The matrix A is called *singular*. Hint: you will have a 4 x 4 linear system to solve; let $B = \begin{bmatrix} a & b \\ c & d \end{bmatrix}$ and set AB = I.

21. Repeat #20 with $A = \begin{bmatrix} 1 & 2 \\ 3 & 4 \end{bmatrix}$; show that $B = \dfrac{-1}{2} \begin{bmatrix} 4 & -2 \\ -3 & 1 \end{bmatrix}$.

22. Show that the matrix $A = \begin{bmatrix} 1 & 2 \\ 2 & 1 \end{bmatrix}$ satisfies the equation

$$p(A) = A^2 - 2A - 3I = \begin{bmatrix} 0 & 0 \\ 0 & 0 \end{bmatrix}$$

23. Find all 2 by 2 matrices $A = \begin{bmatrix} a & b \\ c & d \end{bmatrix}$ that $A^t = A$. These matrices are called *symmetric* matrices. ans. $A = \begin{bmatrix} a & b \\ b & c \end{bmatrix}$

24. Find conditions on a, b, c and d such that $A \begin{bmatrix} 1 & 1 \\ 1 & 1 \end{bmatrix} = \begin{bmatrix} 0 & 0 \\ 0 & 0 \end{bmatrix}$ if $A = \begin{bmatrix} a & b \\ c & d \end{bmatrix}$. A does NOT have to be the zero matrix; this means that under matrix multiplication there exists matrices that satisfy AB = zero matrix even though neither A nor B is the zero matrix. Such matrices are called in algebra *divisors of zero*. In the arithmetic of real numbers, if xy = 0 then either x = 0 or y = 0. You use this result often in real arithmetic; it is not true in general for matrices. ans. a = -b, c = -d

25. Repeat #24 by trying to find A such that $A \begin{bmatrix} 3 & -2 \\ 4 & -3 \end{bmatrix} = \begin{bmatrix} 0 & 0 \\ 0 & 0 \end{bmatrix}$; here the only way this can happen is if a = b = c = d = 0.

26. Given the matrix $A = \begin{bmatrix} -2 & -1 \\ 3 & 1 \end{bmatrix}$; show that $A^3 = I_2$. This means that A is *of order 3*.

27. Show that if $A' = P^{-1} A P$ (ie. A' is *similar* to A in # 26) also is of order 3.

28. Given $\exp(A) = I + A + \dfrac{1}{2!} A^2 + \dfrac{1}{3!} A^3 + \dots$ is the matrix exponential of A; find a quadratic (degree two) approximation for exp(A) if $A = \begin{bmatrix} 1 & 0 \\ 0 & 3 \end{bmatrix}$.

Part D. The matrix $A = \begin{bmatrix} a & b & c \\ c & a & b \\ b & c & a \end{bmatrix}$ is called a (3 by 3) *circulant matrix*. Note how **row 2** is formed from row 1, row 3 from row 2, and row 1 from row 3.

29. Show that the product of two circulant matrices is a circulant matrix.

30. Show that circulant matrices commute under multiplication.

31. Is the sum of two circulant matrices a circulant matrix? Prove or disprove. ans. YES

Part E. Given the matrix $M = \begin{bmatrix} 0.5 & 0.6 \\ 0.5 & 0.4 \end{bmatrix}$; this is a *Markov matrix* (square with non-negative entries such that the sum of the entries of each column = 1)

32. Determine whether M^2 is a Markov matrix or not.

33. Given $X_0 = \begin{bmatrix} 30 \\ 70 \end{bmatrix}$; show that if $X_1 = MX_0$, then the sum of the entries in X_1 is the same as the sum of the entries in X_0. $X_1 = \begin{bmatrix} 57 \\ 43 \end{bmatrix}$

34. Using X_1(from #33) find X_2 if $X_2 = MX_1$. Is the sum of the entries the same as that of X_0?

35. Find all X such that MX = X.

ans. $X = \begin{bmatrix} 12t \\ 10t \end{bmatrix}$

$$x = 1$$

Part F. Given the system $2x + y = 3$;

$$4x + 3y = 8$$

36. write it in the form AX = B and show that it is inconsistent.

37. multiply both sides by A transpose and solve the resulting system. It will be consistent. It is called the "least squares solution". ans. $\{(6/7, 3/2)\}$

Part G. Airline problem:

38. Write down the incidence matrix for "Happy Flier" Airline Co. which has the following 4 cities on its routes (use order Erie-Pittsburgh-Cleveland-Buffalo):

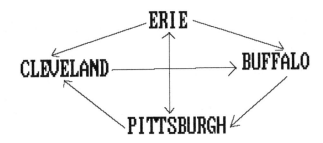

39. Find the matrix which tells the passenger how he can fly from city i to city j in exactly two

flights. ans. $A^2 = \begin{bmatrix} 1 & 1 & 1 & 1 \\ 0 & 1 & 1 & 2 \\ 0 & 1 & 0 & 0 \\ 1 & 0 & 1 & 0 \end{bmatrix}$

40. Find the matrix which tells the passenger how he can fly from city i to city j in two flights or less.

41. Find the matrix which tells the passenger how he can fly from city 1 to city j in exactly 3 flights.

ans. $A^3 = \begin{bmatrix} 1 & 2 & 2 & 2 \\ 1 & 2 & 1 & 1 \\ 1 & 0 & 1 & 0 \\ 0 & 1 & 1 & 2 \end{bmatrix}$

$$x_1 + 2x_2 - x_3 + 11x_4 = 1$$

Part H. Given the system   ;

$$-x_1 + 3x_2 + 6x_3 - x_4 = 4$$

42. Write the system in the form AX = B, where A is the coefficient matrix.

43. Solve the system using Gaussian elimination, writing the answer in the form of a column

matrix. ans. $X = \begin{bmatrix} -1+3s-7t \\ 1-s-2t \\ s \\ t \end{bmatrix}$

44. Prove that your solution works by doing one simple matrix multiplication.

45. Write the solution is the form $X = X_h + X_p$ (where X_h is the solution of the corresponding homogeneous system); show $AX_h = \theta$ & $AX_p = B$.

ans. $X = \begin{bmatrix} 3s-7t \\ -s-2t \\ s \\ t \end{bmatrix} + \begin{bmatrix} -1 \\ 1 \\ 0 \\ 0 \end{bmatrix}$

46. Take X_h from the answer to #45 and segregate the s & t scalars to product two different column matrices, call them X_1 and X_2.

47. Show that both X_1 & X_2 separately satisfy the corresponding homogeneous system $AX = \theta$; show also that X_p does NOT.

ans. $X_1 = \begin{bmatrix} 3 \\ -1 \\ 1 \\ 0 \end{bmatrix}$, $X_2 = \begin{bmatrix} -7 \\ -2 \\ 0 \\ 1 \end{bmatrix}$

Part I. Given the matrix A = $A = \dfrac{1}{5}\begin{bmatrix} 1 & 2 \\ 2 & 4 \end{bmatrix}$;

48. Show A satisfies the equation $A^2 = A$ so A is *idempotent*.

49. If B = I - A where $I = \begin{bmatrix} 1 & 0 \\ 0 & 1 \end{bmatrix}$ show that B is also idempotent.

50. Amazingly enough, show that AB = the zero matrix so both are divisors of zero. Neither A nor B has an inverse.

51. Show that AB = BA so that they commute.

52. Let U = 2A - I; show that $U^2 = I$. This means that U is its own inverse; it is called *unipotent*.

53. Show that U commutes with its idempotent "creator" A.

Part J. LU decomposition: given the system AX = B where $\begin{bmatrix} 1 & 0 & 2 \\ 0 & 3 & 3 \\ 1 & 3 & 6 \end{bmatrix}\begin{bmatrix} x \\ y \\ z \end{bmatrix} = \begin{bmatrix} 1 \\ 2 \\ 3 \end{bmatrix}$:

54. Show that A can be factored as $\begin{bmatrix} 1 & 0 & 0 \\ 0 & 1 & 0 \\ 1 & 1 & 1 \end{bmatrix}\begin{bmatrix} 1 & 0 & 2 \\ 0 & 3 & 3 \\ 0 & 0 & 1 \end{bmatrix} = LU$

(L - unit *lower* triangular, U - **upper** triangular).

55. Then show that the system LU X = B can be solved via a two step procedure:
a) using *forward substitution* if UX = Y, solve LY = B for Y
b) then using **back-substitution**, solve UX = Y for X. ans X = (1,2/3,0)

Part K. Calculus and matrices: define the derivative of a matrix

thus: $Y = \begin{bmatrix} y_1 & y_2 \\ y_3 & y_4 \end{bmatrix} \Rightarrow \dfrac{dY}{dt} = \begin{bmatrix} y_{1'} & y_{2'} \\ y_{3'} & y_{4'} \end{bmatrix}$ ie. to differentiate a matrix, just differentiate each entry in

the matrix.

56. Show that if $A = \begin{bmatrix} 1 & 2 \\ 0 & 3 \end{bmatrix}$, $Y = \begin{bmatrix} e^t & e^{3t} - e^t \\ 0 & e^{3t} \end{bmatrix}$ then dY/dt = AY.

57. Show that if $X = \begin{bmatrix} x \\ y \end{bmatrix}$, $C = \begin{bmatrix} c_1 \\ c_2 \end{bmatrix}$, then the solution to dX/dt = AX is $X = \begin{bmatrix} e^t & e^{3t} - e^t \\ 0 & e^{3t} \end{bmatrix}C$.

Hint: just multiply out the right side and do the calculus on the left

Part L. Given $\begin{aligned} \dfrac{dx}{dt} &= y \\ \dfrac{dy}{dt} &= x \end{aligned}$;

58. Show that this system of differential equations can be written in the form dX/dt = AX where $A = \begin{bmatrix} 0 & 1 \\ 1 & 0 \end{bmatrix}$.

59. Show that $X_1 = e^{-t} \begin{bmatrix} 1 \\ -1 \end{bmatrix}$, $X_2 = e^t \begin{bmatrix} 1 \\ 1 \end{bmatrix}$ both satisfy this system of differential equations.

60. Show that $X = c_1 X_1 + c_2 X_2$ satisfies this system too.

61. Show that the solution X from #60 can be written as $X = \begin{bmatrix} 1 & 1 \\ -1 & 1 \end{bmatrix} \begin{bmatrix} e^{-t} & 0 \\ 0 & e^t \end{bmatrix} \begin{bmatrix} c_1 \\ c_2 \end{bmatrix}$

Hint: Multiply the matrices out!

62. Show that $X = \dfrac{1}{2} \begin{bmatrix} e^t + e^{-t} & e^t - e^{-t} \\ e^t - e^{-t} & e^t + e^{-t} \end{bmatrix} \begin{bmatrix} k_1 \\ k_2 \end{bmatrix} = \Phi(t)K$ also satisfies the system of differential equations.

61. Show that the matrix $\Phi(t)$ satisfies $\Phi'(t) = A\Phi(t)$; this means that $\Phi(t) = \exp(At)$ since $\dfrac{d}{dt} \exp(At) = A \exp(At)$. Also show that $\Phi(0) = I$ (cf. $e^0 = 1$).

2.2 Inverses and Square Systems

In the previous section, we mentioned that any square system can be represented in a matrix multiplication format, for example

$$x_1 + 2x_2 = 5$$
$$\Rightarrow \begin{bmatrix} 1 & 2 \\ 3 & 4 \end{bmatrix} \begin{bmatrix} x_1 \\ x_2 \end{bmatrix} = \begin{bmatrix} 5 \\ 6 \end{bmatrix}$$
$$3x_1 + 4x_2 = 6$$

To be as simple as possible, we write this in the form AX = B where

$$A = \begin{bmatrix} 1 & 2 \\ 3 & 4 \end{bmatrix} \quad X = \begin{bmatrix} x_1 \\ x_2 \end{bmatrix} \quad B = \begin{bmatrix} 5 \\ 6 \end{bmatrix}$$

Here A is the *coefficient matrix* of the system and it is square. Now to the question for this section- how do we get the inverse of a square matrix?

Let's start with the simple 2 by 2 case; for this we have a nice algorithm:

$$A = \begin{bmatrix} a & b \\ c & d \end{bmatrix} \quad \Rightarrow \quad A^{-1} = \frac{1}{ad - bc} \begin{bmatrix} d & -b \\ -c & a \end{bmatrix}$$

EXAMPLE 1 Solve the system given above by using the inverse of the coefficient matrix.

$$A^{-1} = \frac{1}{4-6} \begin{bmatrix} 4 & -2 \\ -3 & 1 \end{bmatrix} = \frac{1}{-2} \begin{bmatrix} 4 & -2 \\ -3 & 1 \end{bmatrix} = \begin{bmatrix} -2 & 1 \\ \dfrac{3}{2} & -\dfrac{1}{2} \end{bmatrix}$$

Now multiply both sides of the original system by the inverse of A:

$$AX = B \Rightarrow A^{-1}AX = A^{-1}B \Rightarrow IX = A^{-1}B \Rightarrow X = A^{-1}B$$

Therefore we have for our system:

$$X = \begin{bmatrix} -2 & 1 \\ \dfrac{3}{2} & -\dfrac{1}{2} \end{bmatrix} \begin{bmatrix} 5 \\ 6 \end{bmatrix} = \begin{bmatrix} -4 \\ \dfrac{9}{2} \end{bmatrix}$$

The reader can easily check that this solution satisfies the original system; the system is consistent (ie. there is a *unique* solution).

Terminology

If A is a square matrix, then there may or may not be a square matrix B of the same size which satisfies AB = I. If such a matrix X exists, then $B = A^{-1}$ (read "A inverse"). It is unique and commutes with A; in this case A is called **invertible** (or nonsingular). If no such B exists, then is called noninvertible or *singular*.

Here is a another important difference between matrix algebra and real algebra; in the real number field, only zero has no multiplicative inverse; however there are infinitely-many singular square matrices. For example, any matrix of the form

$$A = \begin{bmatrix} 2 & b \\ \dfrac{30}{b} & 15 \end{bmatrix}$$

is *singular* for any non-zero b since ad - bc = $30 - b(30/b) = 0$.

Below are some important properties of inverse matrices. A and B are square matrices of the same size.

Properties of inverses

1) $(AB)^{-1} = B^{-1} A^{-1}$

Proof:

$$(AB)(B^{-1} A^{-1}) = A(B B^{-1}) A^{-1} = A(I) A^{-1} = A A^{-1} = I$$

thus

$$(AB)^{-1} = B^{-1} A^{-1}$$

2) $(A^{-1})^{-1} = A$

3) $(A^{t})^{-1} = (A^{-1})^{t}$

The inverse of the transpose is the transpose of the inverse.

4) $(kA)^{-1} = \dfrac{1}{k} A^{-1}$

Inverses of 3 by 3 and larger matrices

There is no simple algorithm to find the inverse of a 3 by 3 matrix (or any larger square matrix). However, suppose that the inverse of A exists; call it B. If B is the inverse of A, then the 9 by 9 linear system gotten from setting AB = I (the identity matrix) has a solution which must be unique. This system would be a 9 by 9 linear system! In Section 2.3 we'll be able to prove that the same row operations which take A to its reduced row-echelon form (ie. the identity matrix) also take I to B. On the other hand, if A is not row-equivalent to I, then B does not exist and there is no A^{-1} (ie. A is singular).

EXAMPLE 2 Find the inverse of $A = \begin{bmatrix} 1 & 0 & 3 \\ 0 & 1 & 2 \\ -1 & 0 & 6 \end{bmatrix}$.

$$\begin{bmatrix} 1 & 0 & 3 & 1 & 0 & 0 \\ 0 & 1 & 2 & 0 & 1 & 0 \\ -1 & 0 & 6 & 0 & 0 & 1 \end{bmatrix} row_1 + row_3 \sim \begin{bmatrix} 1 & 0 & 3 & 1 & 0 & 0 \\ 0 & 1 & 2 & 0 & 1 & 0 \\ 0 & 0 & 9 & 1 & 0 & 1 \end{bmatrix} \frac{1}{9} row_3 \sim$$

$$\begin{bmatrix} 1 & 0 & 3 & 1 & 0 & 0 \\ 0 & 1 & 2 & 0 & 1 & 0 \\ 0 & 0 & 1 & \frac{1}{9} & 0 & \frac{1}{9} \end{bmatrix} \begin{matrix} -2\,row_3 + row_2 \\ -3\,row_3 + row_1 \end{matrix} \sim \begin{bmatrix} 1 & 0 & 0 & \frac{2}{3} & 0 & -\frac{1}{3} \\ 0 & 1 & 0 & -\frac{2}{9} & 1 & -\frac{2}{9} \\ 0 & 0 & 1 & \frac{1}{9} & 0 & \frac{1}{9} \end{bmatrix}$$

so we have the inverse of A:

$$A^{-1} = \begin{bmatrix} \frac{2}{3} & 0 & -\frac{1}{3} \\ -\frac{2}{9} & 1 & -\frac{2}{9} \\ \frac{1}{9} & 0 & \frac{1}{9} \end{bmatrix}$$

This can be checked by direct multiplication (ie. $A A^{-1} = I$). If at any time in this process, we produce a <u>row of zeros</u> on the left, this implies that A is *singular* and we would stop.

THEOREM: Inverses and Systems: Given A is an n by n matrix; AX = B has a unique solution iff:
1) A in invertible
2) $A \sim I_n$
3) rank(A) = n
4) $A X = \theta$ has only the solution $X = \theta$
5) nullity(A) = 0

One can see then that knowing whether A is singular or not is important as regards the solution of the corresponding system A X = B. Also it is obvious that actually finding the inverse is a tedious, time-consuming process, since the matrix A must be put in reduced row-echelon form. A simple criterion for determining whether A is singular or not will be given in the next section.

Application: Electrical circuits

In some practical applications, the coefficient matrix remains the same but the constant matrix B changes. We may think of B as an *input* and X as an output- if the input B changes then the output X also must change. A practical example in the form of an electrical circuit is given below:

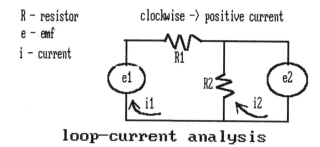

In this application "i" stands for current and I can't change this!

For example, the circuit above leads to the following two equations for the currents[2]:

$$(R_1 + R_2)i_1 - R_2 i_2 = e_1$$

$$- R_2 i_1 + R_2 i_2 = e_2$$

This could be written in matrix multiplication format as:

$$\begin{bmatrix} R_1 + R_2 & - R_2 \\ - R_2 & R_2 \end{bmatrix} \begin{bmatrix} i_1 \\ i_2 \end{bmatrix} = \begin{bmatrix} e_1 \\ e_2 \end{bmatrix}$$

or more succinctly as R [i] = E (almost a matrix version of Ohm's law!). Check out the dimensions- R is a 2 z 2, [i] is a 2 x 1 column matrix and E is a 2 x 1 column matrix (as it should be). As a simple example, let e_1 = 12 volts, R_1 = 2, R_2 = 4, and e_2 = 0 then we have:

$$(2 + 4)i_1 - 4 i_2 = 12$$

$$- 4 i_1 + 4 i_2 = 0$$

---

[2] see *Foundations of Electrical Circuits* by Cogdell, p. 85

In the form R [i] = E this is

$$\begin{bmatrix} 6 & -4 \\ -4 & 4 \end{bmatrix} \begin{bmatrix} i_1 \\ i_2 \end{bmatrix} = \begin{bmatrix} 12 \\ 0 \end{bmatrix}$$

Multiplying both sides by the inverse of R implies:

$$\begin{bmatrix} i_1 \\ i_2 \end{bmatrix} = \frac{1}{8} \begin{bmatrix} 4 & 4 \\ 4 & 6 \end{bmatrix} \begin{bmatrix} 12 \\ 0 \end{bmatrix} = \begin{bmatrix} 6 \\ 6 \end{bmatrix}$$

The advantage of this method is that, if e_1 and e_2 are changed (ie. the input), you don't need to go back and solve the system again. It is already in terms of e_1 and e_2 and the only work that needs to be done is one matrix multiplication, which is pretty easy.

EXERCISES 2.2

Part A. Find the inverse if possible; if not possible, write singular.

1. $A = \begin{bmatrix} 1 & 2 \\ 3 & 4 \end{bmatrix}$ ans. $A = \begin{bmatrix} -2 & 1 \\ \dfrac{3}{2} & \dfrac{-1}{2} \end{bmatrix}$

2. $A = \begin{bmatrix} 1 & 2 \\ 2 & 4 \end{bmatrix}$

3. $A = \begin{bmatrix} 1 & 2 & 0 \\ 0 & 3 & 4 \\ 1 & 1 & 1 \end{bmatrix}$ ans. $A = \dfrac{1}{7} \begin{bmatrix} -1 & -2 & 8 \\ 4 & 1 & -4 \\ -3 & 1 & 3 \end{bmatrix}$

4. $A = \begin{bmatrix} 2 & 2 & 0 \\ 2 & 4 & 2 \\ 1 & 1 & 1 \end{bmatrix}$

5. $A = \begin{bmatrix} 1 & 2 & 0 \\ 0 & 3 & 4 \\ 1 & 5 & 4 \end{bmatrix}$ ans. singular

6. $A = \begin{bmatrix} 1 & 2 & 0 & 0 \\ 0 & 1 & 0 & 2 \\ 0 & 0 & 2 & 2 \\ 1 & 1 & 0 & 1 \end{bmatrix}$

7. $A = \begin{bmatrix} 5 & -4 & 0 \\ 6 & -5 & 0 \\ 0 & 0 & 1 \end{bmatrix}$ ans. $A^{-1} = A$

8. $A = \begin{bmatrix} -3 & 4 & 0 \\ -2 & 3 & 0 \\ 0 & 0 & 1 \end{bmatrix}$

Part B. Solve each system by writing the system in the form AX = B and multiplying both sides by the inverse of A to get X.

9. $\begin{bmatrix} 1 & 2 \\ 3 & 4 \end{bmatrix} \begin{bmatrix} x \\ y \end{bmatrix} = \begin{bmatrix} 5 \\ 6 \end{bmatrix}$ ans. $X = \begin{bmatrix} -4 \\ \dfrac{9}{2} \end{bmatrix}$

10.
$$x_1 + x_2 \quad\;\; = 0$$
$$-x_1 + 2x_2 + x_3 = 1$$
$$2x_1 + 4x_2 + 6x_3 = 3$$

11. $\begin{bmatrix} 1 & 2 & 0 \\ 0 & 3 & 4 \\ 1 & 5 & 5 \end{bmatrix} \begin{bmatrix} x \\ y \\ z \end{bmatrix} = \begin{bmatrix} 5 \\ 10 \\ 15 \end{bmatrix}$ ans. $X = \begin{bmatrix} -\dfrac{5}{3} \\ \dfrac{10}{3} \\ 0 \end{bmatrix}$

12. $\begin{bmatrix} 1 & 2 & 0 & 0 \\ 0 & 1 & 0 & 2 \\ 0 & 0 & 2 & 2 \\ 1 & 1 & 0 & 1 \end{bmatrix} \begin{bmatrix} x \\ y \\ z \\ w \end{bmatrix} = \begin{bmatrix} 1 \\ 4 \\ 4 \\ 8 \end{bmatrix}$

13. $\begin{bmatrix} 1 & 2 & 3 & 4 \\ 0 & 1 & 0 & 2 \\ 0 & 0 & 4 & 4 \\ 0 & 0 & 0 & 1 \end{bmatrix} \begin{bmatrix} x \\ y \\ z \\ w \end{bmatrix} = \begin{bmatrix} 1 \\ 4 \\ 4 \\ 2 \end{bmatrix}$ ans. $X = \begin{bmatrix} -4 \\ 0 \\ -1 \\ 2 \end{bmatrix}$

14.
$$x_1 + x_2 \quad = 1$$
$$-x_1 + 2x_2 + x_3 = 2$$
$$4x_2 + 6x_3 = -4$$

Part C. Matrix algebra: Solve for A; assume all matrices are square of the same size.

15. $A B^t + D = A$ ans. $D(I - B^t)^{-1}$

16. $(B A)^{-1} + C = A^{-1}$

17. $P^{-1} A P = B + 4C$ ans. $P(B + 4C)P^{-1}$

18. $(P A^{-1})^t = B + C$

Part D. Let $A = \begin{bmatrix} a & b \\ c & d \end{bmatrix}$;

19. show directly that $(A^{-1})^t = (A^t)^{-1}$ for non-singular matrices A.

20. Show directly that $(k A)^{-1} = \frac{1}{k} A^{-1}$.

Part E. If $B = \begin{bmatrix} 1 & 2 \\ 3 & 4 \end{bmatrix}$;

21. Show that $(B^2)^{-1} = (B^{-1})^2$.

22. Determine for what lambda $B - \lambda \begin{bmatrix} 1 & 0 \\ 0 & 1 \end{bmatrix}$ is singular.

23. Show $B B^t$ and $B^t B$ are both symmetric and invertible.

24. Show that $B + B^t$ is also symmetric. Is it invertible?

Part F. Using the theorem from this section, for each n by n matrix A given

a) find the rank of A

b) if rank(A) < n show that $A X = \theta$ has infinitely many solutions

c) if rank(A) = n show that the inverse of A exists by finding it

25. $A = \begin{bmatrix} 2 & 4 \\ 0 & 5 \end{bmatrix}$ ans. rank(A) = 2, $A^{-1} = \begin{bmatrix} \dfrac{1}{2} & -\dfrac{2}{5} \\ 0 & \dfrac{1}{5} \end{bmatrix}$

26. $A = \begin{bmatrix} 1 & 2 & 0 \\ 0 & 3 & 4 \\ 1 & 5 & 4 \end{bmatrix}$

27. $A = \begin{bmatrix} 1 & 2 & 0 \\ 0 & 3 & 4 \\ 1 & 5 & 4 \end{bmatrix}$ ans. rank(A) = 2, X = t(8/3,-4/3,1), A is singular

Part G. Given $A = \begin{bmatrix} -2 & 1 & 0 \\ -1 & 2 & -1 \\ 0 & 1 & -2 \end{bmatrix}$ $B = \begin{bmatrix} 10 \\ 10 \\ 10 \end{bmatrix}$; find:

28. inverse of A
29. solution of AX = B using the inverse of A ans. X = (-5,0,-5)
30. solution of $A^{-1} X = B$
31. solution of $A^t X = B$ using your answer to #28. ans. X = (-15,20,-15)

Part H. Given the matrix $A = \begin{bmatrix} 1 & 2 \\ 0 & 3 \end{bmatrix}$;

32. Show that A is a solution of the matrix polynomial equation $p(A) = A^2 - 4A + 3I = \begin{bmatrix} 0 & 0 \\ 0 & 0 \end{bmatrix}$.

33. Use this equation to find the inverse of A. Hint: isolate I and then multiply both sides by the inverse of A. ans. $A^{-1} = \dfrac{1}{3}(4I - A)$

34. If $B = P^{-1} A P$ (A,B,P square of same size), then B is *similar to A*. Show that if A is a solution of the polynomial equation in #32, then so is B.

Part I. Given the matrix $A = \begin{bmatrix} a & b \\ c & d \end{bmatrix}$;

35. Let $X = \begin{bmatrix} x_1 & x_3 \\ x_2 & x_4 \end{bmatrix}$; set $AX = \begin{bmatrix} 1 & 0 \\ 0 & 1 \end{bmatrix}$ and write down the augmented matrix.

66

ans. $\begin{bmatrix} a & b & 0 & 0 & 1 \\ 0 & 0 & a & b & 0 \\ c & d & 0 & 0 & 0 \\ 0 & 0 & c & d & 1 \end{bmatrix}$

36. Solve for X (which is the inverse of A); show the result agrees with the algorithm given in the text.

Part J. Application: The **steady state temperature** at each point in metal plate is the average of the temperatures at the 4 points surrounding it (West, North, East, South)

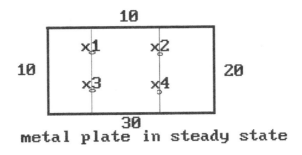

metal plate in steady state

37. Set up the system of equations for each point, starting with x_1. Hint:
equ1:= $x_1 = \frac{1}{4}(10 + 10 + x_2 + x_3)$. Write the system in the form AX = B where

$X = (x_1, x_2, x_3, x_4)$.

38. Solve the corresponding system for X using the inverse of A.
ans. (205/48,25/6,65/12,235/48)

Part K. Given the matrix $A = \begin{bmatrix} 0 & 1 & 0 \\ 0 & 0 & 1 \\ 2 & -5 & 4 \end{bmatrix}$; this matrix (called a companion matrix) is a solution of

the matrix equation $p(A)= A^3 - 4 A^2 + 5 A - 2 I = \theta$.

39. Use this polynomial equation to isolate I. ans. $I = \frac{1}{2}(A^3 - 4 A^2 + 5 A)$

40. Notice that each term on the other side has a factor of A; now multiply both sides by A inverse to find the inverse of A in terms of powers of A. Actually carry out the calculations to get A^{-1}. Verify your result by multiplication.

Part L. Given the matrix $B(t)=\begin{bmatrix} 1 & t & \dfrac{t^2}{2} \\ 0 & 1 & t \\ 0 & 0 & 1 \end{bmatrix}$;

41. Show that B(t) is nonsingular for all t by finding its inverse. ans. $\begin{bmatrix} 1 & -t & \dfrac{t^2}{2} \\ 0 & 1 & -t \\ 0 & 0 & 1 \end{bmatrix}$

42. Find B(4) and its inverse.

43. Show that $B'(t)=\begin{bmatrix} 0 & 1 & 0 \\ 0 & 0 & 1 \\ 0 & 0 & 0 \end{bmatrix} B(t)$.

Part M. Using the circuit given in the text, solve for the currents i_1 and i_2 if:
44. R1 = 3, R2 = 3, e1 = 12, e2 = 0
45. R1 = 3, R2 = 3, e1 = 6, e2 = 12
ans. $i_1=6, i_2=10$

Part N. Given the system of differential equations

$$\begin{bmatrix} \dfrac{dx}{dt} \\ \dfrac{dy}{dt} \end{bmatrix} = \begin{bmatrix} x+2y \\ 2x+y \end{bmatrix} = \begin{bmatrix} 1 & 2 \\ 2 & 1 \end{bmatrix}\begin{bmatrix} x \\ y \end{bmatrix} \Rightarrow \dfrac{dX}{dt} = AX \text{ where } A = \begin{bmatrix} 1 & 2 \\ 2 & 1 \end{bmatrix} \text{and } X = \begin{bmatrix} x \\ y \end{bmatrix}$$

which is called *coupled* since at least one equation contains both x and y terms;
let X = PY where $P = \begin{bmatrix} 1 & 1 \\ -1 & 1 \end{bmatrix}$. Knowing that d/dt(PY) = P dY/dt (since P is constant):

46. Make the given replacement for X on both sides of dX/dt = AX.
47. Show that the "new system" in Y is of the form dY/dt = DY where

$$D = P^{-1} A P = \begin{bmatrix} -1 & 0 \\ 0 & 3 \end{bmatrix}$$

48. Thus the solution of the (new) system in Y is of the form

$$Y = \exp(\,D\,t\,)\,C \ \ \text{where} \exp(\,D\,t\,) = \begin{bmatrix} e^{-1t} & 0 \\ 0 & e^{3t} \end{bmatrix} \quad \& \quad C = \begin{bmatrix} C_1 \\ C_2 \end{bmatrix}$$

What is the solution of the original system $dX/dt = AX$? Hint: remember $X = PY$.

49. Find the solution of the original system which satisfies $X(\,0\,) = \begin{bmatrix} x(\,0) \\ y(\,0\,) \end{bmatrix} = \begin{bmatrix} 30 \\ 10 \end{bmatrix}$.

Verify this by replacing X in the original system $dX/dt = AX$ and then show that $dx/dt = x + 2y$ and $dy/dt = 2x + y$.

ans. $X = \begin{bmatrix} x \\ y \end{bmatrix} = \begin{bmatrix} 1 & 1 \\ -1 & 1 \end{bmatrix} \begin{bmatrix} e^{-t} & 0 \\ 0 & e^{3t} \end{bmatrix} \begin{bmatrix} 10 \\ 20 \end{bmatrix} = \begin{bmatrix} 10\,e^{-t} + 20\,e^{3t} \\ -10\,e^{-t} + 20\,e^{3t} \end{bmatrix}$

2.3 Elementary Matrices

An elementary matrix is a first cousin of the identity matrix I; it is a square matrix which is one step away from the identity matrix I.

DEFINITION	An **elementary matrix** is a matrix E formed from the identity matrix I by performing one elementary row operation on I.

This means that we can form an elementary matrix in one of three ways:
1) multiplying any one row of I by a nonzero constant
2) multiplying row i of I by a (nonzero) constant k and adding that row to row j
3) interchanging any two rows of I

EXAMPLE 1 Suppose we take the identity for 2 x 2 matrices and multiply row one by 2; we get

$$I_2 = \begin{bmatrix} 1 & 0 \\ 0 & 1 \end{bmatrix} \underset{2\,row_1}{\sim} \begin{bmatrix} 2 & 0 \\ 0 & 1 \end{bmatrix} = E_1$$

Now if we were to perform some elementary row operation on E_1 we would, in general, not produce an elementary matrix (it would be two steps away from the identity-not a first cousin!).
The utility of these matrices can be seen by doing some multiplication; let's pick a particular 2 x 2 matrix, call it A:

$$A = \begin{bmatrix} 4 & 3 \\ 2 & 1 \end{bmatrix}$$

If we multiply A on the left by E_1, we get:

$$E_1 A = \begin{bmatrix} 2 & 0 \\ 0 & 1 \end{bmatrix} \begin{bmatrix} 4 & 3 \\ 2 & 1 \end{bmatrix} = \begin{bmatrix} 8 & 6 \\ 2 & 1 \end{bmatrix} = B$$

So we produced a second matrix B which could be gotten (from A) by doing the same elementary row operation on A as we did to create E_1 (ie. multiply row 1 by 2). Now suppose we would interchange the rows of B; we'd produce a second matrix row equivalent to A. Let's create a second elementary matrix E_2:

$$I_2 = \begin{bmatrix} 1 & 0 \\ 0 & 1 \end{bmatrix} \sim \begin{bmatrix} 0 & 1 \\ 1 & 0 \end{bmatrix} = E_2$$

Now let's multiply B on the left by E_2:

$$E_2 B = \begin{bmatrix} 0 & 1 \\ 1 & 0 \end{bmatrix} \begin{bmatrix} 8 & 6 \\ 2 & 1 \end{bmatrix} = \begin{bmatrix} 2 & 1 \\ 8 & 6 \end{bmatrix} = C$$

By now we should be able to see what is going on; if we create an elementary matrix by doing a particular row operation on I and then multiply some matrix A on the **left** by this elementary matrix,

we get the same result as if we had done this very same elementary row operation on A itself.
That is, elementary matrices can by used to "keep track" of elementary row operations (we just need to remember to multiply on the left!). So at this point, we could write down a matrix equation relating all the matrices in the game so far:

$$E_1 A = B \quad and \quad E_2 B = C \implies E_2(E_1 A) = C$$

If we wanted to do more row operations on A, we could just "pile on" the matrix multiplications which would keep track of these row operations on A by successive elementary matrices. Since these elementary matrices are so close to the identity, getting their inverses is very easy, in fact it can be done on sight:

elementary matrix	corresponding inverse
multiply row i by k	multiply row i by 1/k
multiply row i by k and add to row j	multiply row i by -k and add to row j
interchange rows i and j	interchange rows i and j

Remember that you are doing these operations always to the identity; otherwise the result would not be an elementary matrix. Of course, this means that the inverse of an elementary matrix is also an elementary matrix.

EXAMPLE 2 Find the inverses of the following three elementary matrices:

$$E_1 = \begin{bmatrix} 2 & 0 \\ 0 & 1 \end{bmatrix} \quad E_2 = \begin{bmatrix} 1 & 0 \\ 5 & 0 \end{bmatrix} \quad E_3 = \begin{bmatrix} 0 & 1 \\ 1 & 0 \end{bmatrix}$$

First check out that all three are elementary, then ask yourself how they were formed. The first is just formed by multiplying row 1 on the identity by 2 so its inverse is found by multiplying row 1 of I_2 by 1/2:

$$I_2 = \begin{bmatrix} 1 & 0 \\ 0 & 1 \end{bmatrix} \underset{\frac{1}{2} row_1}{\sim} \begin{bmatrix} \frac{1}{2} & 0 \\ 0 & 1 \end{bmatrix} = E_1^{-1}$$

The second is formed by multiplying row 1 by 5 and adding to row 2:

$$I_2 = \begin{bmatrix} 1 & 0 \\ 0 & 1 \end{bmatrix} \underset{-5 \, row_1 + row_2}{\sim} \begin{bmatrix} 1 & 0 \\ -5 & 0 \end{bmatrix} = E_2^{-1}$$

The third is formed by interchanging rows 1 and 2; this one is the easiest of all- it is its own inverse!

$$I_2 = \begin{bmatrix} 1 & 0 \\ 0 & 1 \end{bmatrix} \underset{row_1 \leftrightarrow row_2}{\sim} \begin{bmatrix} 0 & 1 \\ 1 & 0 \end{bmatrix} = E_3^{-1}$$

The reader is invited to actually multiply each matrix by its alleged inverse to see that indeed we have found the correct inverse.

Now for the main use of the elementary matrices, a theorem.

Correspondence theorem

If one elementary row operation is performed on a matrix A (of size m x n) to form the matrix B and if the matrix E is formed from the identity matrix I_m by performing the very same row operation, then $EA = B$.

For example, let

$$A = \begin{bmatrix} 1 & -1 & 5 \\ 2 & 0 & -3 \end{bmatrix}$$

and add twice row 1 to row 2:

$$A = \begin{bmatrix} 1 & -1 & 5 \\ 2 & 0 & -3 \end{bmatrix} \underset{2\,row_1 + row_2}{\sim} \begin{bmatrix} 1 & -1 & 5 \\ 4 & -2 & 7 \end{bmatrix} = B$$

Now create E by doing this very same row operation to the identity matrix I_2 (since A is a 2 x 3 matrix):

$$I_2 = \begin{bmatrix} 1 & 0 \\ 0 & 1 \end{bmatrix} \underset{2\,row_1 + row_2}{\sim} \begin{bmatrix} 1 & 0 \\ 2 & 1 \end{bmatrix} = E \qquad EA = \begin{bmatrix} 1 & 0 \\ 2 & 1 \end{bmatrix}\begin{bmatrix} 1 & -1 & 5 \\ 2 & 0 & -3 \end{bmatrix} = \begin{bmatrix} 1 & -1 & 5 \\ 4 & -2 & 7 \end{bmatrix} = B$$

The point of the theorem is that we can keep track of row operations using elementary matrices via left-multiplication. This allows us to prove that our algorithm for finding the inverse of a matrix is valid. For example, suppose it takes us 3 row operations to take the matrix A to the identity. Each of these can be kept track of via an elementary matrix:

$$E_3(E_2(E_1 A))= I \tag{1}$$

Now in order to get A we need to left-multiply both sides first by E_3^{-1}, then by E_2^{-1} and then by E_1^{-1} which leaves us with

$$A = E_1^{-1} E_2^{-1} E_3^{-1} I = E_1^{-1} E_2^{-1} E_3^{-1}$$

Thus to get the inverse of A we finally have:

$$A^{-1} = (E_1^{-1} E_2^{-1} E_3^{-1})^{-1}$$

But inverting a product means we have to invert each factor and reverse the order:

$$A^{-1} = E_3(E_2(E_1 I)) \tag{2}$$

Comparing equ(1) and equ(2), we see that the same elementary matrices which take A to I also take I to A^{-1}.

Factor theorem

From Section 2.2, we know that to find the inverse of a matrix, we reduce that matrix to the identity using elementary row operations. But from this section, we now know that these row operations can be kept track of using elementary matrices. Let's see how this works for a particular invertible matrix.

EXAMPLE 3 Show that A can be factored as a product of elementary matrices by reducing A to the identity matrix I if $A = \begin{bmatrix} 1 & 2 \\ 0 & 3 \end{bmatrix}$.

To find the inverse, we know that we use row operations on A until we get I:

$$\begin{bmatrix} 1 & 2 \\ 0 & 3 \end{bmatrix} \underset{\frac{1}{3} row_1}{\overset{\sim}{}} \begin{bmatrix} 1 & 2 \\ 0 & 1 \end{bmatrix} \underset{-2\, row_2\, +row_1}{\overset{\sim}{}} \begin{bmatrix} 1 & 0 \\ 0 & 1 \end{bmatrix}$$

The corresponding elementary matrices are:

$$E_1 = \begin{bmatrix} 1 & 0 \\ 0 & \dfrac{1}{3} \end{bmatrix} \qquad E_2 = \begin{bmatrix} 1 & -2 \\ 0 & 1 \end{bmatrix}$$

Their inverses are easy to find:

$$E_1^{-1} = \begin{bmatrix} 1 & 0 \\ 0 & 3 \end{bmatrix} \qquad E_2^{-1} = \begin{bmatrix} 1 & 2 \\ 0 & 1 \end{bmatrix}$$

Thus from the correspondence theorem we have:

$$E_2\, E_1\, A = I$$

Left-multiplying both sides first by E_2^{-1} and then by E_1^{-1} means we get:

$$E_2^{-1}(\, E_2\, E_1\, A\,) = E_2^{-1} I \Rightarrow (\, E_2^{-1} E_2\,) E_1\, A = E_2^{-1} I \Rightarrow E_1\, A = E_2^{-1} I \Rightarrow$$

$$(\, E_1^{-1} E_1\,) A = E_1^{-1} E_2^{-1} I \Rightarrow A = E_1^{-1} E_2^{-1} I$$

so that A can be written as

$$A = \begin{bmatrix} 1 & 0 \\ 0 & 3 \end{bmatrix} \begin{bmatrix} 1 & 2 \\ 0 & 1 \end{bmatrix}$$

This procedure can be carried out on any *invertible* square matrix A. Thus we have:

THEOREM (FACTOR TH) Any invertible square matrix A can be factored as the product of elementary matrices[3].

Algorithm for finding the inverse

Let's use the fact that elementary matrices keep track of row operations to justify the algorithm from Section 2.2 for finding the inverse of a square matrix A. We used the idea that the same row operations taking A to I take I to A inverse. For convenience suppose it takes three row operations ; keep track of them by multiplying A on the left in succession by three elementary matrices; do the same to the identity matrix I :

$$E_3 E_2 E_1 A = I \qquad E_3 E_2 E_1 I = B$$

Solve for A on the left and realize the identity matrix factor I is unnecessary:

$$A = E_1^{-1} E_2^{-1} E_3^{-1} \qquad E_3 E_2 E_1 = B$$

If we multiply A and B we have:

$$A B = (E_1^{-1} E_2^{-1} E_3^{-1})(E_3 E_2 E_1) = (E_1^{-1} E_2^{-1})(E_3^{-1} E_3)(E_2 E_1) = E_1^{-1} (E_2^{-1} E_2) E_1 = E_1^{-1} E_1 = I$$

This proves that B must be in fact the inverse of A and shows the underlined statement above is true.

EXERCISES 2.3

Part A. Determine if A is an elementary matrix; if so, write down the row operation that was used on I$_n$ to produce it.

1. $A = \begin{bmatrix} 1 & 0 \\ 2 & 1 \end{bmatrix}$ ans. 2row1 + row2

2. $A = \begin{bmatrix} 0 & 1 \\ 1 & 0 \end{bmatrix}$

3. $A = \begin{bmatrix} 1 & 0 \\ 2 & 4 \end{bmatrix}$ ans. NOT

---

[3] The order of the factors may differ.

$4.\ A = \begin{bmatrix} 3 & 0 \\ 0 & 1 \end{bmatrix}$

$5.\ A = \begin{bmatrix} 1 & 0 & 0 \\ 0 & 3 & 0 \\ 0 & 0 & 1 \end{bmatrix}$ ans. 3Rrow2

Part B. Perform the indicated row operation on the matrix A and call the result B; then find an elementary matrix E which accounts for this row operation. Multiply A on the left by E and show that the result is B.

$6.\ A = \begin{bmatrix} 1 & 0 & 3 \\ 2 & 4 & 8 \end{bmatrix}$, 3 row1 + row2

$7.\ A = \begin{bmatrix} 1 & 0 \\ 2 & 2 \\ 5 & 6 \end{bmatrix}$, -2 row2 + row3 ans. $E = \begin{bmatrix} 1 & 0 & 0 \\ 0 & 1 & 0 \\ 0 & -2 & 1 \end{bmatrix}$

$8.\ A = \begin{bmatrix} 1 & 0 & 0 \\ 0 & 3 & 6 \\ 2 & 2 & 4 \end{bmatrix}$, interchange rows 1 and 3

Part C. Given the matrix A; determine what row operations took A to C; find two corresponding elementary matrices E1 and E2 so that E1 A = B and then E2 B = C.

$9.\ A = \begin{bmatrix} 1 & 0 & 3 \\ 2 & 1 & -4 \end{bmatrix} \sim \begin{bmatrix} 2 & 1 & -4 \\ 1 & 0 & 3 \end{bmatrix} = B \sim \begin{bmatrix} 2 & 1 & -4 \\ 5 & 0 & 15 \end{bmatrix} = C$ ans. $E_1 = \begin{bmatrix} 0 & 1 \\ 1 & 0 \end{bmatrix}$, $E_2 = \begin{bmatrix} 1 & 0 \\ 0 & 5 \end{bmatrix}$

$10.\ A = \begin{bmatrix} 1 & 0 & 3 \\ 0 & 0 & 2 \\ 0 & 5 & 0 \end{bmatrix} \sim \begin{bmatrix} 1 & 0 & 3 \\ 2 & 0 & 8 \\ 0 & 5 & 0 \end{bmatrix} = B \sim \begin{bmatrix} 1 & 0 & 3 \\ 2 & 0 & 8 \\ 6 & 5 & 24 \end{bmatrix} = C$

Part D. Given the matrix A; carefully use Gauss-Jordan elimination on A to reduce A to the identity I. For each row operation you perform, write down the elementary matrix Ei which affects this very same row operation by left-multiplication. Then write A as the product of elementary matrices. Check your answer by multiplication!

$11.\ A = \begin{bmatrix} 1 & 2 \\ 2 & 5 \end{bmatrix}$ ans. $A = \begin{bmatrix} 1 & 0 \\ 2 & 1 \end{bmatrix}\begin{bmatrix} 1 & 2 \\ 0 & 1 \end{bmatrix}$

12. $A = \begin{bmatrix} 1 & 0 & 0 \\ 0 & 0 & 2 \\ 0 & 4 & 0 \end{bmatrix}$

13. $A = \begin{bmatrix} 2 & 0 & 4 \\ 0 & 4 & 0 \\ 0 & 0 & 6 \end{bmatrix}$ ans. $\begin{bmatrix} 1 & 0 & 0 \\ 0 & 4 & 0 \\ 0 & 0 & 1 \end{bmatrix}\begin{bmatrix} 1 & 0 & 0 \\ 0 & 1 & 0 \\ 0 & 0 & 6 \end{bmatrix}\begin{bmatrix} 2 & 0 & 0 \\ 0 & 1 & 0 \\ 0 & 0 & 1 \end{bmatrix}\begin{bmatrix} 1 & 0 & 2 \\ 0 & 1 & 0 \\ 0 & 0 & 1 \end{bmatrix}$

14. $A = \begin{bmatrix} 0 & 0 & 3 \\ 3 & 3 & 3 \\ 6 & 0 & 0 \end{bmatrix}$

15. $A = \begin{bmatrix} 1 & 0 & 0 \\ -4 & 0 & 1 \\ 3 & 2 & 0 \end{bmatrix}$ ans. $\begin{bmatrix} 1 & 0 & 0 \\ 0 & 0 & 1 \\ 0 & 1 & 0 \end{bmatrix}\begin{bmatrix} 1 & 0 & 0 \\ 3 & 1 & 0 \\ 0 & 0 & 1 \end{bmatrix}\begin{bmatrix} 1 & 0 & 0 \\ 0 & 2 & 0 \\ 0 & 0 & 1 \end{bmatrix}\begin{bmatrix} 1 & 0 & 0 \\ 0 & 1 & 0 \\ -4 & 0 & 1 \end{bmatrix}$

16. $A = \begin{bmatrix} 1 & 2 \\ 2 & 5 \end{bmatrix}$

17. $A = \begin{bmatrix} 1 & 0 & 0 \\ 0 & 0 & 2 \\ 0 & 4 & 0 \end{bmatrix}$ ans. $\begin{bmatrix} 1 & 0 & 0 \\ 0 & 0 & 1 \\ 0 & 1 & 0 \end{bmatrix}\begin{bmatrix} 1 & 0 & 0 \\ 0 & 4 & 0 \\ 0 & 0 & 1 \end{bmatrix}$

18. $A = \begin{bmatrix} 0 & 0 & 3 \\ 3 & 3 & 3 \\ 6 & 0 & 0 \end{bmatrix}$

19. $A = \begin{bmatrix} 1 & 2 & 0 \\ 0 & 1 & 0 \\ 2 & 8 & 1 \end{bmatrix}$ ans. $\begin{bmatrix} 1 & 0 & 0 \\ 0 & 1 & 0 \\ 2 & 0 & 1 \end{bmatrix}\begin{bmatrix} 1 & 2 & 0 \\ 0 & 1 & 0 \\ 0 & 0 & 1 \end{bmatrix}\begin{bmatrix} 1 & 0 & 0 \\ 0 & 1 & 0 \\ 0 & 4 & 1 \end{bmatrix}$

20. $A = \begin{bmatrix} 0 & 2 & 0 \\ 3 & 0 & 0 \\ 0 & 0 & 4 \end{bmatrix}$

21. $A = \begin{bmatrix} 0 & 0 & 3 \\ 0 & 4 & 0 \\ 5 & 0 & 0 \end{bmatrix}$ ans. $A = \begin{bmatrix} 0 & 0 & 1 \\ 0 & 1 & 0 \\ 1 & 0 & 0 \end{bmatrix} \begin{bmatrix} 5 & 0 & 0 \\ 0 & 1 & 0 \\ 0 & 0 & 1 \end{bmatrix} \begin{bmatrix} 1 & 0 & 0 \\ 0 & 4 & 0 \\ 0 & 0 & 1 \end{bmatrix} \begin{bmatrix} 1 & 0 & 0 \\ 0 & 1 & 0 \\ 0 & 0 & 3 \end{bmatrix}$

22. From #21, find the inverse of A by evaluating the product of the inverses of each elementary factor of A in the ans. to #21, in the *reverse* order.

2.4 Determinants

There are two important matrix topics in which determinants play an important role; one is in distinguishing invertible from singular matrices and the second is solving the eigenvalue problem (cf. Chapter 6). In this section, I have only two goals:

1) to be able to calculate the determinants of small matrices
2) to introduce some properties of the determinant function

Since there are many computer programs which can calculate the determinants of very large matrices in a picosecond, I don't expect the student to calculate by hand the determinant of any matrix larger than a 4 by 4. But can the student see when the determinant of a matrix might be zero without doing any calculations? This might be an important observation...

The value in knowing how to calculate a determinant lies in the connection between inverses and determinants. Recall the algorithm for getting the inverse of a 2 by 2 matrix:

$$A = \begin{bmatrix} a & b \\ c & d \end{bmatrix} \quad \Rightarrow \quad A^{-1} = \frac{1}{ad-bc} \begin{bmatrix} d & -b \\ -c & a \end{bmatrix}$$

The *scalar multiplier* $1/(ad - bc)$ is in fact the reciprocal of the *determinant* of the 2 by 2 matrix A.

DEFINITION	Given the matrix $A = \begin{bmatrix} a & b \\ c & d \end{bmatrix}$; the **determinant** of A is ad - bc. We write

$$\det(A) = ad - bc$$

An alternate notation is

$$|A| = \begin{vmatrix} a & b \\ c & d \end{vmatrix} = ad - bc$$

Be careful not to confuse the determinant of A(which is a real number) with the matrix A. We clearly see then that if det(A) = 0, the matrix A is **singular**. This is true also for larger matrices.

THEOREM: A is singular iff det(A) = 0
This theorem is the most important reason for knowing about determinants. There is a corollary to the theorem.
Corollary: A is invertible iff $\det(A) \neq 0$.
Thus a nonzero determinant is the hallmark of an invertible matrix.

EXAMPLE 1 Find det(A) if

$$A = \begin{bmatrix} 3 & 4 \\ 5 & 10 \end{bmatrix}$$

det(A) = 30 - 20 = 10 which means A is invertible.

Algorithm for a 3 by 3

For a 3 by 3 matrix, the determinant can also be found with an algorithm:

$$\begin{vmatrix} 1 & 2 & 3 \\ 4 & 5 & 6 \\ 7 & 8 & 10 \end{vmatrix} = (\ 1 \cdot 5 \cdot 10 + 4 \cdot 8 \cdot 3 + 7 \cdot 6 \cdot 2\ ) - (\ 3 \cdot 5 \cdot 7 + 6 \cdot 8 \cdot 1 + 10 \cdot 4 \cdot 2\ ) =$$

$$230 - 233 = -3$$

One starts at the <u>upper left</u> and moves from *left to right* to get the *positive* products; the first one is the entries along the main diagonal. Then we imitate this maneuver, starting with the "4" and sweeping along the subdiagonal, always needing three factors in each product. There are three positive products and three negative ones; the negative ones are found by starting with the "3" on the minor diagonal and sweeping down from right to left, mimicking what was done to get the positive products. Each entry is hit twice; once in the positive sense and once in the negative sense. A word of caution - this does **NOT work for a 4 by 4 or larger!**

Properties of determinants

A few of the important properties can be obtained by playing a little with 2 by 2 determinants. Basically we wish to know the effect of doing row operations on a matrix and what this does to the value of the determinant. Here I will use my favorite matrix A = $\begin{bmatrix} 1 & 2 \\ 3 & 4 \end{bmatrix}$; we know that

det(A) = -2.
Suppose we interchange two rows of the matrix A; what happens to det(A)?

$$B = \begin{bmatrix} 3 & 4 \\ 1 & 2 \end{bmatrix} \Rightarrow \det(\ B\ ) = 2 \cdot 3 - 4 \cdot 1 = 6 - 4 = 2 = -\det(\ A\ )$$

Interchanging two rows means that the value of the determinant is *opposite* the value of the original. Now suppose we multiply one row by a scalar, say k = 10:

$$C = \begin{bmatrix} 10 & 20 \\ 3 & 4 \end{bmatrix} \Rightarrow \det(C\ ) = \begin{vmatrix} 10 & 20 \\ 3 & 4 \end{vmatrix} = 40 - 60 = -20 = 10\ (\ -2\ ) = 10\det(A\ )$$

Very simply the value of the determinant is multiplied by 10.

Now for the most popular row operation- adding a multiple of one row to another. Suppose we multiply row 1 by 7 and add to row 2 and then get the value of the determinant:

$$D = \begin{bmatrix} 1 & 2 \\ 10 & 18 \end{bmatrix} \Rightarrow \det(D) = 18 - 20 = -2 = \det(A)$$

Nothing happens! The value of the determinant is the same as the original. Let's summarize our results:

row operation	value *of determinant*
interchange two rows	det(new) = -det(old)
multiply one row by a constant k	det(new) = k det(old)
add a multiple of one row to another	det(new) = det(old)

So now we know the effect of doing row operations on a matrix and the value of the determinant of that new matrix. But the question remains- why would we want to do row operations on a matrix in order to evaluate its determinant? The answer is the following theorem:

THEOREM: If A is a triangular matrix, then det(A) = product of elements on the main diagonal[4].

So what we wish to do is perform row operations on the matrix in order to put it in triangular form (and we are used to doing that, since a square matrix in row-echelon form is also a triangular matrix). We account for the row operations on the value of the det(A) and then we get the determinant of the triangular matrix easily using the theorem.

EXAMPLE 2 Evaluate the following determinant using "row operations".

$$A = \begin{bmatrix} 5 & 10 & 15 & 0 \\ 0 & 2 & 4 & 6 \\ -1 & -2 & 5 & 1 \\ 0 & 2 & 12 & 10 \end{bmatrix} \begin{array}{c} \\ \\ \frac{1}{5}row_1 \\ \\ \end{array} \sim \begin{bmatrix} 1 & 2 & 3 & 0 \\ 0 & 2 & 4 & 6 \\ -1 & -2 & 5 & 1 \\ 0 & 2 & 12 & 10 \end{bmatrix} \begin{array}{c} \\ \\ row_1 + row_3 \\ \\ \end{array} \sim$$

---

[4] This theorem is proved using the Laplace expansion of a determinant. This involves multiplying elements of any one row (or column) by their corresponding cofactors.

$$\begin{bmatrix} 1 & 2 & 3 & 0 \\ 0 & 2 & 4 & 6 \\ 0 & 0 & 8 & 1 \\ 0 & 2 & 12 & 10 \end{bmatrix} \underset{-row_2+row_4}{\sim} \begin{bmatrix} 1 & 2 & 3 & 0 \\ 0 & 2 & 4 & 6 \\ 0 & 0 & 8 & 1 \\ 0 & 0 & 8 & 4 \end{bmatrix} \underset{-row_3+row_4}{\sim} \begin{bmatrix} 1 & 2 & 3 & 0 \\ 0 & 2 & 4 & 6 \\ 0 & 0 & 8 & 1 \\ 0 & 0 & 0 & 3 \end{bmatrix} = B$$

We need to "compensate" for the row operations that we did; adding a multiple of one row to another does NOT change the value of the determinant so we need account for the multiplication of the row1 of A by 1/5 so we get det(A) = 5 det(B)= $5 \cdot (1 \cdot 2 \cdot 8 \cdot 3)= 240$ since B is triangular. To my mind, this is much easier than the minors and cofactors expansion and much less prone to error. Besides which you are already good at putting matrices in triangular form.

EXAMPLE 3 Find det(A) if $A = \begin{bmatrix} 0 & 0 & 6 & 20 \\ 0 & 0 & 1 & 2 \\ 0 & 10 & 20 & 10 \\ 4 & 0 & 6 & 4 \end{bmatrix}$.

$$A = \begin{bmatrix} 0 & 0 & 6 & 20 \\ 0 & 0 & 1 & 2 \\ 0 & 10 & 20 & 10 \\ 4 & 0 & 6 & 4 \end{bmatrix} \underset{row_1 \leftrightarrow row_4}{\sim} = \begin{bmatrix} 4 & 0 & 6 & 4 \\ 0 & 0 & 1 & 2 \\ 0 & 10 & 20 & 10 \\ 0 & 0 & 6 & 20 \end{bmatrix} \underset{row_2 \leftrightarrow row_3}{\sim}$$

$$\begin{bmatrix} 4 & 0 & 6 & 4 \\ 0 & 10 & 20 & 10 \\ 0 & 0 & 1 & 2 \\ 0 & 0 & 6 & 20 \end{bmatrix} \underset{-6\,row_3+row_4}{\sim} \begin{bmatrix} 4 & 0 & 6 & 4 \\ 0 & 10 & 20 & 10 \\ 0 & 0 & 1 & 2 \\ 0 & 0 & 0 & 8 \end{bmatrix} = B$$

Thus det(B) = (-1)(-1)(4)(10)(1)(8) = 320.

Other properties of the determinant function

1) $\det (k A)= k^n \det (A)$

2) $\det (A^t)= \det (A)$

3)*** $\det (A^{-1})= \dfrac{1}{\det (A)}$ (This reminds you that the determinant of a **singular** matrix is **zero**!)

4) $\det (A B)= \det (A) \det (B)$ (One says the determinant function is *multiplicative*.)

So much for determinants. Again keep in mind that the main reason for knowing about determinants is that the *determinant of a nonsingular matrix is nonzero*.

Fundamental theorem

For the square n by n matrix A, the following are equivalent:
1) A is *invertible*
2) A is row equivalent to the identity matrix I
3) $\det(A) \neq 0$
4) $AX = \theta$ has only the trivial solution $X = \theta$
5) $AX = B$ has a unique solution
6) rank(A) = n
7) nullity(A) = 0

EXAMPLE 4 Given the matrix $A = \begin{bmatrix} 1 & 2 & 0 & 3 \\ 2 & 4 & 2 & 4 \\ 1 & 2 & 1 & -1 \\ 4 & 8 & 3 & 6 \end{bmatrix}$; use the fundamental theorem to write several

important facts about A without doing any calculations.

Observing that the fourth row is just the sum of the first three, when putting A in echelon form, the last row would disappear and thus we have a matrix whose determinant is **zero** so A is *singular*. This means that

1) A is not row-equivalent to I
2) $AX = \theta$ has infinitely many solutions
3) $AX = B$ does not have a unique solution (the system will either be inconsistent or consistent and dependent) for any B
4) rank(A) < 4
5) nullity(A) $\neq 0$

EXERCISES 2.4

Part A. Find det(A) using an algorithm:

1. $A = \begin{bmatrix} 1 & 2 \\ 3 & 6 \end{bmatrix}$ ans. 0

2. $A = \begin{bmatrix} 1 & 2 & 3 \\ 0 & 0 & 1 \\ 1 & 2 & 4 \end{bmatrix}$

3. $A = \begin{bmatrix} 1 & 0 & 2 \\ 3 & 3 & 6 \\ 0 & 4 & 4 \end{bmatrix}$ ans. 12

4. $A = \begin{bmatrix} 0 & 2 & 3 \\ 4 & 4 & 4 \\ -2 & 3 & 5 \end{bmatrix}$

$5. A = \begin{bmatrix} 1 & 2 & 3 \\ 4 & 5 & 6 \\ 7 & 8 & 9 \end{bmatrix}$ ans. 0

Part B. Find det(A) for every matrix A; look for any hints that the determinant may be zero. For a 4 by 4 or larger matrix, find a matrix B (row equivalent to A) which is in *upper triangular* form. By compensating for your row operations on A, find det(A) using the value of det(B).

$6. A = \begin{bmatrix} 1 & 2 & 3 \\ 3 & 3 & 3 \\ 4 & 5 & 7 \end{bmatrix}$

$7. A = \begin{bmatrix} 1 & 0 & 0 \\ 2 & 3 & 0 \\ 5 & 5 & 5 \end{bmatrix}$ ans. 15

$8. A = \begin{bmatrix} 1 & 2 & 3 \\ 0 & 4 & 5 \\ 0 & 0 & 8 \end{bmatrix}$

$9. A = \begin{bmatrix} 1 & 2 & 3 & 4 \\ 4 & 5 & 6 & 7 \\ 7 & 8 & 9 & 10 \\ 10 & 11 & 12 & 13 \end{bmatrix}$ ans. 0

$10. A = \begin{bmatrix} 1 & 2 & 3 & 4 \\ 4 & 5 & 6 & 7 \\ 7 & 8 & 9 & 10 \\ 12 & 15 & 18 & 21 \end{bmatrix}$

11. $A = \begin{bmatrix} -1 & 2 & -3 & 4 \\ 0 & 3 & 1 & 0 \\ 0 & 0 & -7 & 5 \\ 0 & 0 & 0 & 10 \end{bmatrix}$ ans. 210

12. $A = \begin{bmatrix} 1 & 2 & 3 & 1 \\ 2 & 4 & 6 & 2 \\ 3 & 6 & 9 & 3 \\ 6 & 12 & 18 & 6 \end{bmatrix}$

13. $A = \begin{bmatrix} 2 & 2 & 3 & 4 \\ -4 & -4 & -4 & -8 \\ 10 & 10 & 20 & 20 \\ -1 & -2 & -3 & -4 \end{bmatrix}$ ans. 0

14. $A = \begin{bmatrix} -1 & 2 & -3 & 4 \\ 1 & 0 & 1 & 0 \\ 0 & 2 & -2 & 5 \\ 0 & 4 & -5 & 10 \end{bmatrix}$

15. $A = \begin{bmatrix} 1 & 0 & 1 & 0 \\ 0 & 2 & 0 & 2 \\ 3 & 0 & 3 & 6 \\ 10 & 10 & 30 & 20 \end{bmatrix}$ ans. -240

16. $A = \begin{bmatrix} 1 & 0 & 1 & 0 \\ 0 & 3 & 0 & 3 \\ 4 & 0 & 4 & 4 \\ 6 & 6 & 6 & 6 \end{bmatrix}$

17. $A = \begin{bmatrix} 2 & 2 & 0 & 2 \\ 0 & 4 & 0 & 4 \\ 3 & 3 & 3 & 3 \\ 0 & 5 & 0 & 6 \end{bmatrix}$ ans. 24

18. $A = \begin{bmatrix} 1 & 0 & 1 & 0 & 1 \\ 0 & 2 & 0 & 2 & 2 \\ 3 & 0 & 3 & 6 & 6 \\ 0 & 4 & 0 & 4 & 6 \\ 1 & 2 & 3 & 4 & 5 \end{bmatrix}$

Part C. Given $A = \begin{bmatrix} a & b \\ c & d \end{bmatrix}$ and $B = \begin{bmatrix} e & f \\ g & h \end{bmatrix}$ which are non-singular; show that:

19. $\det (A + B) \neq \det (A) + \det (B)$
20. $\det(AB) = \det(A) \det(B)$
21. $\det((kA)B) = k^2 \det(A) \det(B)$
22. $\det (A B) = \det (B A)$
23. $\det ((A B)^t) = \det (A B)$
24. $\det (A B^t) = \det (A) \det (B)$
25. $\det (A B^{-1}) = \dfrac{\det (A)}{\det (B)}$

26. $\det ((A B)^{-1}) = \dfrac{1}{\det (A) \det (B)}$

27. Find two 2 by 2 matrices such that det(A) = det(B) yet A is not equal to B.

ans. $A = \begin{bmatrix} 1 & 2 \\ 3 & 4 \end{bmatrix}$ $B = \begin{bmatrix} 1 & 2 \\ 6 & 10 \end{bmatrix}$, det(A) = det(B) = -2 (there are many other correct answers)

Part D. Given A and B are square matrices of the same size; use the determinant function to prove:
28. If AB is singular, then either A is singular or B is singular.
29. If $A B = \theta$, then either A is singular or B is singular.
30. If $A^k = \theta$ for some positive integer k, then A is *nilpotent*. Prove that the determinant of a nilpotent matrix must be zero.
31. If $A^2 = A$ then det(A) = 0 or det(A) = 1. NB. A is called *idempotent*.
32. If $B = P^{-1} A P$ then det(A) = det(B). NB. B is said to be *similar* to A.
33. If $A^{-1} = A^t$ then $\det (A) = \pm 1$. NB. A is called an *orthogonal* matrix.

Part E. Given the 3 by 3 matrices of the form $A = \begin{bmatrix} a & b & c \\ c & a & b \\ b & c & a \end{bmatrix}$; such matrices are called *circulants*.

34. Find det(A).

35. Show that if a + b + c = 0, then det(A) = 0.

36. If a + b + c = 0, solve for c and then replace c in the matrix A. Calculate det(A) using the algorithm and show that its determinant is ZERO. Then give a particular example of such a singular circulant.

37. What is $\det(A^{-1})$? ans. $\dfrac{1}{a^3 + b^3 + c^3 - 3\,a\,b\,c}$

Part F. Given the matrix $V = \begin{bmatrix} 1 & a & a^2 \\ 1 & b & b^2 \\ 1 & c & c^2 \end{bmatrix}$, which is a *Vandermonde* matrix;

38. Find det(V).

39. Show that a,b and c are nonzero and distinct implies V is invertible.
ans. det(V) = (a - b)(b - c)(c - a) is nonzero if a,b & c satisfy the given criteria

NB. The matrix V arises from trying to find the parabola $y = a + b\,x + c\,x^2$ which passes through 3 distinct points.

Part G. Given the matrix $A = \begin{bmatrix} e^t & e^{2t} \\ 0 & e^{3t} \end{bmatrix}$;

40. Find det(A) and show it is never zero.

41. Find A^{-1}. ans. $\begin{bmatrix} e^{-t} & -e^{-2t} \\ 0 & e^{-3t} \end{bmatrix}$

Part H. Given the matrix $A = \begin{bmatrix} 1 & 1 \\ 4 & 1 \end{bmatrix}$;

42. Let $X = \begin{bmatrix} 5 \\ 10 \end{bmatrix}$ show that AX = 3 X

43. Find all values of k for which A - kI is singular. Hint: what is should det(A - kI) be?
ans. k = -1,3

Part I. Given the matrix $A = \begin{bmatrix} 3 & -2 & 0 \\ 1 & 0 & 0 \\ 0 & 1 & 0 \end{bmatrix}$;

44. Find all values of k for which A - kI is *singular*.

86

45. For each value of k, solve the corresponding homogenous system $(A-kI)X = \theta$ (which must be consistent and dependent!).

ans. k = 0 & $X = r \begin{bmatrix} 0 \\ 0 \\ 1 \end{bmatrix}$, k = 1 & $X = s \begin{bmatrix} 1 \\ 1 \\ 1 \end{bmatrix}$, k = 2 & $X = t \begin{bmatrix} 4 \\ 2 \\ 1 \end{bmatrix}$

46. Show that $(A-kI)X = \theta \Rightarrow AX = kX$; then demonstrate that for each value of k found in #45, this equation holds. Thus AX is just a scalar multiple of X for each k.

Part J. Given the matrix $M = \begin{bmatrix} a & b \\ 1-a & 1-b \end{bmatrix}$;

47. Show M - I is singular

48. Show MX = X has infinitely many solutions.
Hint: rewrite by bring X carefully over to the left side; you'll get a homogeneous system!

Part K. Given twice-differentiable functions f, g & h; the determinant of the matrix

$W = \begin{bmatrix} f & g & h \\ f' & g' & h' \\ f'' & g'' & h'' \end{bmatrix}$ is called the *Wrońskian* of the functions f,g & h.

49. Find the Wrońskian for $f = 1, g = x, h = x^2$. ans. det(W) = 2

50. Find the Wrońskian for $f = e^x, g = \sin(x), h = \cos(x)$.

51. Show that det(W) is NOT zero for any x using the functions from #50. ans. $-2e^x \neq 0$

Part L. A square matrix A is of order n if $A^n = I$ for some positive integer n.

52. Show that $A = \begin{bmatrix} 3 & -2 \\ 4 & -3 \end{bmatrix}$, then A is of order 2. A is *unipotent*.

53. If A is of order n, what must be true about det(A)? ans. $\det(A) = \pm 1$

54. Show that if $A = \begin{bmatrix} 1 & 2 \\ 2 & 4 \end{bmatrix}$ (which is singular) no such n exists (ie. A has infinite order).

55. Show that even if det(A) = 1, A may have infinite order. Hint: use a matrix like $A = \begin{bmatrix} 1 & 2 \\ 0 & 1 \end{bmatrix}$.

56. Show that if A = $A = \begin{bmatrix} \dfrac{1}{5} & \dfrac{2}{5} \\ \dfrac{2}{5} & \dfrac{4}{5} \end{bmatrix}$, then $A^2 = A$ so all powers of A higher than two reduce to A. A is

idempotent.

87

2.5 Special Matrices

This section is devoted to analyzing certain types of matrices from an algebraic point of view. Certainly matrices can be classified according to many different types of properties, for example a symmetric matrix is one which is the same as its transpose: $A = A^t$

Consider the symmetric matrices A and B:

$$A = \begin{bmatrix} 1 & 2 \\ 2 & 5 \end{bmatrix} \quad B = \begin{bmatrix} 2 & 3 \\ 3 & 4 \end{bmatrix} \quad C = A\,B = \begin{bmatrix} 8 & 11 \\ 19 & 26 \end{bmatrix}$$

$$A = \begin{bmatrix} 1 & 2 \\ 2 & 5 \end{bmatrix} \quad B = \begin{bmatrix} 2 & 3 \\ 3 & 4 \end{bmatrix} \quad C = A + B = \begin{bmatrix} 3 & 5 \\ 5 & 9 \end{bmatrix}$$

We see that the product of two symmetric matrices here is NOT symmetric. We say that the set of symmetric matrices is not closed under multiplication. However, if two symmetric matrices are *added*, the sum is a symmetric matrix; the set of symmetric matrices is closed under addition. What we want to define is something called a **group**- it is one of the simplest algebraic structures yet very important.

DEFINITION

A **group** is a set of elements (say G) and a binary operation on G (we write $* | G \, x \, G \rightarrow G$) and use the symbol * to be generic) which satisfies the following:
1) associativity: (a*b)*c = a*(b*c)
2) identity element: G has an identity element e such that a*e = e*a = a for all a in G
2) inverses: for all $a \in G$ there exists $a^{-1} \in G$ such that

$$a * a^{-1} = a^{-1} * a = e$$

You already are familiar with many examples of groups. The set of reals $\Re$ under *addition* – we write $(\Re,+)$, is a group. So is the set of all nonzero reals under multiplication. Zero must be excluded because under multiplication it has no inverse. The set of positive integers in not a group under either multiplication or addition. Under multiplication, no element (except 1) has an inverse (since 1/a is NOT an integer); under addition, no element has an inverse, since the additive inverse of a positive integer is a negative integer. I think you get the idea.

Let's consider symmetric 2 x 2 matrices under addition. We need two arbitrary 2 x 2 symmetric matrices:

$$A = \begin{bmatrix} a & b \\ b & c \end{bmatrix} \quad B = \begin{bmatrix} d & e \\ e & f \end{bmatrix} \Rightarrow A + B = \begin{bmatrix} a+d & b+e \\ b+e & c+f \end{bmatrix}$$

adding we get another *symmetric matrix* so the set of all symmetric matrices is *closed* under addition.

The zero matrix is the identity, which is certainly symmetric. What about the inverse under addition? Clearly

$$-A = \begin{bmatrix} -a & -b \\ -b & -c \end{bmatrix}$$

so the additive inverse is also symmetric. Thus (G,+) where G = {all symmetric 2 by 2 matrices} is a group under addition.

General Linear Group

Matrix multiplication is more exciting than matrix addition. If we want to get a set of square matrices to be a group under multiplication, we need to reject those matrices which are singular. The determinant function is a good tool to use here, since we know that if det(A) = 0 then A is singular. Let's now define the **general linear group**:

$$GL(n,\Re) = \{ A : A \text{ is an } n \text{ by } n \text{ matrix st. } \det(A) \neq 0 \}$$

Since det(AB) = det(A)det(B), if A and B are in this set, then det(AB) is also nonzero. Thus
1) the product AB is in this set
2) the identity matrix I is in the set, since det(I) = 1
3) A inverse is also in the set, since then $\det(A^{-1}) = \dfrac{1}{\det(A)} \neq 0$

For example, $GL(2,\Re)$ is the set of all 2 by 2 invertible matrices. It is a non-commutative group.

Subgroups

A fundamental concept in group theory is the idea of a subgroup. If G is a group and H is a subset of G, then a natural question is whether H itself is a group wrt. the binary operation defined on G. If H in its own right is a group, it is called a **subgroup** of G. We are mainly interested in *proper* subgroups of G (ie. G itself and the subset containing only the identity are obviously always subgroups of G – they are not proper subgroups).

For example, let's consider the set of all 2 by 2 upper triangular matrices:

$$H = \{ \begin{bmatrix} a & b \\ 0 & c \end{bmatrix} : a,b,c \in \Re \}$$

Clearly the sum of two such matrices is an upper triangular matrix so the set is closed under addition. Also, the additive inverse of such a matrix also has zero below the main diagonal so the set is closed under the taking of additive inverses. And the zero matrix is upper triangular. Therefore, under addition, the set forms a group. (H,+) is a group. What about under multiplication? Well, one easily sees that the product of two upper triangular matrices is an upper triangular matrix. But, if we try to get the inverse of such a matrix, and there is a *zero* on the main diagonal, the determinant will be **zero** so the matrix will be singular. Therefore, we must restrict the main diagonal entries- none can be zero. Under this restriction, the matrix will be invertible (the determinant will not be zero) and this subset of the general linear group is *closed* under multiplication.

The inverse of an upper triangular matrix is upper triangular; and the identity matrix I is upper triangular. Thus we have observed that the invertible matrices in H form a subgroup of the general linear group. Is there a shorter way to show that a subset is a subgroup? The answer is given in the theorem below.

THEOREM If H is a (nonempty) subset of a group G, then H is a *subgroup* of G if:

1) a*b is in H whenever $a, b \in H$ (ie. closure) and 2) $a^{-1} \in H$ whenever $a \in H$

Some authors put these two criteria together into one; suppose A and B are in H, then if $A B^{-1}$ is in H, H is a subgroup of G.

EXAMPLE 1 Show that the given subset H of $GL(2, \Re)$ is a subgroup.

$$H = \{ \begin{bmatrix} a & b \\ b & a \end{bmatrix} \} \subseteq GL(2, \Re)$$

We need two matrices from H:

$$A = \begin{bmatrix} a & b \\ b & a \end{bmatrix} \quad B = \{ \begin{bmatrix} c & d \\ d & c \end{bmatrix}$$

Now multiply:

$$AB = \begin{bmatrix} a & b \\ b & a \end{bmatrix} \begin{bmatrix} c & d \\ d & c \end{bmatrix} = \begin{bmatrix} ac+bd & ad+bc \\ ad+bc & ac+bd \end{bmatrix}$$

so the product is in H (ie. the child resembles the parents!). Now for inverses:

$$A^{-1} = \frac{1}{a^2 - b^2} \begin{bmatrix} a & -b \\ -b & a \end{bmatrix}$$

There is a problem here if a = -b or a = b but H is a subset of the general linear group so all matrices in H are already invertible. Since H is a subset of $GL(2, \Re)$, we have proved that this subset is also a sub*group*. By the way, H consists of 2 by 2 *circulant* matrices; it is a commutative subgroup (this is unusual).

EXAMPLE 2 Show the set $H = \{ A = \begin{bmatrix} a & b \\ c & d \end{bmatrix} : \det(A) = 2 \} \subseteq GL(2, \Re)$ is not a subgroup under multiplication.

Suppose A and B lie in H; then det(A) = 2 and det(B) = 2. But det(AB) = det(A)det(B) = 4 so $AB \notin H$ thus H is *not closed* under multiplication. H cannot be a subgroup.

Alternative: if we choose to use the two-in-one approach, we'd pick A and B in H and form the product $A B^{-1}$ where det(A) = 2 and det(B) = 2. But then

$$\det(\ B^{-1}\ ) = \frac{1}{\det(\ B\ )} = \frac{1}{2} \Rightarrow \det(\ A\ B^{-1}\ ) = \det(\ A\ )\det(\ B^{-1}\ ) = (\ 2\ )(\ \frac{1}{2}\ ) = 1$$

so this product is NOT in H.

Below is a chart containing some special matrices discussed in this chapter.

Some special matrices:

Type of matrix	form
upper triangular	$\begin{bmatrix} a & b \\ 0 & c \end{bmatrix}$
Diagonal	$\begin{bmatrix} a & 0 \\ 0 & b \end{bmatrix}$
Scalar	$\begin{bmatrix} a & 0 \\ 0 & a \end{bmatrix}$
Symmetric	$\begin{bmatrix} a & b \\ b & c \end{bmatrix}$
skew-symmetric	$A^t = -A$
general linear	$\begin{bmatrix} a & b \\ c & d \end{bmatrix} \quad a\,d - b\,c \neq 0$
Orthogonal	$A^{-1} = A^t$
Vandermonde	$\begin{bmatrix} 1 & a & b \\ 1 & a^2 & b^2 \\ 1 & a^3 & b^3 \end{bmatrix}$
Circulant	$\begin{bmatrix} a & b & c \\ c & a & b \\ b & c & a \end{bmatrix}$

EXERCISES 2.5

Part A. Determine whether the set G is a group under **addition**:

1. G = {all 2 by 2 matrices diagonal matrices}

ans. yes

2. G = {all 2 by 2 matrices A such that $A = A^t$ }

3. G = {all 2 by 2 invertible matrices}

ans. no

4. G = {A: A is a 2 by 2 matrix with det(A) = 2}

5. G = {A: A is a 2 by 2 matrix such that sum of all entries = 0}
ans. yes

6. G = {A: A is 2 by 2 such that tr(A) = 0} NB. recall that the **trace** of A is the sum of the diagonal entries, written tr(A)

7. G = {A: A is 2 by 2 & det(A) = 0} ans. no

8. G = {A: A is 2 by 2 & det(A) = 1}

9. G = {A: A is a 3 by 3 matrix such that $A^t = -A$ } NB. such matrices are called skew-symmetric
ans. yes

10. G = {A: A is a 3 by 3 circulant matrix}

Part B. Determine if the set G is a group under **multiplication**:

11. G = {A: all 2 by 2 invertible matrices} ans. yes

12. G = {A: all 2 by 2 matrices with det(A) = 1} NB. this is called the *special linear group*

13. G = {A: all 2 by 2 matrices with det(A) = -1} ans. no (not closed)

14. G = {A: all 2 by 2 upper triangular matrices with no zero entries on the main diagonal}

15. G = {A: all 2 by 2 symmetric matrices} ans. no (not closed)

16. $G = \{ A = \begin{bmatrix} 1 & 0 & a \\ 0 & 1 & 0 \\ 0 & 0 & 1 \end{bmatrix}, a, b, c \in \Re \}$

17. $G = \{ A = \begin{bmatrix} 1 & a & 0 \\ 0 & 1 & b \\ 0 & 0 & 1 \end{bmatrix}, a, b \in \Re \}$ ans. no (not closed)

18. $G = \{ A = \begin{bmatrix} 1 & a & b \\ 0 & 1 & c \\ 0 & 0 & 1 \end{bmatrix}, a, b, c \in \Re \}$

19. G = {A: $A^{-1} = A^t$ where A is 2 by 2}. NB. These are *orthogonal*
matrices. ans. yes

20. $G = \{ A = \begin{bmatrix} a & b \\ -b & a \end{bmatrix}, a, b \in \Re \}$ (NB. a and b not *simultaneously* zero)

21. G = {A: A is a 2 by 2 formed from I by doing row switches} ans. yes
22. G = {A: A is a 3 by 3 formed from I by doing row switches}. NB. These matrices are called *permutation* matrices; there should be 6 of them! Here G is a *finite group* of order 6.

Part C. Let G = {3 by 3 circulant matrices}
23. Pick any two 3 by 3 circulant matrices with nonzero entries. Show the product is a circulant matrix.

24. Use one of the two circulants from #23 (with non-zero determinant!) and show that its inverse is also a circulant. Realizing that the identity is also a circulant, what important fact might you conjecture about the set of 3 by 3 invertible circulants?

25. Let A be the 3 by 3 arbitrary circulant in the "special matrices" chart; find det(A). ans. $\det(A) = a^3 + b^3 + c^3 - 3abc$

Part D: Let $H = \{ A : A = \begin{bmatrix} a & b \\ 0 & 1 \end{bmatrix}, a \neq 0, b \in \Re \}$;

26. Show that H is closed under multiplication and taking of inverses.

27. Show that the inverse of any A in H is also in H. Thus H is a subgroup of $GL(2, \Re)$ so H is a group in its own right.

28. Let K = {f: f(x) = ax + b} with "a" nonzero; show under *composition* that K is closed. Hint: use g(x) = cx + d.
29. Realizing that K behaves exactly like H does in Part D, find $f^{-1}(x)$; prove that it works.

One says that H and K are *isomorphic* groups. ans. $f^{-1}(x) = \frac{1}{a}x - \frac{b}{a}$

Part E. Let $H = \{ A = \begin{bmatrix} a & b \\ -b & a \end{bmatrix}, a, b \in \Re \}$;

30. Show that if A,B are in H then AB = BA. Thus H is a commutative subgroup.

31. Let $J = \begin{bmatrix} 0 & 1 \\ -1 & 0 \end{bmatrix}$; show that any matrix in G is of the form A = aI + bJ.

32. Show that $A^2 = (a^2 - b^2)I + 2abJ$.

33. If $B = \begin{bmatrix} c & d \\ -d & c \end{bmatrix}$ find AB; thus the matrices in G behave like complex numbers under

multiplication. ans. $AB = \begin{bmatrix} ac - bd & ad + bc \\ -ad - bc & ac - bd \end{bmatrix}$

Part F. Let H = {A: det(A) = 1} where $H \subseteq GL(2, \Re)$;

34. Show that H is a subgroup of G; it is called the special linear group.

35. Write down two matrices A and B in H; show that det(AB) = 1.

36. Using A from #35, find its inverse and verify that the value of its determinant is one.

37. Using $A = \begin{bmatrix} a & b \\ c & d \end{bmatrix}$, force det(A) = 1 to show how to create a matrix A which will always be in

H. ans. $A = \begin{bmatrix} \dfrac{bc+1}{d} & b \\ c & d \end{bmatrix}$

38. Let B be any matrix in $GL(2, \Re)$; if A is in H, show that $\det(B^{-1}AB) = 1$ so that $B^{-1}AB$ winds up in H. This is an important property of the special linear group $SL(n, \Re)$.

Part G. Let $H = \{ A : A^{-1} = A^t \} \subseteq GL(2, \Re)$; such matrices are called *orthogonal*.

39. Show that H is closed under multiplication.

40. Show that if A is in H, so is its inverse. Thus conclude that H is a subgroup of G.

41. Show that any matrix of the form $A = \begin{bmatrix} \dfrac{a}{\sqrt{a^2+b^2}} & \dfrac{-b}{\sqrt{a^2+b^2}} \\ \dfrac{b}{\sqrt{a^2+b^2}} & \dfrac{a}{\sqrt{a^2+b^2}} \end{bmatrix}$ is orthogonal.

42. Prove that if A is orthogonal, then det(A) = +/- 1. Hint: multiply the defining equation by something simple!

Part H. Let $H = \{ M : M = \begin{bmatrix} a & b \\ 1-a & 1-b \end{bmatrix} \}$ where $0 < a \le 1$ and $0 < b \le 1$ so M is a *Markov* matrix;

43. Show H is closed under multiplication.

44. But show that the inverse of such a matrix is NOT in H; your conclusion?

2.6 Chapter Two Outline and Review

2.1. Matrix Operations

A matrix is a rectangular array of real numbers $A = [\ a_{ij}\ ]$.

Basic operations: 1) matrix addition 2) scalar multiplication 3) matrix multiplication

Addition: $A + B = [\ a_{ij}\ ] + [\ b_{ij}\ ] = [\ c_{ij}\ ]$ *where* $c_{ij} = a_{ij} + b_{ij}$

Identity matrix under addition (zero matrix): m x n matrix with all zero entries

Properties:

associative: $(A + B) + C = A + (B + C)$

commutative: $A + B = B + A$

additive inverse: $A - B = A + (-B)$

Scalar multiplication: $k\,A = [\ k\,a_{ij}\ ]$

Matrix multiplication: $A\,B = C$ where A is **m** x p and B is p x r (C is then **m** x r)

entries of C are found by multiplying **ROWS** of A by <u>COLUMNS</u> of B

Properties:

associative: $(AB)C = A(BC)$

distributive: $A(B + C) = AB + AC, \ (A + B)C = AC + BC$

*** not *commutative*

Other operations:

1) transpose- A^t ; switch rows and columns so if $A = [\ a_{ij}\ ]$ then $A^t = [\ a_{ji}\ ]$

2) trace- tr(A) = sum of elements on main diagonal (only for square matrices)

2.2 Inverses and Square Systems

If A is a square matrix and there exists a matrix of the same size B such that AB = BA = I, then $B = A^{-1}$ and A is called <u>invertible</u>; if no such B exists, then A is *singular* (non-invertible).

System AX = B has (unique) solution $X = A^{-1}B$ if A is invertible

Algorithm for 2 by 2: $A = \begin{bmatrix} a & b \\ c & d \end{bmatrix} \Rightarrow A^{-1} = \dfrac{1}{\det(\ A\ )} A = \begin{bmatrix} d & -b \\ -c & a \end{bmatrix}$ *if* $\det(\ A\ ) \neq 0$

Finding the inverse of larger matrices- use Gauss-Jordan elimination to change $[\ A : I\ ] \sim [\ I : B\ ]$. If this is possible, then $B = A^{-1}$. If not possible, then A is singular.

Properties of inverses:

1) $(\ A^{-1}\ )^{-1} = A$

***2) $(\ A\,B\ )^{-1} = B^{-1}\,A^{-1}$

3) $(\ k\,A\ )^{-1} = \dfrac{1}{k}\,A^{-1}$

4) $(\ A^t\ )^{-1} = (\ A^{-1}\ )^t$

<u>THEOREM</u> An n by n matrix A is invertible iff rank(A) = n (ie. echelon form of A has n nonzero rows)

2.3 Elementary Matrices

An **elementary** matrix is one formed from the identity matrix I by doing one (and only one) row operation. There are three types (since there are 3 elementary row operations) and the inverse of an elementary matrix E is also elementary.

<u>CORRESPONDENCE THEOREM</u> If B is formed from A by doing one ERO and E is formed from I by doing exactly the same ERO then B = EA (ie. E keeps track of the row operation done to A via a left multiplication on the matrix A.

<u>FACTOR THEOREM</u> If A is invertible, then A can be written as the product of elementary matrices.

2.4 Determinants

Know algorithm for 2 by 2's and 3 by 3's

Determinants and row operations: if A ~ B

1) switching rows implies det(B) = -det(A)

2) multiplying one row by k implies det(B) = k det(A)

***3) adding a multiple of one row to another implies det(B) = det(A)

determinant of triangular matrix = product of elements on main diagonal

If A is 4 by 4 (or larger): reduce A to (upper) triangular form using row operations, being careful to use above three properties; if a row disappears, then det(A) = 0 (and A is singular)

*** <u>THEOREM</u> If $\det(A)\neq 0$ then A is **invertible** (if det(A) = 0, A is singular)

Properties:

1) det(AB) = det(A)det(B)

2) $\det(A^{-1}) = \dfrac{1}{\det(A)}$

3) $\det(A) = \det(A^{t})$

4) $\det(kA) = k^{n} \det(A)$ (for n x n matrix A)

2.5 Special Matrices

Group - a set of elements G and a binary operation (call it "*") on G which satisfies the following:

1) associativity: (a*b)*c = a*(b*c)

2) identity element: G has an element e such that a*e = e*a = a *for all a* in G

3) inverses:

$$a\in G \Rightarrow a^{-1}\in G \text{ such that } a * a^{-1} = a^{-1} * a = e$$

G may be a group under "+" or "multiplication"(or some other binary operation);

Most important example: $G L(n, \Re)$ (the general linear group) is the set of all *invertible* n x n matrices (under multiplication)

Special matrices: upper triangular, diagonal, symmetric, scalar, orthogonal

Subgroup: if H is a subset of group (G,*) and (H,*) itself is a group, then H is a *subgroup* of G.

If H is a subset of G = $G L(n, \Re)$ and $A B^{-1}$ is in H for all A,B in H, then H is a *subgroup* of G.

Review Problems

Part A: Given $A = \begin{bmatrix} 0 & 1 & 3 \\ 2 & 0 & 4 \end{bmatrix}$ $B = \begin{bmatrix} 0 & 1 \\ 3 & 2 \\ 0 & 4 \end{bmatrix}$; find (if possible):

1. AB ans. $\begin{bmatrix} 3 & 14 \\ 0 & 18 \end{bmatrix}$

2. BA ans. $\begin{bmatrix} 2 & 0 & 4 \\ 4 & 3 & 17 \\ 8 & 0 & 16 \end{bmatrix}$

3. $(AB)^t$ ans. $A = \begin{bmatrix} 3 & 0 \\ 14 & 18 \end{bmatrix}$

4. $(BA)^t$ ans. $\begin{bmatrix} 2 & 4 & 8 \\ 0 & 3 & 0 \\ 4 & 17 & 16 \end{bmatrix}$

5. $2A + (-3B)^t$ ans. $\begin{bmatrix} 0 & -7 & 6 \\ 1 & -6 & -4 \end{bmatrix}$

6. $B + A^t$ ans. $\begin{bmatrix} 0 & 3 \\ 4 & 2 \\ 3 & 8 \end{bmatrix}$

7. solve AX = B if $B = \begin{bmatrix} 6 \\ 7 \end{bmatrix}$ ans. $X = \begin{bmatrix} \dfrac{7}{2} - 2t \\ 6 - 3t \\ t \end{bmatrix}$

8. solve BX = C if $C = \begin{bmatrix} 2 \\ 5 \\ 4 \end{bmatrix}$ ans. inconsistent

9. $A A^t$ ans. $\begin{bmatrix} 10 & 12 \\ 12 & 20 \end{bmatrix}$

10. $B B^t$ ans. $\begin{bmatrix} 1 & 2 & 4 \\ 2 & 17 & 8 \\ 4 & 8 & 16 \end{bmatrix}$

Part B: Given $A = \begin{bmatrix} 1 & 0 & 1 \\ 0 & 1 & 2 \\ -1 & -1 & 1 \end{bmatrix}$ find:

11. A^{-1} ans. $\dfrac{1}{4} \begin{bmatrix} 3 & -1 & -1 \\ -2 & 2 & -2 \\ 1 & 1 & 1 \end{bmatrix}$

12. A^2 ans. $\begin{bmatrix} 0 & -1 & 2 \\ -2 & 1 & 4 \\ -2 & -2 & -2 \end{bmatrix}$

13. A^{-2} ans. $\dfrac{1}{8} \begin{bmatrix} 5 & -3 & -1 \\ -6 & 2 & -2 \\ 1 & 1 & -1 \end{bmatrix}$

14. $(A^t)^{-1}$ ans. $\dfrac{1}{4} \begin{bmatrix} 3 & -2 & 1 \\ -1 & 2 & 1 \\ -1 & -2 & 1 \end{bmatrix}$

15. solution of $AX = B$ for $B = \begin{bmatrix} 1 \\ 2 \\ 3 \end{bmatrix}$ using the inverse of A ans. $\begin{bmatrix} \dfrac{-1}{2} \\ -1 \\ \dfrac{3}{2} \end{bmatrix}$

16. det(A) ans. 4
17. solution of $A X = \theta$ without doing any work ans. θ
$$x_1 + 2 x_2 + \quad + x_4 = 4$$

Part C: Given the system $-x_1 + 2 x_2 + 4 x_3 - 3 x_4 = 2$;

$$-x_1 + 2 x_2 - x_3 + x_4 = 2$$

18. Write the system in the form $AX = B$. ans. $\begin{bmatrix} 1 & 2 & 0 & 1 \\ -1 & 2 & 4 & -3 \\ -1 & 2 & -1 & 1 \end{bmatrix} \begin{bmatrix} x_1 \\ x_2 \\ x_3 \\ x_4 \end{bmatrix} = \begin{bmatrix} 4 \\ 2 \\ 2 \end{bmatrix}$

19. Solve the system using Gaussian elimination.

ans. $\begin{bmatrix} 1 - \dfrac{2}{5}t \\[2mm] \dfrac{3}{2} - \dfrac{3}{10}t \\[2mm] \dfrac{4}{5}t \\[2mm] t \end{bmatrix}$

20. Write the answer as a column matrix in the form $X = X_h + X_p$.

ans. $\begin{bmatrix} -\dfrac{2}{5}t \\[2mm] -\dfrac{3}{10}t \\[2mm] \dfrac{4}{5}t \\[2mm] t \end{bmatrix} + \begin{bmatrix} 1 \\[2mm] \dfrac{3}{2} \\[2mm] 0 \\[2mm] 0 \end{bmatrix}$

21. Check your answer by a matrix multiplication; show that

$$A X_h = \theta \quad \& \quad A X_p = B$$

Part D: If $A = \begin{bmatrix} 0 & 1 \\ 2 & 0 \end{bmatrix}$;

22. Find two elementary matrices such that $E_2 E_1 A = I_2$. ans.

$$E_1 = \begin{bmatrix} 0 & 1 \\ 1 & 0 \end{bmatrix}, \ E_2 = \begin{bmatrix} \dfrac{1}{2} & 0 \\ 0 & 1 \end{bmatrix}$$

23. Write A as the product of elementary matrices using the answer to #22. ans.

$$\begin{bmatrix} 0 & 1 \\ 1 & 0 \end{bmatrix} \begin{bmatrix} 2 & 0 \\ 0 & 1 \end{bmatrix}$$

24. Show $\det(A) = \det(E_1^{-1}) \det(E_2^{-1})$.

25. Write A^{-1} as the product of elementary matrices. ans. $\begin{bmatrix} \dfrac{1}{2} & 0 \\ 0 & 1 \end{bmatrix} \begin{bmatrix} 0 & 1 \\ 1 & 0 \end{bmatrix}$

Part E: Find the determinant of each matrix; if the matrix is larger than 3 by 3, reduce the matrix to an upper triangular matrix and use a theorem to evaluate. Be aware of the effect of your row operations on the value of det(A) !

26. $A = \begin{bmatrix} 1 & 0 & 1 \\ 0 & 1 & 2 \\ 1 & 3 & 6 \end{bmatrix}$ ans. -1

27. $A = \begin{bmatrix} 1 & 0 & 1 & 2 \\ 0 & 1 & 2 & -2 \\ -1 & -1 & 4 & 1 \\ 0 & 0 & 3 & 6 \end{bmatrix}$ ans. 39

28. $A = \begin{bmatrix} 1 & 0 & 1 & 2 \\ 0 & 1 & 2 & -2 \\ -1 & -1 & 4 & 1 \\ 0 & 0 & 7 & 1 \end{bmatrix}$ ans. 0

Part F: Given A is a 3 by 3 matrix such that det(A) = 10; find det(B) if:

29. A~B and B is formed from A by performing the row operation 2 row1 + row 3
ans. 10

30. $B = A^2$ ans. 100

31. $B = A^t$ ans. 10

32. $B = A^{-1}$ ans. 1/10

33. A~B and B is formed from A by performing the row operation row2 <-> row3
ans. -10

Part G: Determine if each set G is a group under the indicated operation.

34. $G = \{ \begin{bmatrix} 1 & a \\ 0 & 1 \end{bmatrix}, \ a \in \Re \}$ under addition ans. NO

35. $G = \{ \begin{bmatrix} 1 & a \\ 0 & 1 \end{bmatrix}, \ a \in \Re \}$ under multiplication ans. YES

36. $G = \{ \begin{bmatrix} a & b \\ 0 & c \end{bmatrix}, a,b,c \in \Re \}$ under addition ans. YES

37. $G = \{ \begin{bmatrix} a & b \\ 0 & c \end{bmatrix}, a,b,c \in \Re \}$ under multiplication ans. NO

38. $G = \{ \begin{bmatrix} a & b \\ c & d \end{bmatrix}, ad - bc = 0 \}$ under addition ans. NO

39. $G = \{ \begin{bmatrix} 1 & a & b \\ 0 & 1 & c \\ 0 & 0 & 1 \end{bmatrix} : a,b,c \in \Re \}$ under multiplication ans. YES

40. G = {invertible 2 x 2 symmetric matrices} under multiplication ans. NO

41. G = {2 x 2 symmetric matrices} under addition ans. YES

42. Suppose $G = \{ A = \begin{bmatrix} a & b \\ -b & a \end{bmatrix} : A^t = A^{-1} \}$; show that for this to hold, $a^2 + b^2 = 1$.

43. Using G from #42, show that G is a group under multiplication. NB. These matrices are called *orthogonal* matrices

44. $G = \{ \begin{bmatrix} a & b \\ c & -a \end{bmatrix}, a,b,c \in \Re \}$ under addition ans. YES

45. $G = \{ \begin{bmatrix} a & b \\ c & -a \end{bmatrix}, a,b,c \in \Re \}$ under multiplication ans. NO

46. $G = \{ A = \begin{bmatrix} a & b \\ b & a \end{bmatrix} : A\ invertible \}$ under multiplication. ans.YES

47. $G = \{ \begin{bmatrix} 1 & 0 \\ 0 & 1 \end{bmatrix}, \begin{bmatrix} 1 & 0 \\ 0 & -1 \end{bmatrix} \begin{bmatrix} -1 & 0 \\ 0 & 1 \end{bmatrix} \begin{bmatrix} -1 & 0 \\ 0 & -1 \end{bmatrix} \}$ under multiplication. ans. YES

NB. This finite subgroup[5] of $GL(2,\Re)$ has many unusual properties; each matrix is its own inverse and the multiplication is commutative!

48. G = {A: dim(A) = 2 by 2 and det(A) > 0}. ans. YES

49. G = {A: dim(A) = 2 by 2 and det(A) <0 0}. ans. NO

---

[5]This subgroup is isomorphic to Klein's Viergruppe

Part H: Given the flight schedule for "GREAT LAKES AIR"; use the flight diagram below and the order Erie-Pittsburgh-Cleveland-Buffalo to answer the following:

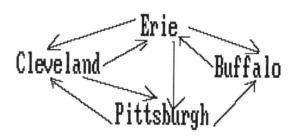

50. Find the matrix which represents the possibilities for flying from city i to city j in one flight.

ans. $A = \begin{bmatrix} 0 & 1 & 1 & 1 \\ 0 & 0 & 1 & 1 \\ 1 & 1 & 0 & 0 \\ 1 & 0 & 0 & 0 \end{bmatrix}$

51. Find the matrix which represents the possibilities for flying from city i to city j in exactly two

flights.ans. $\begin{bmatrix} 2 & 1 & 1 & 1 \\ 2 & 1 & 0 & 0 \\ 0 & 1 & 2 & 2 \\ 0 & 1 & 1 & 1 \end{bmatrix}$

52. Find the matrix which represents the possibilities for flying from city i to city j in two flights or

less.ans. $\begin{bmatrix} 2 & 2 & 2 & 2 \\ 2 & 1 & 1 & 1 \\ 1 & 2 & 2 & 2 \\ 1 & 1 & 1 & 1 \end{bmatrix}$

Part I: Given the singular matrix $A = \begin{bmatrix} 1 & 3 \\ 4 & 12 \end{bmatrix}$;

53. Let $B = \begin{bmatrix} a & b \\ c & d \end{bmatrix}$; find the product AB. ans. $\begin{bmatrix} a+3c & b+3d \\ 4a+12c & 4b+12d \end{bmatrix}$

54. Set AB equal to the zero matrix; solve the corresponding system for a,b,c and d.
ans. a = -3c, b = -3d

55. You thus showed that there are infinitely-many nonzero matrices B such that $AB = \begin{bmatrix} 0 & 0 \\ 0 & 0 \end{bmatrix}$;

these matrices are called *divisors of zero*. Give an example of at least one and check that it satisfies AB = the zero matrix. ans. $B = \begin{bmatrix} -3 & -6 \\ 1 & 2 \end{bmatrix}$

56. For any two square matrices of the same size A and B, if AB = the zero matrix, then det(AB) = 0; what does this mean about either A or B?
ans. either A or B is singular, either det(A) = 0 or det(B) = 0

Part J: Given $A = \begin{bmatrix} e^t & 0 & 1 \\ -1 & 1 & 0 \\ 0 & 0 & e^{-t} \end{bmatrix}$;

57. find det(A) ans. 1
58. find the reduced echelon form of A ans. I_3
59. determine for what t the matrix A is invertible ans. for all t

60. find the inverse of A ans. $\begin{bmatrix} e^{-t} & 0 & -1 \\ e^{-t} & 1 & -1 \\ 0 & 0 & e^t \end{bmatrix}$

Part J: Given $J = \begin{bmatrix} k & 1 & 0 \\ 0 & k & 1 \\ 0 & 0 & k \end{bmatrix}$;

61. Show that if $K = \begin{bmatrix} k & 0 & 0 \\ 0 & k & 0 \\ 0 & 0 & k \end{bmatrix}$ $L = \begin{bmatrix} 0 & 1 & 0 \\ 0 & 0 & 1 \\ 0 & 0 & 0 \end{bmatrix}$, then J = K + L and $J^2 = K^2 + 2KL + L^2$

62. Show J and K commute, then show $J^3 = K^3 + 3K^2 L + 3KL^2$.

Part K: Given $k_1(1) + k_2(x) + k_3(x^2) = 0$;

63. Differentiate this equation twice, thus producing 3 equations in $k_1, k_2, \& k_3$.

64. Write this 3 by 3 system in the form $WK = \theta$ $K = \begin{bmatrix} k_1 \\ k_2 \\ k_3 \end{bmatrix}$, where W contains 1, x and x^2 in

the first row, the first derivatives in the second row, etc.
65. Show that rank(W) = 3, so W is invertible.
66. Find det(W); observe that this is NOT zero; thus the only solution of $WK = \theta$ is $K = \theta$.
ans. det(W) = 2

Part L: Given $A = \begin{bmatrix} 1 & 1 \\ 4 & 1 \end{bmatrix}$, $P = \begin{bmatrix} 1 & 1 \\ -2 & 2 \end{bmatrix}$, $D = \begin{bmatrix} -1 & 0 \\ 0 & 3 \end{bmatrix}$;

67. show AP = PD.

68. solve for A. ans. $A = P D P^{-1}$

69. find A^2 in terms of P & D. ans. $A^2 = P D^2 P^{-1}$

70. find A^3 using the factorization in #68. ans. $A^3 = P D^3 P^{-1}$

Part M: Given $A = \begin{bmatrix} 1 & 2 \\ 0 & 3 \end{bmatrix}$;

71. if $X = \begin{bmatrix} e^t & e^{3t} \\ 0 & e^{3t} \end{bmatrix}$ find dX/dt ans. $\dfrac{dX}{dt} = \begin{bmatrix} e^t & 3 e^{3t} \\ 0 & 3 e^{3t} \end{bmatrix}$

72. show that dX/dt = AX

Part N: Leontief model of economy: use the matrix equation CX + D = X where
X - price vector, C - consumption matrix, D - demand;

73. Solve the Leontief equation for X (the price vector). ans. $X = (I - C)^{-1} D$

74. The equation has a realistic solution only if $(I - C)^{-1}$ exists and has non-negative entries; determine if this is the case for $C = \begin{bmatrix} 0.7 & 0.2 \\ 0.2 & 0.5 \end{bmatrix}$. ans. YES

75. If $D = \begin{bmatrix} 100 \\ 200 \end{bmatrix}$, find the price vector X. ans. $X = \dfrac{1}{11} \begin{bmatrix} 9000 \\ 8000 \end{bmatrix}$

Part O: Given the system of differential equations $\dfrac{dx}{dt} = 2x + y$

$\dfrac{dy}{dt} = x + 2y$

$x(0) = 3, y(0) = 4$;

76. If the system is transformed using Laplace transforms, the result is of the form

$s \bar{x} - 3 = 2 \bar{x} + \bar{y}$

$s \bar{y} - 4 = \bar{x} + 2 \bar{y}$

By putting the transformed variables (ie. $\bar{x}$ & $\bar{y}$) on the left and the constants on the right, write the system in the form AX = B. ans. $\begin{bmatrix} s - 2 & -1 \\ -1 & s - 2 \end{bmatrix} \begin{bmatrix} \bar{x} \\ \bar{y} \end{bmatrix} = \begin{bmatrix} 3 \\ 4 \end{bmatrix}$

77. Solve this system for X using the inverse of A. ans.

$$X = \begin{bmatrix} \bar{x} \\ \bar{y} \end{bmatrix} = \frac{1}{s^2 - 4s + 3} \begin{bmatrix} 3s - 2 \\ 4s - 5 \end{bmatrix}$$

78. For what values of s is this solution NOT valid??? ans. s = 1, 3

Part P: Suppose x = number of city dwellers and y = number of country dwellers in the state of Calivania; further suppose that 70% of city dwellers move to the country while 50% of country dwellers move to the city. This means that

$$0.30 \, x_0 + 0.50 \, y_0 = x_1$$

$$0.70 \, x_0 + 0.50 \, y_0 = y_1$$

Let $X_1 = \begin{bmatrix} x_1 \\ y_1 \end{bmatrix}$ & $X_0 = \begin{bmatrix} x_0 \\ y_0 \end{bmatrix}$;

79. Write this system in the form of a matrix multiplication; show that the coefficient matrix (call it M) is a Markov matrix. $\begin{bmatrix} x_1 \\ y_1 \end{bmatrix} = \begin{bmatrix} 0.3 & 0.5 \\ 0.7 & 0.5 \end{bmatrix} \begin{bmatrix} x_0 \\ y_0 \end{bmatrix} = M \, X_0$

80. Find X_1 if x_0 = 1100 and y_0 = 100. ans. $X_1 = \begin{bmatrix} 380 \\ 820 \end{bmatrix}$

81. If $X_{n+1} = M \, X_n$ find X_3. ans. $X_3 = \begin{bmatrix} 495 \\ 705 \end{bmatrix}$

82. Show that no one moves if x0 = 500 and y0 = 700; this is called the *steady state*.

$$\frac{d\,y}{d\,t} = v$$

Part Q: Given the system of differential equations   with y(0) = 1 and v(0) = 2; if this

$$\frac{d\,v}{d\,t} = 6\,y + v$$

system is transformed using Laplace transforms, one gets

$$s\,\bar{y} - 1 = \bar{v}$$

$$s\,\bar{v} - 2 = 6\,\bar{y} + \bar{v}$$

83. Write this system in the form of a matrix multiplication, with $Y = \begin{bmatrix} \bar{y} \\ \bar{v} \end{bmatrix}$.

ans. $\begin{bmatrix} s & -1 \\ -6 & s-1 \end{bmatrix} \begin{bmatrix} \bar{y} \\ \bar{v} \end{bmatrix} = \begin{bmatrix} 1 \\ 2 \end{bmatrix}$

84. Solve for Y using the inverse of the coefficient matrix. ans.

$$Y = \begin{bmatrix} \bar{y} \\ \bar{v} \end{bmatrix} = \frac{1}{s^2 - s - 6} \begin{bmatrix} s-1 & 6 \\ 1 & s \end{bmatrix} \begin{bmatrix} 1 \\ 2 \end{bmatrix}$$

85. For what s-values is the coefficient matrix singular? ans. s = -2,3

Part R: Given the matrix $A = \begin{bmatrix} a & b \\ c & d \end{bmatrix}$;

86. Find B and C if $B = A + A^t$ and $C = A - A^t$ .ans.

$$B = \begin{bmatrix} 2a & b+c \\ b+c & 2d \end{bmatrix}, \quad C = \begin{bmatrix} 0 & b-c \\ c-b & 0 \end{bmatrix}$$

87. Show that B is symmetric (ie. $B = B^t$) and C is skew-symmetric (ie. $C = -C^t$).
88. Show that A = 1/2(B + C).
89. Find B and C for $A = \begin{bmatrix} 1 & 2 \\ 3 & 4 \end{bmatrix}$. ans. $B = \begin{bmatrix} 2 & 5 \\ 5 & 8 \end{bmatrix}$, $C = \begin{bmatrix} 0 & -1 \\ 1 & 0 \end{bmatrix}$

Part S: Let $A = \frac{1}{5} \begin{bmatrix} 1 & 2 \\ 2 & 4 \end{bmatrix}$;

90. Show that $A^2 = A$ so A is *idempotent*.

91. If U = 2A - I and A is idempotent, show in general that $U^2 = \begin{bmatrix} 1 & 0 \\ 0 & 1 \end{bmatrix} = I$. U is called

unipotent. Verify this for the particular matrix A given.

92. Show in general that U is *its own inverse*; verify this for the matrix U formed from the given

matrix A. ans. $U = \frac{1}{5} \begin{bmatrix} -3 & 4 \\ 4 & 3 \end{bmatrix}$

CHAPTER 3

VECTOR SPACES

CHAPTER OVERVIEW

This chapter involves the understanding of what a **vector space** is; our model is the way vectors behave in 2-space or 3-space. We only need two operations- vector *addition* and *scalar multiplication*. We observe what properties these operations have in 2-space or 3-space and use these as our **axioms** for the concept of a vector space-any set of elements where vector additions and scalar multiplication behave precisely the same as in 2-space or 3-space. We will think of polynomials, matrices and even functions as vectors. If a subset of a vector space is itself a vector space, then this will be called a **subspace** and when solving systems of equations, we actually have been dealing with subspaces even in chapter one.

Thinking of the vector X = (3,4) as 3(1,0) + 4(0,1) makes us think of (3,4) as a **linear combination** of (1,0) and (0,1); since k1(1,0) + k2(0,1) = (0,0) only if k1 = k2 = 0, then we will say that the set {(1,0),(0,1)} is an **independent** set of vectors. Taking it further, since 2-space is formed from this very same set, we will say that $\Re^2 = span\{ (1,0),(0,1) \}$ which means that {(1,0),(0,1)} is a **basis** for 2-space. Also 2-space is called "2-space" since it takes two independent vectors (no more and no less) to span 2-space. This means that 2-space is *two*- dimensional.

When solving any system of equations AX = B, we will show that we thereby produce 3 subspaces formed from the matrix A- the **row space** of A, the *column space* of A and the **null** space of A. The row echelon form of A will be a key ingredient in determining the sizes of these 3 subspaces.

Finally, we discuss the **coordinates** of a vector. Precisely that when we write down a vector like X = (3,4) we automatically are thinking of a *standard basis* (1,0) and (0,1) and that the numbers (scalars) 3 and 4 are the multipliers (in order) which when combined with the basis vectors **i** = (1,0) and **j** = (0,1) produce the given vector (3,4). Also we will determine what effect a *change of basis* will have on the coordinates of a given vector.

3.1 Vector Spaces

Recall that when we solve linear systems, the solution is an n-tuple (a point or set of points in n-space, that is, having n components). We can think of such an object as a vector, by which the physicists means some quantity which has both a magnitude (size) and a direction. In 2-space or 3-space, we can think of a vector as an arrow drawn from the origin. In 5-space of course we cannot draw any arrows. We wish to generalize the algebraic structure of vectors in n-space, which we call $\mathfrak{R}^n$. Essentially we need to be able to add two vectors and to multiply a vector by a scalar. In 3-space these operations can be visualized with diagrams but mathematicians don't wish to rely on diagrams for proofs. We need to be able to prove things algebraically. For notation purposes, we will feel free to represent a vector in n-space horizontally with the traditional mathematical notation: $X = (\ x_1, x_2, ..., x_n\ )$ or vertically with the column matrix notation: $X = \begin{bmatrix} x_1 \\ x_2 \\ ... \\ x_n \end{bmatrix}$. Sometimes things that you don't see easily in a horizontal format are easily seen in a vertical format so feel free to use whichever format you like best.

As far as the two operations of addition and scalar multiplication are concerned, the following definitions are the ones which are standard:

$$X + Y = (\ x_1, x_2, ..., x_n\ ) + (\ y_1, y_2, ..., y_n\ ) = (\ x_1 + y_1, x_2 + y_2, ..., x_n + y_n\ )$$

In other words, addition is simply done component-wise. Scalar multiplication is just as easy:

$$k\,X = (\ k\,x_1, k\,x_2, ..., k\,x_n\ )$$

These definitions correspond to the standard engineering or physics definitions when one writes vectors in the i-j-k format in 3-space:

$$\vec{X} = x_1 \vec{i} + x_2 \vec{j} + x_3 \vec{k} \Rightarrow c\,\vec{X} = c\,x_1 \vec{i} + c\,x_2 \vec{j} + c\,x_3 \vec{k}$$

As far as the algebraic properties of these two operations, it would be very easy (although somewhat tedious) to prove the following fundamental properties:

1) $X + Y = Y + X$ $\qquad\qquad$ $X + (Y + Z) =)X + Y) + Z$
(addition is both commutative and associative)

2) existence of additive identity $\theta \in \mathfrak{R}^n$ (zero vector theta) which is the unique vector satisfying $X + \theta = \theta + X = X$

3) $X \in \mathfrak{R}^n \Rightarrow - X \in \mathfrak{R}^n$ *such that* $X + (- X\ ) = \theta$ (existence of additive inverses)

4) for all scalars $k \in \mathfrak{R}$ and all vectors $X \in \mathfrak{R}^n$ the scalar multiple $kX \in \mathfrak{R}^n$

If $X, Y \in \mathfrak{R}^n$ then for all scalars k and j:

5) $k(X + Y) = KX + kY$

6) $(k + j)X = kX + jX$

7) $(jk)X = j(kX)$

8) $1X = X$

We will consider these to be the *fundamental properties of how vectors should behave* and generalize these so that they become axioms- that is, any set of elements having these two operations (vector addition and scalar multiplication) behaves exactly like vectors in n-space and will be called a vector space. The definition for a vector space is thus the following:

DEFINITION

Given any (non-empty) set of elements V and scalars $k \in \Re$; V is called a **vector space** if the following axioms are satisfied:

---

1) addition[1] is a binary operation on V which is both commutative and *associative*
2) there exists a **zero vector** $\theta \in V$ such that $X + \theta = \theta + X = X$ for all X in V
3) for all X in V, there exists -X in V such that $X + (- X) = \theta$
4) for all scalars k and all vectors X in V, kX is in V

5) *scalar multiplication properties* (for all $X, Y \in V$ & $k, j \in \Re$):
a) k(X + Y) = kX + kY
b) (k + j)X = kX + jX
c) (jk)X = j(kX)
d) 1X = X

NB. We say V is a vector space over the field $\Re$, since the scalars are *real* numbers.
If "+" is a binary operation on V, this means that $+ | V \ x \ V \to V$ ie. addition takes two elements in V and produces a third vector *also in V* (the child resembles the parents). This means that V must be "**closed under addition**". The set consisting of the zero vector alone trivially satisfies all these axioms and is called the *trivial* (vector) *space*.

EXAMPLE 1 Show that the set of all 2 by 2 matrices

$$V = \left\{ \begin{bmatrix} a & b \\ c & d \end{bmatrix} : a, b, c, d \in \Re \right\}$$

forms a vector space.
1) We already know that matrix addition is commutative and associative
2) The zero vector is the zero matrix $\theta = \begin{bmatrix} 0 & 0 \\ 0 & 0 \end{bmatrix} \in V$.

3) the additive inverse for any $M \in V$ is $-M = \begin{bmatrix} -a & -b \\ -c & -d \end{bmatrix}$ which is also in V

4) clearly multiplying a 2 by 2 matrix by a scalar produces another 2 by 2 matrix:

---

[1] The first 3 axioms could be replaced by the simple statement that V forms a commutative group under addition.

$$kM = k\begin{bmatrix} a_1 & b_1 \\ c_1 & d_1 \end{bmatrix} = \begin{bmatrix} k\,a_1 & k\,b_1 \\ k\,c_1 & k\,d_1 \end{bmatrix} = \begin{bmatrix} a_2 & b_2 \\ c_2 & d_2 \end{bmatrix} \in V$$

5) scalar multiplication properties
a)

$$k(\,M+N\,) = k\left(\begin{bmatrix} a_1 & b_1 \\ c_1 & d_1 \end{bmatrix} + \begin{bmatrix} a_2 & b_2 \\ c_2 & d_2 \end{bmatrix}\right) = k\left(\begin{bmatrix} a_1+a_1 & b_1+b_2 \\ c_1+c_2 & d_1+d_2 \end{bmatrix}\right) =$$

$$\begin{bmatrix} k(\,a_1+a_2\,) & k(\,b_1+b_2\,) \\ k(\,c_1+c_2\,) & k(\,d_1+d_2\,) \end{bmatrix} = \begin{bmatrix} k\,a_1+k\,a_2 & k\,b_1+k\,b_2 \\ k\,c_1+k\,c_2 & k\,d_1+k\,d_2 \end{bmatrix} =$$

$$\begin{bmatrix} k\,a_1 & k\,b_1 \\ k\,c_1 & k\,d_1 \end{bmatrix} + \begin{bmatrix} k\,a_2 & k\,b_2 \\ k\,c_2 & k\,d_2 \end{bmatrix} = k\begin{bmatrix} a_1 & b_1 \\ c_1 & d_1 \end{bmatrix} + k\begin{bmatrix} a_2 & b_2 \\ c_2 & d_2 \end{bmatrix} = k\,M+k\,N$$

The other scalar multiplication properties are proved similarly; one may also use previous properites of matrices (cf. Section 2.1) to claim that all the scalar multiplication axioms hold in V.

EXAMPLE 2 Show that the set of functions $V = \{\ f : f\ [\ 0,2\ ] \rightarrow \Re,\ f\ continuous\ \}$ is a vector space.
First, we need to know how to add two functions. Well, if we want to add f(x) and g(x), we define
$(f + g)(x) = f(x) + g(x)$; also $(kf)(x) = k\,f(x)$ for all x in [0,2].
1) clearly functional addition is both commutative and associative, since real addition is both commutative and associative:
$(f + g)x = f(x) + g(x) = g(x) + f(x) = (g + f)(x)$ and

$((f + g) + h)(x) = (f + g)(x) + h(x) = (f(x) + g(x)) + h(x) = f(x) + (g(x) + h(x)) = (f + (g + h))(x)$
2) the zero vector is the function $\theta(\,x\,) = 0$ which is in V since a constant function is always continuous
3) additive inverse of f(x) is –f(x), which is continuous if f(x) is continuous and therefore is in V
4) closure under scalar multiplication: $k \in \Re\ and\ f(\,x\,) \in V \Rightarrow (\,k\,f\,)(\,x\,) \in C[\,0,2\,]$

(ie. a scalar multiple of a continuous function is a continuous function)
5) scalar multiplication properties:
a) $k[\,f(\,x\,) + g(\,x\,)] = k\,f(\,x\,) + k\,g(\,x\,) = (\,k\,f\,)(\,x\,) + (\,k\,g\,)(\,x\,)$
b) $(\,k+j\,)\,f(\,x\,) = k\,f(\,x\,) + j\,f(\,x\,) = (\,k\,f\,)(\,x\,) + (\,j\,f\,)(\,x\,)$
c) $(\,k\,j\,)\,f(\,x\,) = k(\,j\,f(\,x\,)\,) = k(\,j\,f\,)(\,x\,)$
d) $1 \cdot f(\,x\,) = (\,1 \cdot f\,)(\,x\,) = f(\,x\,)$

Therefore the set V of functions continuous on [0,2] is a vector space and we can think of functions then as vectors. The interval [0,2] is in no way special and we can say in general that functions continuous on [a,b] form a vector space, which we denote C[a,b]. This is a very important vector space; this type of vector space is called a *function space*, since its elements are really functions[2].

EXAMPLE 3 Determine if the set $V = \{\ p(\ x\ ): p(\ x\ ) = 3 + k_1\ x,\ \ k_1 \in \Re\ \ \}$ is a vector space.

1) Certainly addition of polynomials is both commutative and associative, since real addition is both commutative and associative. However, if we add two polynomials in V, we get

$$p(\ x\ ) = 3 + k_1\ x,\ \ q(\ x\ ) = 3 + k_2\ x \Rightarrow p(\ x\ ) + q(\ x\ ) = 6 + (\ k_1 + k_2\ )\ x$$

and this polynomial is NOT in V. Thus V is NOT a vector space. Thus to show that a set of elements does not form a vector space, we need find *only one axiom* that fails.

EXAMPLE 4 Show that the set of polynomials

$V = \{\ p(\ x\ ): p(\ x\ ) = a\ x^2 + b\ x + c\ , a, b, c \in \Re\ \}$ is a vector space.

1) clearly addition of polynomials is both commutative and associative, since real addition is both commutative and associative. This is very familiar from high school algebra.

2) the zero vector is the polynomial $\theta(\ x\ ) = 0$ (let a = 0, b = 0, c = 0)

3) additive inverse of $p(\ x\ ) = a\ x^2 + b\ x + c$ is $-p(\ x\ ) = -a\ x^2 - b\ x - c$, which is certainly a polynomial of degree 2 or less

4) closure under scalar multiplication:

$$k \in \Re\ and\ p(\ x\ ) \in V \Rightarrow k\ p(\ x\ ) = k\ [\ a\ x^2 + b\ x + c\ ]\ k\ a\ x^2 + k\ b\ x + k\ c = a_1\ x^2 + b_1\ x + c_1 \in P_2$$

(ie. a scalar multiple of a polynomial of degree 2 or less is a polynomial of degree 2 or less)

5) scalar multiplication properties:

a) $k\ [\ p(\ x\ ) + q(\ x\ )] = k\ p(\ x\ ) + k\ q(\ x\ )$

b) $(\ k + j\ )\ p(\ x\ ) = k\ p(\ x\ ) + j\ p(\ x\ )$

c) $(\ k\ j\ )\ p(\ x\ ) = k(\ j\ p(\ x\ ))$

d) $1 \cdot p(\ x\ ) = p(\ x\ )$

Therefore the set V of polynomials of degree 2 or less is a vector space and we can think of these polynomials as *vectors*. The fact that we used degree 2 was in no way special and we can say in general that the set of all polynomials of degree n or less forms a vector space, which we denote P_n.

For reference, here is a list of important vector spaces; you should know the notation and what type of vectors are in each vector space.

<u>Classic Vector Spaces</u>

Vector Space	Notation
n-dimensional space	$\Re^n$
polynomials of degree n or less	P_n

---

[2] Functions will still be denoted by lowercase letters, since using capital letters would cause too much confusion.

111

polynomial of any degree	P
m by n matrices	$M_{m,n}$
continuous functions on [a,b]	$C[a,b]$
continuous functions on $\Re$	C(-oo,oo)

EXERCISES 3.1

Part A. Given the set $V = \{\, p : p = a + b\,x^2, a, b \in \Re \,\}$ where addition is just ordinary addition of polynomials. Show that V is a vector space by proving the listed axioms hold:
1. axiom 1
2. axiom 2
3. axiom 3
4. axiom 4
5. axiom 5 (parts a and b)
6. axiom 5 (parts c and d)

Part B. Using V from EX1, prove:
7. axiom 5 b)
8. axiom 5 c)
9. axiom 5 d)

Part C. Given the set $V = \{\, A : A = \begin{bmatrix} a & 0 \\ 0 & b \end{bmatrix}, a, b \in \Re \,\}$; show that V is a vector space by proving the listed axioms:
10. axiom 1
11. axiom 2
12. axiom 3
13. axiom 4
14. axiom 5 (parts a and b)
15. axiom 5 (parts c and d)

Part D. Given the set $V = \{\, A = \begin{bmatrix} a & b \\ 0 & c \end{bmatrix}, a, b, c \in \Re, a + c = 0 \,\}$; show that V is a vector space by proving the listed axioms:
16. axiom 1
17. axiom 2
18. axiom 3
19. axiom 4
20. axiom 5 (parts a and b)

Part E. Given the set $V = \{ f : \int_a^b f \, dx \text{ exists} \}$; show

V is vector space by proving the listed axioms:
21. axiom 1
22. axiom 2
23. axiom 3
24. axiom 4
25. axiom 5 (parts a and b)

Part F. Given the set $V = \{ t(1,2,3) : t \in \Re \}$; show V is a vector space by proving the listed axioms:
26. axiom 1
27. axiom 2
28. axiom 3
29. axiom 4
30. axiom 5 (parts a and b)
NB. geometrically V is a line through the origin

Part G. Show that the set V does NOT form a vector space by finding the first axiom that fails.
31. $V = \{ f(x) : f(x) = 1, \; f \mid \Re \to \Re \}$ ans. axiom 1(sum of two such functions not in V)
32. $V = \{ p(x) : p(x) = 2 + k_1 x^2, \; k_1 \in \Re \}$
33. $V = \{ (x, y) : x, y \in \Re \; \& \; x \geq 0 \}$ ans. axiom 3 ((x,y) has no additive inverse in V since
x cannot be < 0)
34. $V = \left\{ \begin{bmatrix} a & 1 \\ 0 & b \end{bmatrix} : a, b \in \Re \right\}$

Part H. Given the set $V = \{ (x, y, z) : 2x - 3y + z = 0, \; x, y, z \in \Re \}$; show V is a vector space by proving the listed axioms: Hint: solve for z first and rewrite V.
35. axiom 1
36. axiom 2
37. axiom 3
38. axiom 4
39. axiom 5 (parts a and b)

Part I. Given the set $K = \{ a + b\sqrt{2} : a, b \in Q \}$; K is a vector space over Q (the field of rational numbers).
40. Show K is closed under addition and scalar multiplication (scalars here are elements of Q).
41. Find the additive inverse of $k = a + b\sqrt{2}$ and show it is an element of K.
42. Show that axioms 5 (parts a and b) hold.
NB. This type of vector space is used when studying *extension fields* (needed when solving polynomial equations over Q)
43. Show that K is closed under vector multiplication (this is NOT a vector space axiom!!!)
44. Show that the multiplicative inverse of $k = a + b\sqrt{2}$ is in K (where $a \neq 0, b \neq 0$).

NB. A vector space which is closed under MULTIPLICATION of vectors (together with a few other properties) is called an *algebra*.

45. P_2 could NOT form an algebra (why??).

ans. product of two degree two polynomials is of degree FOUR(and thus NOT in P2).

3.2 Subspaces

Suppose we analyze a subset W of a given vector space V. A logical question arises: can it be that W itself is a vector space? For example, in $\Re^3$ consider the set

$$W = \{ \ (\ x,0,z \ ) : x \in \Re, y \in \Re \ \}$$

Clearly if we add two vectors with y component zero, we get a third vector with y-component zero. If we multiply (x,0,z) by k, we get another vector with zero y-component. Since W is already contained in $\Re^3$ (a known vector space), all the axioms of a vector space would hold by inheritance. Therefore W itself is a vector space but since $W \subseteq V$, we call W a **subspace**. The following theorem is very helpful:

THEOREM Given W is a (non-empty) subset of the vector space V; W is a **subspace** of V (and therefore a vector space in its own right) iff W is *closed* under
1) *vector addition* and 2) *scalar multiplication*

NB. If W consists of just the zero vector or W = V, then clearly W is automatically a subspace of V. We are mainly interested in *proper subspaces* of V.

EXAMPLE 1 Show that any solution of a homogeneous linear equation forms a subspace of $\Re^n$. Any linear equation is of the form

$$a_1 x_1 + a_2 x_2 + \dots + a_n x_n = 0$$

Let

$$W = \{ \ (\ x_1, x_2, \dots, x_n \ ) : a_1 x_1 + a_2 x_2 + \dots + a_n x_n = 0 \, , a_i \in \Re \ \}$$

and suppose two solutions of this equation are:

$$X = (\ x_1, x_2, \dots, x_n \ ) \quad Y = (\ y_1, y_2, \dots, y_n \ )$$

Then we have:

$$a_1 x_1 + a_2 x_2 + \dots + a_n x_n = 0 \ \& \ a_1 y_1 + a_2 y_2 + \dots + a_n y_n = 0 \Rightarrow$$

$$(\ a_1 x_1 + a_2 x_2 + \dots + a_n x_n \ ) + (\ a_1 y_1 + a_2 y_2 + \dots + a_n y_n \ ) = 0$$

$$a_1(\ x_1 + y_1 \ ) + a_2(\ x_2 + y_2 \ ) + \dots + a_n(\ x_n + y_n \ ) = 0$$

so X + Y $= (\ x_1 + y_1, x_2 + y_2, \dots, x_n + y_n \ )$ is a solution and lies in W so the set is closed under addition.

Closure under scalar multiplication:

$$a_1 x_1 + a_2 x_2 + ... + a_n x_n = 0 \Rightarrow k(a_1 x_1 + a_2 x_2 + ... + a_n x_n) = 0$$

But

$$k(a_1 x_1 + a_2 x_2 + ... + a_n x_n) = a_1(k x_1) + a_2(k x_2) + ... + a_n(k x_n) = 0$$

so that

$$(k x_1, k x_2, ..., k x_n) = k X \in W$$

for any scalar k so W is closed under scalar multiplication and therefore it is a subspace. In fact it is a *hyperplane* through the origin.

EXAMPLE 2 Consider the set W = {(x,y): $x \geq 0$, $y \geq 0$} which is essentially the set of vectors in quadrant I (or its boundary). Determine if W is a subspace of V = $\Re^2$.

Let X = (x_1, y_1) and Y = (x_2, y_2) be in W; then clearly W is closed under addition since

$$X + Y = (x_1, y_1) + (x_2, y_2) = (x_1 + y_1, x_2 + y_2) \qquad x_1 + y_1 \geq 0, \ x_2 + y_2 \geq 0$$

Now for scalar multiplication:

$$k X = k(x_1, y_1) = (k x_1, k y_1)$$

but kX does NOT lie in W if k < 0. Thus W is not a vector subspace of V. Note that by limiting vectors to quadrant I, we have neither additive inverses for any vector in W nor closure under scalar multiplication.

EXAMPLE 3 Consider the set W of matrices $W = \left\{ \begin{bmatrix} a & b \\ c & d \end{bmatrix} : ad - bc = 0 \right\}$. Thus W consists of singular 2 by 2 matrices. Let's A and B be two particular matrices in W:

$$A = \begin{bmatrix} 1 & 2 \\ -2 & -4 \end{bmatrix} \qquad B = \begin{bmatrix} 2 & 3 \\ 4 & 6 \end{bmatrix} \qquad \det(A) = 0 = \det(B)$$

But if we add A and B we have:

$$A + B = \begin{bmatrix} 3 & 5 \\ 2 & 2 \end{bmatrix} \Rightarrow \det(A + B) = -4$$

Thus for this A and B, det(A + B) is not zero so the sum is NOT in W. One counter-example shows that W cannot be a subspace. This shows that two matrices with determinant zero can add up to a third matrix whose determinant is NOT zero (ie. the child does not resemble the parents).

EXAMPLE 3 Let $W = \{ A = \begin{bmatrix} a & b \\ c & d \end{bmatrix} : A = A^t \} \subseteq M_{2,2}$; show W is a subspace of $V = M_{2,2}$. Let

A & B be two matrices in W; then $A^t = A$ & $B^t = B$.

To show closure under addition:

$$(A + B)^t = A^t + B^t = A + B$$

so A + B is in W whenever A and B are.

Now for closure under scalar multiplication:

$$(k A)^t = k A^t = k A$$

so that kA is in W whenever A is. Thus W is closed under both vector addition and scalar multiplication so W is a subspace of V. Note that W is just the set of all 2 x 2 symmetric matrices so $W = \{ A = \begin{bmatrix} a & b \\ b & d \end{bmatrix} \}$.

Calculus considerations

Consider the set of continuous functions C[a,b]; from calculus we know that a function f in C[a,b] is always integrable on [a,b]; that is $\int_a^b f (x) \, dx$ exists. Therefore the set of continuous functions is a proper subset of the set of all integrable functions on [a,b]. Let's call

$$I [a,b] = \{ f (x) : \int_a^b f (x) \, dx \ \ exists \}$$

We know that I[a,b] is in fact a vector space (cf. Section 3.1, Exercises, part E)- a vector space of integrable functions. Some of them are not continuous on [a,b] . But C[a,b] is a *subset* of I[a,b].
Question: if we knew nothing about C[a,b] except that it is a subset of I[a,b], do we have to prove all 5 axioms in order to determine whether C[a,b] is also a vector space? The answer is no- we use the theorem to prove that C[a,b] is **closed** under vector addition and scalar multiplication. Then, since C[a,b] is a subset of the vector space I[a,b]) and C[a,b] is closed under addition and scalar multiplication, it is a **subspace** of I[a,b]. Thus it is a vector space in its own right. Moral of the story- you can save yourself lots of time if you know that W is a subset of a known vector space V.

EXAMPLE 4 Let D[a,b] = {f: df/dx exists on [a,b]}; is D[a,b] a subspace of C[a,b]?
From calculus we know that if a function is differentiable, it is automatically continuous so clearly D[a,b] is a proper subset of C[a,b]. Let f, g be two functions in D[a,b]; then
d/dx[f + g] = df/dx + dg/dx implies f + g is in D[a,b] so it is closed under addition. We also know that if h = kf then dh/dx = k df/dx ; thus D[a,b] is also closed under scalar mulitplication. Thus D[a,b] is a subspace of C[a,b].

EXERCISES 3.2

Part A. In 3-space show that the following are or are not subspaces of 3-space:
1. $W = \{(x,y,z): x - 2y + z = 0\}$ ans. subspace

2. $W = \{(x,y,z): (x,y,z) = t(2,4,5)\}$

3. $W = \{(x,y,z): x - 3y + z = 2\}$ ans. NOT

4. $W = \{(0,0,0)\}$

5. $W = \{(x,y,z): (x,y,z) = s(1,0,1) + t(-2,4,0)$ where s,t are any real numbers$\}$ ans. subspace

6. $W = \{(x,y,z): (x,y,z) = s(1,0,1) + (3,2,1)\}$

Part B. In 4-space, determine if the following are subspaces:

7. $W = \{(x,y,z,w): (x,y,z,w) = s(1,0,1,0) + t(0,2,0,3)\}$ ans. subspace

8. $W = \{(x,y,z,w): (x,y,z,w) = r(1,0,1,0) + s(0,2,0,3) + t(1,2,1,4)\}$

9. $W = \{(x,y,z,w): (x,y,z,w) = t(1,0,1,0) + (1,0,3,4)\}$ ans. NOT

Part C. In $M_{2,2}$ determine if the following are subspaces:

10. $W = \left\{ \begin{bmatrix} a & b \\ a+b & c \end{bmatrix} a,b,c \in \Re \right\}$

11. $W = \left\{ \begin{bmatrix} a & -a \\ b & 2b \end{bmatrix} a,b \in \Re \right\}$ ans. YES

12. $W = \left\{ \begin{bmatrix} a & -b \\ b-a & 2a \end{bmatrix} a,b \in \Re \right\}$

13. $W = \left\{ \begin{bmatrix} a & b \\ -a & c+1 \end{bmatrix} a,b,c \in \Re \right\}$ ans. NOT

14. $W = \left\{ \begin{bmatrix} -a & b \\ a & c+b \end{bmatrix} a,b,c \in \Re \right\}$

15. $W = \left\{ \begin{bmatrix} a & -b \\ b & a \end{bmatrix} a,b \in \Re \right\}$ ans. YES

16. $W = \left\{ \begin{bmatrix} a & b+c \\ b-c & -a \end{bmatrix} a,b,c \in \Re \right\}$

17. $W = \{ A : A + A^t = \begin{bmatrix} 0 & 0 \\ 0 & 0 \end{bmatrix}, \text{ where } A \in M_{2,2}, A = \begin{bmatrix} a & b \\ c & d \end{bmatrix} \}$

Hint: $(A + B)^t = A^t + B^t$ ans. YES

Part D. Given the set W is a subset of P_3; determine whether W is a subspace or not:

18. $W = \{ p : deg(p) \leq 2 \}$

19. $W = \{ p : p = a + b x^3, a, b \in \Re \}$ ans. YES

20. $W = \{ p : p = a + b x^2 + c x^3, a + b + c = 0 \}$

21. $W = \{ p : p = a + b x^2 + c x^3, a + b + c = 1 \}$ ans. NO

Part E. In $\Re^4$ determine whether the subset W is a subspace:

22. $W = \{$ all solutions of the system $\quad \begin{aligned} x_1 + 2x_2 - x_3 - x_4 &= 0 \\ 2x_1 - x_2 + x_3 + x_4 &= 0 \end{aligned} \}$

23. $W = \{$ all solutions of the system $\begin{aligned} x_1 + 2x_2 - x_3 - x_4 &= 0 \\ 2x_1 - x_2 + x_3 + x_4 &= 0 \}$ ans. YES

24. $W = \{$ all solutions of the equation $x_1 + 2x_2 - x_3 - x_4 = 0 \}$

25. $W = \{(x,y,z): x - y + z = 1, 2x + 3y + 4z = 2\}$ ans. NO

26. $W = \{(x,y,z,w): w - 2x + y - 3z = 0\}$

27. $W = \{(x,y,z,w): (x,y,z,w) = t(1,2,3,4)\}$ ans. YES

Part F. Given the vector space $V = F[a,b]$ of *all functions* defined on [a,b] is a vector space; determine whether W (a subset of V) is a subspace of V

28. $W = \{ f : \int_a^b f \, dx \text{ exists} \}$

29. $W = \{\ f : \dfrac{d\,f}{d\,x}$ on $[\ a,b\ ]$ exists$\}$ ans. subspace

30. $W = \{\ f : \dfrac{d^2 f}{d\,x^2}$ exists on $[\ a,b\ ]\ \}$

31. $W = \{\ f : \displaystyle\int_a^b f\,d\,x$ exists$\}$ ans. subspace

32. $W = \{\ f : \displaystyle\int_a^b f\,d\,x = 0\ \}$

33. $W = \{\ f : \displaystyle\int_a^b f\,d\,x = 1\ \}$ ans. NOT

34. $W = \{\ f : f(a\ ) = 0\ \}$

35. $W = \{\ f : f''(\ a\ ) = 0\ \}$ ans. subspace

36. $W = \{\ f : f'(\ a\ ) = 1\ \}$

Part G. Let W be the set of all solutions of the given differential equation. Determine whether W is a subspace of C[0,4], the space of all continuous functions on [0,4].

37. dy/dx - 2y = 0; $W = \{\ y : y = C\,e^{2x}\ \}$ ans. YES

38. y" + y = 0; $W = \{\ y : y = C_1 \cos(\ x\ ) + C_2 \sin(\ x\ )\ \}$

39. dy/dx + 3y = 6; $W = \{\ y : y = C\,e^{-3x} + 2\ \}$ ans. NO
40. y" - 4y = 0; $W = \{\ y : y = C_1 \cosh(\ 2\,x\ ) + C_2 \sinh(\ 2\,x\ )\ \}$

41. dy/dx + 2y = 6x + 11; $W = \{y: y = c \exp(-2x) + 3x + 4\}$ ans. NO

3.3 Linear Independence and Spanning

In this section we need to define some very basic terminology associated with vector spaces. First we need the concept of linear combination.

DEFINITION

Given a set of vectors $S = \{ X_1, X_2, ..., X_n \}$ in a vector space V; if there exist corresponding scalars $k_1, k_2, ..., k_n$ such that

$$X = k_1 X_1 + k_2 X_2 + ... + k_n X_n$$

then we say that the vector X is a **linear combination** of the vectors in S.

EXAMPLE 1 Suppose $S = \{(1,2,3,4),(0,1,-4,5)\}$. Determine if the vector $X = (2,7,-6,23)$ is a linear combination of these two vectors.
If so, there exist two scalars k_1, and k_2 such that

$$k_1(1,2,3,4) + k_2(0,1,-4,5) = (2,7,-6,23)$$

This implies

$$k_1 = 2$$

$$2 k_1 + k_2 = 7$$

$$3 k_3 - 4 k_2 = -6$$

$$4 k_1 + 5 k_2 = 23$$

$$\begin{bmatrix} 1 & 0 & 2 \\ 2 & 1 & 7 \\ 3 & -4 & -6 \\ 4 & 5 & 23 \end{bmatrix} \sim \begin{bmatrix} 1 & 0 & 2 \\ 0 & 1 & 3 \\ 0 & 0 & 0 \\ 0 & 0 & 0 \end{bmatrix}$$

It should not take you long to figure out that $k_1 = 2$ and $k_2 = 3$ and this is the only solution to the system. This means that X is indeed a linear combination of the two vectors (1,2,3,4) and (0,1,-4,5).
NB. The solution need not be unique; all we care is that at least one solution to the system exists!
EXAMPLE 2 Suppose

$$S = \{ 1 + x^2, 2 + 4x - x^3, -3 + x - 4x^2 \} \subseteq P_3$$

Is the polynomial $p(x) = 5x - 3x^2 + x^3$ a linear combination of the vectors in S?
If so, there exist three scalars k_1, k_2, and k_3 such that

$$k_1(\, 1 + x^2 \,) + k_2(\, 2 + 4 \, x - x^3 \,) + k_3(\, -3 + x - 4 \, x^2 \,) = 5 \, x - 3 \, x^2 + x^3$$

Grouping like terms implies:

$$(\, k_1 + 2 \, k_2 - 3 \, k_3 \,) + (4 \, k_2 + k_3 \,) \, x + (\, k_1 - 4 \, k_3 \,) \, x^2 + (\, -k_2 \,) \, x^3 = 5 \, x - 3 \, x^2 + x^3$$

This leads to the augmented matrix

$$\begin{bmatrix} 1 & 2 & -3 & 0 \\ 0 & 4 & 1 & 5 \\ 1 & 0 & -4 & -3 \\ 0 & -1 & 0 & 1 \end{bmatrix}$$

whose row-echelon form is

$$\begin{bmatrix} 1 & 2 & -3 & 0 \\ 0 & 1 & 0 & -1 \\ 0 & 0 & 1 & 5 \\ 0 & 0 & 0 & 1 \end{bmatrix}$$

The last row indicates that the system is *inconsistent*, so the polynomial p(x) is NOT a combination of the given three polynomials. Moral of the story: consistent system implies given vector IS a combination of the vectors in S, *inconsistent* system means that the given vector is NOT a combination of the vectors in S.

The next idea builds on the idea of linear combination; what happens if two linear combinations (of vectors from a set S) are added? Well, the sum is another combination of vectors from S. If a linear combination of vectors from S is multiplied by a scalar, the result is another linear combination of vectors from S. The set of all linear combinations of a given set S of vectors is closed under addition and scalar multiplication. Hence we have formed a vector subspace of V by taking all possible linear combinations of vectors from a given set S.

THEOREM: The set of *all linear combinations* of a given set S of vectors from a vector space V is a subspace of V, called the **span** of the set S.
Notation: we write span(S) for this subspace; if we let W = span(S), we can also say that S *spans* W (or S is a spanning set for W).
Proof: Let $X, Y \in W$ and $S = \{ \, X_1, X_2, ..., X_n \, \}$; then

$$X = k_1 X_1 + k_2 X_2 + ... + k_n X_n \qquad Y = j_1 X_1 + j_2 X_2 + ... + j_n X_n$$

Then if we add X and Y we have:

$$X + Y = (\, k_1 X_1 + k_2 X_2 + ... + k_n X_n \,) + (\, j_1 X_1 + j_2 X_2 + ... + j_n X_n \,)$$

$$X + Y = (\, k_1 + j_1 \,) X_1 + (\, k_2 + j_2 \,) X_2 + ... + (\, k_n + j_N \,) X_n \in W$$

122

So W is closed under *addition*.
Now for scalar multiplication:

$$j\,X = j\,(\;k_1\,X_1 + k_2\,X_2 + \ldots + k_n\,X_n\;)$$

$$j\,X = (\;j\,k_1\;)\,X_1 + (\;j\,k_2\;)\,X_2 + \ldots + (\;j\,k_n\;)\,X_n = l_1\,X_1 + l_2\,X_2 + \ldots + l_n\,X_n$$

Thus jX is a linear combination of the vectors in S; therefore W is closed under *scalar multiplication* so W = span(S) is a subspace of V.

EXAMPLE 3 Describe the span(S) if $S = \{\;1 + x, 1 - x, x^2\;\} \subseteq P_2$.

Let W = span(S). A polynomial p(x) is in W iff there exist scalars $k_1\,k_2\;\&\;k_3$ such that:

$$p(\;x\;) = k_1(\;1 + x\;) + k_2(\;1 - x\;) + k_3\,x^2$$

But any polynomial $p \in P_2$ is of the form $p(\;x\;) = a + b\,x + c\,x^2$ so this means that

$$p(\;x\;) = k_1(\;1 + x\;) + k_2(\;1 - x\;) + k_3\,x^2 = a + b\,x + c\,x^2$$

This implies that$(\;k_1 + k_2\;) + (\;k_1 - k_2\;)\,x + k_3\,x^2 = a + b\,x + c\,x^2$ so

$$k_1 + k_2 = a$$

$$k_1 - k_2 = b$$

$$k_3 = c$$

Things become clearer sometimes by looking at the augmented matrix:

$$\begin{bmatrix} 1 & 1 & 0 & a \\ 1 & -1 & 0 & b \\ 0 & 0 & 1 & c \end{bmatrix} \sim \begin{bmatrix} 1 & 1 & 0 & a \\ 0 & 1 & 0 & \dfrac{b-a}{-2} \\ 0 & 0 & 1 & c \end{bmatrix}$$

This means *for any a,b, and c* we can find $k_1, k_2, \;\&\;k_3$ so that

$$p(\;x\;) = a + b\,x + c\,x^2 \in span(\;S\;)$$

Thus S spans P_2.

EXAMPLE 4 Describe the span(S) if $S = \{\;(\;1, 0, 2, 3\;), (\;0, 1, -3, 4\;), (\;0, 0, 1, 1\;)\;\} \in \Re^4$.

Let W = span(S). A vector in 4-space is in W iff there exist scalars k_1, k_2, and k_3 such that:

$$k_1(\ 1,0,2,3) + k_2(\ 0,1,-3,4\ ) + k_3(\ 0,0,1,1\ ) = (\ x, y, z, w\ )$$

Adding:

$$(\ k_1, k_2, 2\,k_1 - 3\,k_2 + k_3, 3\,k_1 + 4\,k_2 + k_3\ ) = (\ x, y, z, w\ )$$

This implies a 4 by 3 system:

$$\begin{bmatrix} 1 & 0 & 0 & x \\ 0 & 1 & 0 & y \\ 2 & -3 & 1 & z \\ 3 & 4 & 1 & w \end{bmatrix}$$

Getting the row-echelon form of this augmented matrix:

$$\begin{bmatrix} 1 & 0 & 0 & x \\ 0 & 1 & 0 & y \\ 0 & 0 & 1 & z-2\,x+3\,y \\ 0 & 0 & 1 & w-3\,x-4\,y \end{bmatrix} \sim \begin{bmatrix} 1 & 0 & 0 & x \\ 0 & 1 & 0 & y \\ 0 & 0 & 1 & z-2\,x+3\,y \\ 0 & 0 & 0 & w-x-7\,y-z \end{bmatrix}$$

This means that x + 7y + z - w = 0 ; this is then the equation of the span of the three given vectors . It is a linear equation in four unknowns and therefore represents a *hyperplane* (which goes through the origin) in 4-space. Note that the coordinates of each vector in S satisfy this equation (a good check on your work!). Here S does NOT span $\Re^4$. For example, (1,1,0,7) is not in span(S).

EXAMPLE 5 Describe the span of the set S if $S = \{\ 1+x, x^2\ \}$.
Suppose

$$p(\ x\ ) = a + b\,x + c\,x^2 \in span(\ S_1\ )$$

then there exist two scalars k_1 and k_2 such

$$p(\ x\ ) = a + b\,x + c\,x^2 = k_1(\ 1 + x\ ) + k_2\,x^2 \Rightarrow\ = a + b\,x + c\,x^2 = k_1 + k_1\,x + k_2\,x^2$$

The augmented matrix of this system is:

$$\begin{bmatrix} 1 & 0 & a \\ 1 & 0 & b \\ 0 & 1 & c \end{bmatrix} \sim \begin{bmatrix} 1 & 0 & a \\ 0 & 1 & c \\ 0 & 0 & b-a \end{bmatrix}$$

from which it is clear that a = b so that p(x) if of the form $p(\ x\ ) = b + b\,x + c\,x^2$.

EXAMPLE 6 Describe the span of the set S if $S = \{\ 1+x, x^2, 3+3\,x+4\,x^2\ \}$.

Suppose $p(x) = a + bx + cx^2 \in span(S_2)$; then there exist three scalars $k_1, k_2,$ and k_3 such that

$$k_1(1+x) + k_2 x^2 + k_3(3 + 3x + 4x^2) = a + bx + cx^2 = p(x)$$

$$(k_1 + 3k_3) + (k_1 + 3k_3)x + (k_2 + 4k_3)x^2 = a + bx + cx^2$$

whose augmented matrix is

$$\begin{bmatrix} 1 & 0 & 3 & a \\ 1 & 0 & 3 & b \\ 0 & 1 & 4 & c \end{bmatrix} \sim \begin{bmatrix} 1 & 0 & 3 & a \\ 0 & 1 & 4 & c \\ 0 & 0 & 0 & b-a \end{bmatrix}$$

which means (again!) that a = b so that p(x) is of the form $p(x) = b + bx + cx^2$. Why did it not matter in EX6 that S had one extra vector compared to the set S in EX5 yet the span of S was the same in each example? The reason is that the third vector in EX6 was a *linear combination of the first two vectors* and therefore span(S) is the same in both examples. The relationship among the vectors in EX5 is different from the relationship among the vectors in S in EX6. A definition is in order.

Linear Independence

DEFINITION

A set of vectors $S = \{X_1, X_2, ..., X_n\} \subseteq V$ is **linearly independent** iff $k_1 X_1 + k_2 X_2 + ... + k_n X_n = \theta \Rightarrow k_1 = 0, k_2 = 0, ..., k_n = 0$.

EXAMPLE 7 Show that the set $S = \{1, -1 + 3x^2, -3x + 5x^3\}$ is independent.
First, let's set up the independence equation:

$$k_1 1 + k_2(-1 + 3x^2) + k_3(-3x + 5x^3) = 0$$

Gathering like terms:

$$(k_1 - k_2) \cdot 1 + k_3(-3x) + k_2(3x^2) + k_3(5x^3) = (k_1 - k_2) \cdot 1 - 3k_3 x + 3k_2 x^2 + 5k_3 x^3 = 0$$

This implies a homogeneous linear system whose augmented matrix is:

$$\begin{bmatrix} 1 & -1 & 0 & 0 \\ 0 & 0 & -3 & 0 \\ 0 & 3 & 0 & 0 \\ 0 & 0 & 5 & 0 \end{bmatrix} \sim \begin{bmatrix} 1 & -1 & 0 & 0 \\ 0 & 1 & 0 & 0 \\ 0 & 0 & 1 & 0 \\ 0 & 0 & 0 & 0 \end{bmatrix}$$

Using back-substitution, we see rather quickly that $k_3 = 0 \Rightarrow k_2 = 0 \Rightarrow k_1 = 0$.

Thus the only way for an arbitrary linear combination of these three polynomials to be the zero polynomial is if all the scalar multipliers are zero- ie. only the **trivial solution** of the system. This means that the set S is independent. If a set of vectors in not independent, it is *dependent*. This means that there exists a non-trivial solution to the independence equation- at least one of the k_i's does NOT have to be zero.

EXAMPLE 8 Show that the set S = {(1,0,1),(0,2,0),(1,2,1)} is dependent.
Once again, we set up the independence equation
$k_1 X_1 + k_2 X_2 + k_3 X_3 = \theta$.

$$k_1(1,0,1) + k_2(0,2,0) + k_3(1,2,,1) = (0,0,0) \Rightarrow (k_1 + k_3, 2 k_2 + 2 k_3, k_1 + k_3) = (0,0,0)$$

The augmented matrix is of the form (note how the *components* of each vector in S form each column of the coefficient matrix):

$$\begin{bmatrix} 1 & 0 & 1 & 0 \\ 0 & 2 & 2 & 0 \\ 1 & 0 & 1 & 0 \end{bmatrix} \sim \begin{bmatrix} 1 & 0 & 1 & 0 \\ 0 & 1 & 1 & 0 \\ 0 & 0 & 0 & 0 \end{bmatrix} \Rightarrow k_2 = -k_3 = -t \Rightarrow k_1 = -k_3 = -t$$

From chapter 1 we know that column 3 represents a free variable(ie. k_3) and there are infinitely many solutions; hence S is a *dependent* set. For example, if we let t = 1, we have $k_3 = 1, k_2 = k_1 = -1$ so that it must be that

$$- X_1 - X_2 + X_3 = (0,0,0)$$

We thus observe that for this set S,

$$X_3 = X_1 + X_2$$

Therefore if a set S is *dependent*, at least one of the vectors in S is a linear combination of the other vectors in S. Put simply, if a set is dependent, at least one vector in the set is a "lazy" vector, ie. it is a *linear combination* of the other vectors in the set.

THEOREM: A set S of nonzero vectors from a vector space V is *dependent* iff at least one vector in S is a *linear combination* of the other vectors in S.
Proof: $\Rightarrow$ Given $S = \{ X_1, X_2, ..., X_n \} \subseteq V$ where S is dependent; if we set up the independence equation

$$k_1 X_1 + k_2 X_2 + ... + k_n X_n = \theta$$

we know that there must be nontrivial solutions since the set is dependent. Thus at least one of the scalars, say k_i, is not equal to zero, so we have:

$$k_1 X_1 + k_2 X_2 + \ldots + k_i X_i + \ldots + k_n X_n = \theta$$

Isolating X_i means

$$k_i X_i = -k_1 X_1 - k_2 X_2 - \ldots - k_n X_n$$

$$X_i = -\frac{k_1}{k_i} X_1 - \frac{k_2}{k_i} X_2 - \ldots - \frac{k_n}{k_i} X_n = l_1 X_1 + l_2 X_2 + \ldots + l_n X_n$$

which shows that X_i is indeed a combination of the other (n - 1) vectors in the set. QED

EXAMPLE 9 Show that the set $S = \left\{ \begin{bmatrix} 1 & 2 \\ 0 & 0 \end{bmatrix}, \begin{bmatrix} -1 & 3 \\ 0 & 0 \end{bmatrix}, \begin{bmatrix} 1 & 7 \\ 0 & 0 \end{bmatrix} \right\} \subseteq M_{2,2}$ is dependent.

Setting up the independence equation:

$$k_1 \begin{bmatrix} 1 & 2 \\ 0 & 0 \end{bmatrix} + k_2 \begin{bmatrix} -1 & 3 \\ 0 & 0 \end{bmatrix} + k_3 \begin{bmatrix} 1 & 7 \\ 0 & 0 \end{bmatrix} = \begin{bmatrix} k_1 - k_2 + k_3 & 2 k_1 + 3 k_2 + 7 k_3 \\ 0 & 0 \end{bmatrix} = \begin{bmatrix} 0 & 0 \\ 0 & 0 \end{bmatrix}$$

which means that

$$k_1 - k_2 + k_3 = 0$$

$$2 k_1 + 3 k_2 + 7 k_3 = 0$$

Since there are fewer equations than unknowns, there are infinitely many non-trivial solutions of this system so S is a *dependent* set. Alternatively one may have recognized that the third matrix is just twice the first plus the second (ie. it was the "lazy vector" in S).

Corollary: Given the set $S = \{X_1, X_2\}$; S is dependent iff $X_2 = k X_1$ for some real number k.

EXAMPLE 10 Let S = {(1,2,3,4),(0,1,-4,5)}; determine if S is an independent set.
In light of the corollary, it is clear that neither vector here is a multiple of the other so the set MUST be independent. No need to set up the independence equation here!

EXAMPLE 11 Show that the set $S = \{ \cos(x), \sin(x) \}$ is independent.
Since $\sin(x) \neq k \cos(x)$ for any one k, then S is independent. Alternatively, if the *ratio* of these two functions is not a constant, the set is independent.

EXERCISES 3.3

Part A. Determine if X can be written as a linear combination of the vectors in the set S if:
1. X = (1,2,3), S = {(1,0,1), (0,2,2)} ans. YES
2. X = 2 + 3x, S = {4,x, x^2} $S = \{ 4, x, x^2 \}$

3. $X = 1 + x + x^2$, $S = \{ 2, 3 - x, 4x + x^2 \}$ ans. YES

4. $X = (1,-2,2,4)$, $S = \{(1,0,1,0),(0,-2,2,0),(0,0,0,1)\}$

5. $X = (1,-2,3,1)$, $S = \{(1,0,1,0),(0,-2,2,0),(0,0,0,1)\}$ ans. NO

6. $X = \begin{bmatrix} 1 & 2 \\ 0 & 3 \end{bmatrix}$ $S = \{ \begin{bmatrix} 1 & 0 \\ 0 & 0 \end{bmatrix}, \begin{bmatrix} 0 & 1 \\ 0 & 0 \end{bmatrix}, \begin{bmatrix} 0 & 0 \\ 0 & 1 \end{bmatrix} \}$

7. $X = \begin{bmatrix} 1 & 2 & 3 \\ 3 & 1 & 2 \\ 2 & 3 & 1 \end{bmatrix}$ $S = \{ \begin{bmatrix} 1 & 0 & 0 \\ 0 & 1 & 0 \\ 0 & 0 & 1 \end{bmatrix}, \begin{bmatrix} 0 & 1 & 0 \\ 0 & 0 & 1 \\ 1 & 0 & 0 \end{bmatrix}, \begin{bmatrix} 0 & 0 & 1 \\ 1 & 0 & 0 \\ 0 & 1 & 0 \end{bmatrix} \}$

ans. YES

8. $X = \begin{bmatrix} 1 & 2 & 3 \\ 1 & 3 & 2 \\ 2 & 3 & 1 \end{bmatrix}$ $S = \{ \begin{bmatrix} 1 & 0 & 0 \\ 0 & 1 & 0 \\ 0 & 0 & 1 \end{bmatrix}, \begin{bmatrix} 0 & 1 & 0 \\ 0 & 0 & 1 \\ 1 & 0 & 0 \end{bmatrix}, \begin{bmatrix} 0 & 0 & 1 \\ 1 & 0 & 0 \\ 0 & 1 & 0 \end{bmatrix} \}$

Part B. Describe the polynomial p, if it is in the span of S:

9. $S = \{ 1 + x, 2 - x^2 \}$ ans. $p(x) = (b + 2c) + bx + cx^2$

10. $S = \{ 1, 1 + x, x^2 \}$

11. $S = \{ 1 + x, 2 - x^2, x^3 \}$ ans. $p(x) = (b - 2c) + bx + cx^2 + dx^3$

12. $S = \{ 3, 2 - x^2 \}$

13. $S = \{ 1, -1 + 3x^2, -3x + 5x^3 \}$ ans. $p(x) = a + b(x - \frac{5}{3}x^3) + cx^2$

14. $S = \{ x, 5x^3 - 3x \}$

Part C. Determine if the matrix A is in the span of $S = \{ \begin{bmatrix} 1 & 0 \\ 0 & 1 \end{bmatrix}, \begin{bmatrix} 0 & 2 \\ 2 & 0 \end{bmatrix}, \begin{bmatrix} 0 & 0 \\ 0 & 3 \end{bmatrix} \}$:

15. $A = \begin{bmatrix} 1 & 2 \\ 2 & 3 \end{bmatrix}$ ans. NO

16. $A = \begin{bmatrix} 2 & 0 \\ 0 & 5 \end{bmatrix}$

17. $A = \begin{bmatrix} 1 & 2 \\ 2 & 4 \end{bmatrix}$ ans. YES

18. $A = \begin{bmatrix} 0 & 4 \\ 4 & 6 \end{bmatrix}$

19. $A = \begin{bmatrix} 2 & 2 \\ 2 & 5 \end{bmatrix}$ ans. YES

20. $A = \begin{bmatrix} 1 & 2 \\ 3 & 4 \end{bmatrix}$

Part D. Given the set of vectors $S = \{ \begin{bmatrix} 1 & 1 \\ 0 & -1 \end{bmatrix}, \begin{bmatrix} -1 & 1 \\ 0 & 1 \end{bmatrix}, \begin{bmatrix} 0 & 0 \\ 2 & 0 \end{bmatrix} \}$:

21. Determine if S is independent or not. ans. independent

22. Given $A = \begin{bmatrix} a & b \\ c & d \end{bmatrix}$; find a condition on a,b,c and d which guarantees that A is in span(S).

23. Is $B = \begin{bmatrix} 2 & 0 \\ 4 & -2 \end{bmatrix}$ in span(S)? ans. YES, $a + d = 0$

Part E. Determine if the set S is linearly independent or not. If S is dependent, determine which vector(s) in S is(are) combination(s) of the other vectors in S.

24. S = {(1,0,1,0),(2,0,4,0),(5,0,7,0)}

25. S = {(1,0,1,0),(0,2,0,2),(1,2,1,2)} ans. dependent,(1,2,1,2)

26. S = {(1,0,1,0),(0,2,0,2,),(1,2,1,3)}

27. S = {(1,0,1,0),(0,2,0,2),(1,2,1,3),(2,4,2,6)} ans. independent

28. S = {(1,0,2,0),(1,2,3,0),(2,2,5,0),(4,4,10,6)}

29. S = {(1,2,1,2),(-1,0,0,1),(0,2,1,3),(0,2,1,2)} ans. dependent,(0,2,1,3)

30. S = {(1,0,1,0,1),(0,2,0,2,0),(1,2,1,2,2),(2,4,2,5,0)}

31. S = {(1,2,3,4),(0,2,0,2),(0,0,0,0)} ans. dependent,(0,0,0,0)

Part F. Given the set S = {(1,0,1,0),(0,2,0,2),(1,2,1,3)};
32. Determine if the set S is independent or not. ans. independent

33. If X = (x,y,z,w) is in W = span(S), find an equation(s) which guarantees that X *is* in W.
NB. Three independent vectors in 4-space forms a *hyperplane* through the origin.
ans. x - z = 0

34. Does X = (6,14,6,18) lie in W?

Part G. Given the vectors $X_1 = (1,0,1,2)$ and $X_2 = (0,1,3,-1)$;
35. Is $S = \{X_1, X_2\}$ is an independent set? ans. YES

36. Let $X = (x,y,z,w)$ be in $W = \text{span}(\{X_1, X_2\})$; find an equation(s) which guarantee that X is in W.
NB. Two independent vectors in 4-space forms a plane through the origin.

37. Does $X = (3,4,15,2)$ lie in this plane? ans. YES

Part H. Given the set of 2 by 2 matrices W:

$$W = \left\{ \begin{bmatrix} a & b \\ c & -a \end{bmatrix} : a,b,c \in \Re \right\}$$

38. Show that W is a subspace of $M_{2,2}$; these matrices are called *traceless*.
39. Show that any matrix A in W is a linear combination of the matrices in S, where

$$S = \left\{ \begin{bmatrix} 1 & 0 \\ 0 & -1 \end{bmatrix}, \begin{bmatrix} 0 & 1 \\ 0 & 0 \end{bmatrix}, \begin{bmatrix} 0 & 0 \\ 1 & 0 \end{bmatrix} \right\}$$

40. Show that S is linearly independent.
41. Write the vector $A = \begin{bmatrix} 3 & 4 \\ 5 & -3 \end{bmatrix}$ as a combination of the vectors in S.

Part I. Given the set of 2 by 2 matrices W:

$$W = \left\{ \begin{bmatrix} a & b \\ b & c \end{bmatrix} : a,b,c \in \Re \right\}$$

42. Show that W is a subspace of $M_{2,2}$; these matrices are called *symmetric*.
43. Show that any matrix A in W is a linear combination of the matrices in S, where

$$S = \left\{ \begin{bmatrix} 1 & 0 \\ 0 & 0 \end{bmatrix}, \begin{bmatrix} 0 & 1 \\ 1 & 0 \end{bmatrix}, \begin{bmatrix} 0 & 0 \\ 0 & 1 \end{bmatrix} \right\}$$

44. Show that S is linearly independent.
45. Write the vector $A = \begin{bmatrix} 3 & 4 \\ 4 & 5 \end{bmatrix}$ as a combination of the vectors in S.

Part J. Independence and functions spaces: Let $C[a,b] = \{ f : f \mid [a,b] \to \Re \text{ and } f \text{ continuous} \}$
where $\{f,g\}$ are two functions contained in C[a,b] such that f and g are differentiable on [a,b];
46. Show $\{f,g\}$ is independent if for some $k \neq 0 \in \Re$.
47. Show $\{f,g\}$ is dependent if $g = kf$.

48. Show if {f,g} is *dependent* and $W = \begin{bmatrix} f & g \\ f' & g' \end{bmatrix}$ then det(W) = 0 for all x in [a,b]

(ie. det(W) is identically zero on [a,b]).

49. Show $\det(W) \neq 0$ for some x_0 in [a,b] implies {f,g} is independent.

50. Generalize what is in problems # 48 and # 49 for a set of 3 functions {f,g,h} which are continuous and twice differentiable on [a,b].

51. Show that the set of functions $\{ 1, x, x^2 \}$ is independent on $\Re$ by finding det(W) and showing that it's not identically zero. Here $W = \begin{bmatrix} f & g & h \\ f' & g' & h' \\ f'' & g'' & h'' \end{bmatrix}$.

ans. det(W) = 2 for any x

52. For the set S = $\{ x^2, x^3 \}$, det(W) = 0 if x = 0; does this mean that S is dependent? Explain.

Part K. Given a set S $S = \{ X_1, X_2, \dots, X_n \} \subseteq V$; prove:

53. If one vector in S is a combination of the others, then S in dependent.

54. If one vector in S is the zero vector, then S is dependent.

3.4 Basis and Dimension

Now suppose we have set of vectors B which spans a vector space V (ie. span(B) = V) and also B is an independent set. This type of set of vectors is very important in the life of the vector space V- it is called a basis for V.

DEFINITION	In a vector space V, if a set of vectors B satisfies 1) V = span(B) and 2) B is an independent set then B is a **basis** for V.

EXAMPLE 1 Let B = {(1,0,0,0),(0,1,0,0),(0,0,1,0),(0,0,0,1)}; it is easy to show that this set both spans 4-space and is independent. If X is any vector in 4-space, it is of the form $X = (x_1, x_2, x_3, x_4)$; but

$$X = (x_1, x_2, x_3, x_4) = (x_1, 0, 0, 0) + (0, x_2, 0, 0) + (0, 0, x_3, 0) + (0, 0, 0, x_4) =$$

$$x_1(1,0,0,0) + x_2(0,1,0,0) + x_3(0,0,1,0) + x_4(0,0,0,1)$$

so clearly span(B) = 4-space. Now it is impossible for any arbitrary combination of the vectors in B to equal the zero vector unless all the scalar multipliers are zero so B is independent. Therefore B is a basis for 4-space; in fact, it is the basis most often used for 4-space and will be called the *standard basis* for $\Re^4$.

EXAMPLE 2 Show that $B = \{ 1, x, x^2 \} \subseteq P_2$ is a basis for P_2.

1) Let p(x) be any arbitrary polynomial in P_2; we must show that p(x) can be written as a linear combination of vectors in B:

$$p(x) = a + b x + c x^2 = k_1(1) + k_2(x) + k_3(x^2) \Rightarrow k_1 = a, k_2 = b, k_3 = c$$

2) Show that B is a linearly independent set.

$$\theta(x) = 0 = k_1(1) + k_2(x) + k_3(x^2) \Rightarrow k_1 = 0, k_2 = 0, k_3 = 0$$

Since $k_1, k_2, \& k_3$ must all be zero, B is an independent set. Thus B is a basis for P_2. It is not the only basis (in fact there are infinitely many bases for P_2) but we can prove that any basis for P_2 must have three vectors in it? Why? Well, if we had fewer than three, the set could not span P_2; on the other hand, if we had more than three, the set would no longer be independent. So we can think about a basis this way- a basis is both
1) a **minimal spanning set** and
2) a *maximal independent set* for a vector space V.
This means a basis B is a spanning set for the vector space V which is also *independent*.

EXAMPLE 3 Find a basis for W, where S = {(1,0,0,1),(0,1,1,0),(1,1,1,1)} and W = span(S). Since we already have a spanning set, we only to worry about whether the set S is independent or not. Set an arbitrary combination of the three vectors equal to the zero vector:

132

$$k_1(\,1,0,0,1\,) + k_2(\,0,1,1,0\,) + k_3(\,1,1,1,1\,) = (\,0,0,0,0\,)$$

This implies an augmented matrix of the form:

$$\begin{bmatrix} 1 & 0 & 1 & 0 \\ 0 & 1 & 1 & 0 \\ 0 & 1 & 1 & 0 \\ 1 & 0 & 1 & 0 \end{bmatrix} \sim \begin{bmatrix} 1 & 0 & 1 & 0 \\ 0 & 1 & 1 & 0 \\ 0 & 0 & 0 & 0 \\ 0 & 0 & 0 & 0 \end{bmatrix}$$

This system has one free variable (column 3 is a non-pivot column) which means there are infinitely-many solutions and S is NOT independent. The second row gives

$$k_2 + k_3 = 0 \Rightarrow k_2 = -k_3 \qquad k_3 = t \Rightarrow k_2 = -t$$

Similarly equ(1) gives

$$k_1 + k_3 = 0 \Rightarrow k_1 = -k_3 = -t \Rightarrow K = (\,-t,-t,t\,)$$

It is not difficult to observe that the *third vector was a linear combination of the first two* and can be kicked out without affecting the span. Just set t = 1 so K = (-1,-1,1) and upon substitution into the independence equation, the third vector in W is a combination of the first two:

$$(-1)\cdot(\,1,0,0,1\,) + (-1)\cdot(\,0,1,1,0\,) + (\,1\,)\cdot(\,1,1,1,1\,) = (\,0,0,0,0\,)$$

Thus a basis for W is the set {(1,0,0,1),(0,1,1,0)} which is clearly independent since the second vector is not a scalar multiple of the first. Geometrically W is a plane through the origin in 4-space. Any plane needs two independent vectors to span it.

Dimension of a vector space

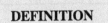

DEFINITION

The dimension of the vector space V is the number of vectors in any basis for V.

For example,

$$P_2 = span(\,B\,)\ where\ B = \{\,1,x,x^2\,\}$$

and B is independent so B forms a basis for P_2 and this means that P_2 is 3-dimensional. If the dimension is a finite non-negative integer, we say that V is **finite-dimensional**. P_2 and $\Re^4$ are finite-dimensional. If no finite basis exists, then V is infinite dimensional. For example, the set C[a,b] of continuous functions on the interval [a,b] has no finite basis and therefore is an infinite dimensional vector space. The trivial space $\{\,\theta\,\}$ has dimension zero.

EXAMPLE 4 Find a basis and also the dimension of the set

$$W = \{ \begin{bmatrix} a & b \\ 0 & c \end{bmatrix} : a, b, c \in \Re \}$$

First, W is a subspace of the vector space $M_{2,2}$ of all 2 by 2 matrices (check out closure under addition and scalar multiplication). If we pick any vector in W, we can write it in the following form:

$$A = \begin{bmatrix} a & b \\ 0 & c \end{bmatrix} = \begin{bmatrix} a & 0 \\ 0 & 0 \end{bmatrix} + \begin{bmatrix} 0 & b \\ 0 & 0 \end{bmatrix} + \begin{bmatrix} 0 & 0 \\ 0 & c \end{bmatrix} = a \begin{bmatrix} 1 & 0 \\ 0 & 0 \end{bmatrix} + b \begin{bmatrix} 0 & 1 \\ 0 & 0 \end{bmatrix} + c \begin{bmatrix} 0 & 0 \\ 0 & 1 \end{bmatrix}$$

Clearly any matrix A in W is thus a combination of the three matrices above so that W is the *span* of these three matrices. The only question is one of independence; setting an arbitrary combination of these three vectors equal to the zero vector gives us:

$$a \begin{bmatrix} 1 & 0 \\ 0 & 0 \end{bmatrix} + b \begin{bmatrix} 0 & 1 \\ 0 & 0 \end{bmatrix} + c \begin{bmatrix} 0 & 0 \\ 0 & 1 \end{bmatrix} = \begin{bmatrix} a & 0 \\ 0 & 0 \end{bmatrix} + \begin{bmatrix} 0 & b \\ 0 & 0 \end{bmatrix} + \begin{bmatrix} 0 & 0 \\ 0 & c \end{bmatrix} = \begin{bmatrix} a & b \\ 0 & c \end{bmatrix} = \begin{bmatrix} 0 & 0 \\ 0 & 0 \end{bmatrix}$$

But clearly the only solution is that a = b = c = 0 so the three vectors are independent and thus form a basis for W. This means that dim(W) = 3.

Dimensions of Classic Vector Spaces

Vector Space	Standard basis	Dimension
$\Re^n$	$\{(1,0,...),(0,1,0..),...(0,0,..,0,1)$	n
P_n	$1, x, x^2, ..., x^n$	n + 1
P	$1, x, x^2, ..., x^n, ...$	oo
$M_{m,n}$	mn matrices consisting of one "1" and mn - 1 zeros, with the "1" starting in upper left corner, moving from left to right, finishing in bottom right corner	mn
$C[a,b]$	no finite basis	oo
C(-oo,oo)	no finite basis	oo

EXAMPLE 5 Find a basis and thus the dimension of V, where $V = M_{2,3}$. Let

$$B = \{ \begin{bmatrix} 1 & 0 & 0 \\ 0 & 0 & 0 \end{bmatrix}, \begin{bmatrix} 0 & 1 & 0 \\ 0 & 0 & 0 \end{bmatrix}, \begin{bmatrix} 0 & 0 & 1 \\ 0 & 0 & 0 \end{bmatrix}, \begin{bmatrix} 0 & 0 & 0 \\ 1 & 0 & 0 \end{bmatrix}, \begin{bmatrix} 0 & 0 & 0 \\ 0 & 1 & 0 \end{bmatrix}, \begin{bmatrix} 0 & 0 & 0 \\ 0 & 0 & 1 \end{bmatrix} \}$$

to show that B spans V, let's show that any 2 x 3 matrix can be written as a linear combination of the vectors in B.

Let $M = \begin{bmatrix} a & b & c \\ d & e & f \end{bmatrix}$; then we must be able to find 6 scalars $k_1, k_2, ..., k_6$ such that

$$k_1 \begin{bmatrix} 1 & 0 & 0 \\ 0 & 0 & 0 \end{bmatrix} + k_2 \begin{bmatrix} 0 & 1 & 0 \\ 0 & 0 & 0 \end{bmatrix} + k_3 \begin{bmatrix} 0 & 0 & 1 \\ 0 & 0 & 0 \end{bmatrix} + k_4 \begin{bmatrix} 0 & 0 & 0 \\ 1 & 0 & 0 \end{bmatrix} + k_5 \begin{bmatrix} 0 & 0 & 0 \\ 0 & 1 & 0 \end{bmatrix} + k_6 \begin{bmatrix} 0 & 0 & 0 \\ 0 & 0 & 1 \end{bmatrix} = M = \begin{bmatrix} a & b & c \\ d & e & f \end{bmatrix}$$

But it is clear that $k_1 = a, k_2 = b, ..., k_6 = f$ so that span(B) = V. We now need to show that B is independent; if we set up the independence equation, we'd have:

$$k_1 \begin{bmatrix} 1 & 0 & 0 \\ 0 & 0 & 0 \end{bmatrix} + k_2 \begin{bmatrix} 0 & 1 & 0 \\ 0 & 0 & 0 \end{bmatrix} + k_3 \begin{bmatrix} 0 & 0 & 1 \\ 0 & 0 & 0 \end{bmatrix} + k_4 \begin{bmatrix} 0 & 0 & 0 \\ 1 & 0 & 0 \end{bmatrix} + k_5 \begin{bmatrix} 0 & 0 & 0 \\ 0 & 1 & 0 \end{bmatrix} + k_6 \begin{bmatrix} 0 & 0 & 0 \\ 0 & 0 & 1 \end{bmatrix} = \begin{bmatrix} 0 & 0 & 0 \\ 0 & 0 & 0 \end{bmatrix}$$

which would produce a 6 by 6 homogeneous system. But without writing out the augmented matrix, it is clear that we must have all k_i's = 0 so the set is independent.

Thus we've shown that $dim(M_{2,3}) = 6$.

Advantages of knowing dim(V)

If we have a vector space V with basis B, where B contains n vectors, then dim(V) = n. The advantage of knowing the dimension of a vector space is that we may be able to answer questions about spanning and independence without doing much work. Suppose $S \subseteq V$ contains n vectors; S has enough vectors to span V, but they need to be independent. So to see if S is a basis for V, all you need to do is to see if S in independent. On the other hand, if we know that span(S) = V, then you just need to determine if S is independent. If it is, then S is a basis.

THEOREM If dim(V) = n and S contains n vectors then:
1) If S spans V, then S is independent.
2) If S is independent, then S spans V.
Proof(1): Suppose S = $\{ X_1, X_2, ..., X_n \}$ where span(S) = V.
Let

$$k_1 X_1 + k_2 X_2 + ... + k_n X_n = \theta$$

and assume S is dependent. Then by a previous theorem, at least one vector in S is a combination of the others; let this vector be X_j (j between 1 and n). Then the set

$$T = \{ X_1, ..., X_{j-1}, X_{j+1}, ..., X_n \}$$

of n-1 vectors spans V (contradiction- since V is n-dimensional). Thus S must be independent.
Proof(2): Suppose S = $\{ X_1, X_2, ..., X_n \}$ and is independent; assume there exists Y in V such that Y is not in span(S).
Then $Y = k_1 X_1 + ... + k_n X_n$ has no solution so $\{ Y, X_1, X_2, ..., X_n \}$ is an independent set larger than S (contradiction- since V is n-dimensional). Thus S spans V.
The gist of the theorem is that if we have n vectors in an *n-dimensional* space, then one needs to

check either spanning OR independence- having ONE of these guarantees that S is a basis for V.

EXAMPLE 6 Given

$$S = \{1 - x^2, 2x, 3 + 4x^2\} \subseteq P_2$$

determine whether S is basis for P_2. We know that $\dim(P_2) = 3$ and S contains three vectors; thus we need only check out whether S is independent or dependent. The independence equation here is:

$$k_1(1 - x^2) + k_2(2x) + k_3(3 + 4x^2) = \theta = 0 \Rightarrow (k_1 + 3k_3) + 2k_2 x + (-k_1 + 4k_3)x^2 = 0$$

The augmented matrix of this system is: $\begin{bmatrix} 1 & 0 & 3 & 0 \\ 0 & 2 & 0 & 0 \\ -1 & 0 & 4 & 0 \end{bmatrix}$ which the reader should be able to show

implies that the k_i's must all be zero so S is *independent* and thus forms a basis for P_2.

Now suppose S is a subset of the n-dimensional space V with basis B (which has n vectors). If S has less than n vectors, it's impossible for S to span V. On the other hand, if S contains more than n vectors then S must be dependent, since any basis for V is a maximal independent set. We thus we have a second theorem:

THEOREM Suppose vector space V has a basis B with n vectors; then
1) any set S with less than n vectors cannot span V
2) any set S with more than n vectors must be dependent.

EXAMPLE 7 Given

$$S = \{1 - x^2, 2x, 3 + 4x^2, 1 + 2x + 3x^2\} \subseteq P_2$$

determine whether S is independent or dependent.
We know that $\dim(P_2) = 3$ and S contains FOUR vectors; since $4 > 3$, the set S must be dependent. No calculations necessary!

EXAMPLE 8 Given S = {(1,0,1,0),(0,2,0,2),(1,2,1,2)}; determine whether S spans $\Re^4$ or not. Since $\dim(\Re^4) = 4$, even if the vectors in S are independent, there are not enough of them to span $\Re^4$, since $3 < 4$.

EXERCISES 3.4

Part A. Given W which is a subspace of some vector space V; determine if the spanning set given for W is linearly independent. If it is dependent, reduce the set of vectors without changing the span to an independent set(ie. a basis for W); then give the dimension of W.
1. W = span{(1,0,2),(0,3,4),(1,3,6)} ans.B = {(1,0,2),(0,3,4)}, 2

2. W = span{(1,0,1,0),(0,1,0,1)}

3. W = span{(1,2,2,1),(2,4,4,2),(3,6,6,6)} ans. B = {(1,2,2,1),(2,4,4,2),(3,6,6,6)}, 3

4. W = span{1 + x, $2 - x^2$ } in P_2

5. W = span{1,1 + x, 1 + x + x^2} in P_2, ans. B = {1,1 + x, 1 + x + x^2}, 3

6. $W = span \left(\left\{ \begin{bmatrix} 1 & 2 \\ 2 & 4 \end{bmatrix}, \begin{bmatrix} 0 & 3 \\ 0 & 6 \end{bmatrix}, \begin{bmatrix} 1 & 5 \\ 2 & 10 \end{bmatrix} \right\} \right)$

7. $W = \left\{ k_1 \begin{bmatrix} 1 & 2 \\ 2 & 4 \end{bmatrix} + k_2 \begin{bmatrix} 0 & 3 \\ 0 & 6 \end{bmatrix} + k_3 \begin{bmatrix} 1 & 5 \\ 2 & 12 \end{bmatrix} : k_1, k_2, k_3 \in \Re \right\}$

ans. $B = \left\{ \begin{bmatrix} 1 & 2 \\ 2 & 4 \end{bmatrix}, \begin{bmatrix} 0 & 3 \\ 0 & 6 \end{bmatrix}, \begin{bmatrix} 1 & 5 \\ 2 & 12 \end{bmatrix} \right\}, 3$

8. W = span{cos(x), sin(x), tan(x)}

9. W = span{cos(x), sin(x), sin(2x)}, ans. B = {cos(x),sin(x),sin(2x)}, 3

10. W = span{cos(x), sin(x), 3cos(x) + 4sin(x)}

11. $W = \left\{ k_1 \begin{bmatrix} 1 & 0 \\ 0 & -1 \end{bmatrix} + k_2 \begin{bmatrix} 0 & 1 \\ 1 & 0 \end{bmatrix} + k_3 \begin{bmatrix} 1 & 1 \\ 1 & -1 \end{bmatrix}, k_1, k_2, k_3 \in \Re \right\}$

ans. $B = \left\{ \begin{bmatrix} 1 & 0 \\ 0 & -1 \end{bmatrix}, \begin{bmatrix} 0 & 1 \\ 1 & 0 \end{bmatrix} \right\}, 2$

Part B. Show that the set B forms a basis for the vector space V by showing that span{B} = V and B is an independent set. Also state dim(V).

12. $B = \left\{ \begin{bmatrix} 1 & 0 \\ 0 & 0 \end{bmatrix}, \begin{bmatrix} 0 & 1 \\ 0 & 0 \end{bmatrix}, \begin{bmatrix} 0 & 0 \\ 1 & 0 \end{bmatrix}, \begin{bmatrix} 0 & 0 \\ 0 & 1 \end{bmatrix} \right\}$ $V = M_{2,2}$

13. $B = \{ 1, x, x^2 \}$ $V = P_2$ ans. 3

14. $B = \{ (1,0,0,0), (0,1,0,0), (0,0,1,0), (0,0,0,1) \}$ $V = \Re^4$

15. $V = \left\{ \begin{bmatrix} a & b & c \\ 0 & d & e \\ 0 & 0 & f \end{bmatrix} \right\}$

$$\begin{bmatrix} 1 & 0 & 0 \\ 0 & 0 & 0 \\ 0 & 0 & 0 \end{bmatrix}, \begin{bmatrix} 0 & 1 & 0 \\ 0 & 0 & 0 \\ 0 & 0 & 0 \end{bmatrix}, \begin{bmatrix} 0 & 0 & 1 \\ 0 & 0 & 0 \\ 0 & 0 & 0 \end{bmatrix}, , \begin{bmatrix} 0 & 0 & 0 \\ 0 & 1 & 0 \\ 0 & 0 & 0 \end{bmatrix} \begin{bmatrix} 0 & 0 & 0 \\ 0 & 0 & 1 \\ 0 & 0 & 0 \end{bmatrix}, , \begin{bmatrix} 0 & 0 & 0 \\ 0 & 0 & 0 \\ 0 & 0 & 1 \end{bmatrix}$$

ans. 6

16. $B = \{ \begin{bmatrix} 0 & 1 & 0 \\ -1 & 0 & 0 \\ 0 & 0 & 0 \end{bmatrix}, \begin{bmatrix} 0 & 0 & 1 \\ 0 & 0 & 0 \\ -1 & 0 & 0 \end{bmatrix}, \begin{bmatrix} 0 & 0 & 0 \\ 0 & 0 & 1 \\ 0 & -1 & 0 \end{bmatrix} \}$

$V = \{ A : A \in M_{3,3}, A = -A^t \}$

NB. these matrices are called *skew-symmetric*

17. $V = \{ \begin{bmatrix} a & b & c \\ c & a & b \\ b & c & a \end{bmatrix}, a,b,c \in \Re \}$ ans. $B = \{ \begin{bmatrix} 1 & 0 & 0 \\ 0 & 1 & 0 \\ 0 & 0 & 1 \end{bmatrix}, \begin{bmatrix} 0 & 1 & 0 \\ 0 & 0 & 1 \\ 1 & 0 & 0 \end{bmatrix}, \begin{bmatrix} 0 & 0 & 1 \\ 1 & 0 & 0 \\ 0 & 1 & 0 \end{bmatrix} \}$

NB. these matrices are called *circulants* ans. 3

18. $B = \{ 1, x, 3x^2 - 1 \}$ $V = P_2$

19. $B = \{ 1, 2x, 4x^2 - 2, 8x^3 - 12x \}$ $V = P_3$ ans. 4

Part C. Given the set S of vectors from V; answer the following:

20. S = {(1,0,2),(0,1,2),(1,1,5),(3,3,3)}, $V = \Re^3$; is S independent?

21. $S = \{ 1 - x, 3x^2 \}, V = P_2$; does S span P_2? ans. NO; 2 vectors cannot span P_2

22. S = {(1,0,2,3),(0,1,2,4),(1,1,4,8)}, $V = \Re^4$; can S span $\Re^4$?

23. S = {(1,0,2,3),(0,1,2,4),(1,1,4,8)}, does S span $\Re^4$? ans. NO; 3 vectors cannot span 4-space

24. Find an equation for span(S) from #22; find dim(span(S)).

Part D. For the given set of matrices W is a *subset* of $M_{2,2}$; determine whether W is a subspace. If W is a subspace, find a basis and dim(W).

25. $W = \{ A : A \in M_{2,2}, tr(A) = 0 \}$

ans. YES, $B = \{ \begin{bmatrix} 1 & 0 \\ 0 & -1 \end{bmatrix}, \begin{bmatrix} 0 & 1 \\ 0 & 0 \end{bmatrix}, \begin{bmatrix} 0 & 0 \\ 1 & 0 \end{bmatrix} \}$, 3; these matrices are called *traceless*

26. $W = \{ \begin{bmatrix} a & b \\ c & d \end{bmatrix} : c = 0 \} \subseteq M_{2,2}$

27. $W = \{ A : A \in M_{2,2}, A = A^t \}$

ans. YES, $B = \{ \begin{bmatrix} 0 & 1 \\ 1 & 0 \end{bmatrix}, \begin{bmatrix} 1 & 0 \\ 0 & 0 \end{bmatrix}, \begin{bmatrix} 0 & 0 \\ 0 & 1 \end{bmatrix} \}$, 3

28. $W = \{ A : A \in M_{2,2}, \det(A) = 0 \}$

29. $W = \{ \begin{bmatrix} a & b+c \\ b-c & -a \end{bmatrix} a,b,c \in \Re \}$ ans. YES, $B = \{ \begin{bmatrix} 1 & 0 \\ 0 & -1 \end{bmatrix}, \begin{bmatrix} 0 & 1 \\ 1 & 0 \end{bmatrix}, \begin{bmatrix} 0 & 1 \\ -1 & 0 \end{bmatrix} \}$, 3

30. $W = \{ \ X : X^t B + B X = \begin{bmatrix} 0 & 0 \\ 0 & 0 \end{bmatrix}, where \ X \in M_{2,2}(\ \Re \ ), B = \begin{bmatrix} 1 & 0 \\ 0 & 2 \end{bmatrix} \}$

31. $W = \{ \begin{bmatrix} a & -b \\ b & 1 \end{bmatrix} : a, b \in \Re \}$ ans. NO

32. $W = \{ \ X : X^t + X = \begin{bmatrix} 0 & 0 \\ 0 & 0 \end{bmatrix}, where \ X = \begin{bmatrix} a & b \\ c & d \end{bmatrix} \}$

Part E. Given the linear system whose **augmented** matrix is $\begin{bmatrix} 1 & 2 & 0 & -1 & 3 \\ 0 & 1 & -1 & 3 & 0 \\ 1 & 3 & -1 & 3 & 3 \end{bmatrix}$;

33. Rewrite the system in the form AX = B; then solve it using Gaussian elimination. Write the answer as a column matrix in the form X = Xh + Xp. Check your answer using matrix multiplication. ans. $X = \begin{bmatrix} -2t \\ t \\ t \\ 0 \end{bmatrix} + \begin{bmatrix} 3 \\ 0 \\ 0 \\ 0 \end{bmatrix}$

34. Show that the set of all solutions of the given (nonhomogeneous) system does NOT form a subspace of 4-space.

35. Show that the set of all solutions of the corresponding *homogeneous* system (ie. X_h) does form a subspace; find a basis and hence its dimension. ans. B = {(-2,1,1,0)}, 1

36. If A is the corresponding *coefficient matrix* of the given augmented matrix, ie.

$A = \begin{bmatrix} 1 & 2 & 0 & -1 \\ 0 & 1 & -1 & 3 \\ 1 & 3 & -1 & 3 \end{bmatrix}$ show that its rows (as elements of $\Re^4$) form an independent set.

37. If A is as in #36, show that its columns (as elements in $\Re^3$) form a dependent set. HINT: just count

Part F. Let $V = \Re^5$;

38. The linear equation $5 x_1 - 4 x_2 + 3 x_3 - 2 x_4 + x_5 = 0$ represents a hyperplane in V; find a basis S of 5-vectors such that span(S) is this hyperplane. Hint: you should need 4 vectors!

39. The span of TWO (independent) vectors in V is a plane; find the equation(s) of the plane whose basis is S = {(1,0,1,-2,3),(0,1,2,-3,4)}. ans. $x_1 + 2 x_2 - x_3 = 0$

40. The span of ONE vector in V is a line; if S = {(1,2,-1,3,4)} is a basis for W = span(S), find a description of the line(which must pass through the origin!) in terms of $X = (\ x_1, x_2, x_3, x_4, x_5 \ )$.

41. Show that the intersection of the two hyperplanes $x_1 + x_3 + x_5 = 0$ and $x_2 + x_3 - x_4 = 0$ is a subspace which is spanned by 3 vectors; find the vectors.
ans. S = {(-1,-1,1,0,0),(0,1,0,1,0),(-1,0,0,0,1)}

Part G. Independence and functions spaces: given {f,g} which are contained in C[a,b] such that f and g are differentiable on [a,b];

42. show {f,g} is independent if $g \neq k\,f$ for some $k \in \Re$.

43. show {f,g} is dependent if g = kf

44. show if {f,g} is dependent and $W(x) = \begin{bmatrix} f & g \\ f' & g' \end{bmatrix}$ then det(W(x)) = 0 for all x in [a,b].

(ie. $\det(W(x)) \equiv 0$ on [a,b])

45. show if $W(x_0) \neq 0$ for some x_0 in [a,b] then {f,g} is independent

46. Given S = {cosh(x), sinh(x)} on [-1,1]; show in two different ways that S is independent.

47. Generalize what is in problems # 44 and # 45 for a set of 3 functions {f,g,h} which are continuous and twice differentiable on [a,b].

48. Given S = {cos(x), sin(x), 2 cos(x) + 3 sin(x)}on $[-\pi,\pi]$; explain why S must be dependent. Then show that det(W) = 0 for any x.

49. Show that the set of functions $\{1, x, x^2\}$ is independent on $\Re$ by finding det(W(x)) and showing that it is *not* identically zero. ans. det(W(x)) = 2

50. For the set S = $\{x^2, x^3\}$, W = 0 if x = 0; does this mean that S is dependent? Explain.

H. Prove: If dim(V) = n and S contains n (nonzero) vectors;

51. any set S with less than n vectors cannot span V.

52. any set S with more than n vectors must be dependent.

3.5 Row and Column Spaces

In this section we wish to discuss several subspaces connected with an m by n matrix A. For example, suppose we wish to know if the set of vectors S = {(1,2,3,4),(0,1,-3,2),(1,3, 0,6)} is independent or not. We could solve the independence equation as we did before or we could create a matrix A whose rows are the components of the vectors in S- in other words use the vectors in S to create a matrix A. Why use *rows*? Well, we are used to doing *row operations* on a matrix and these do not alter the structure of the rows (ie. the span of the rows of A is unchanged by doing elementary row operations). Here we have

$$A = \begin{bmatrix} 1 & 2 & 3 & 4 \\ 0 & 1 & -3 & 2 \\ 1 & 3 & 0 & 6 \end{bmatrix}$$

Let's get the row-echelon form of A:

$$A = \begin{bmatrix} 1 & 2 & 3 & 4 \\ 0 & 1 & -3 & 2 \\ 1 & 3 & 0 & 6 \end{bmatrix} \sim \begin{bmatrix} 1 & 2 & 3 & 4 \\ 0 & 1 & -3 & 2 \\ 0 & 1 & -3 & 2 \end{bmatrix} \sim \begin{bmatrix} 1 & 2 & 3 & 4 \\ 0 & 1 & -3 & 2 \\ 0 & 0 & 0 & 0 \end{bmatrix}$$

So we know now that the third vector was a combination of the first two (so the original rows of A formed a *dependent* set) and in fact a basis for span(S) = {(1,2,3,4),(0,1,-3,2)}. From chapter one, we also know that the rank of A is 2; but now we have a better interpretation of this number- it tells us the **dimension** of the span of the rows of A. Let's agree to call the span of the rows of A the row space of A (and use the abbreviation row(A) for this span).

DEFINITION

The **row space** of an m x n matrix A is the span of the rows of A and is a subspace of $\mathfrak{R}^n$. We denote it by row(A).

From the previous example, row(A) = span{(1,2,3,4),(0,1,-3,2)} $\subseteq \mathfrak{R}^4$ So here the dimension of row(A) = 2 and it is a plane through the origin in 4-space. What about the structure of the columns? Using the matrix A that we have:

$$C_1 = \begin{bmatrix} 1 \\ 0 \\ 1 \end{bmatrix} \quad C_2 = \begin{bmatrix} 2 \\ 1 \\ 3 \end{bmatrix} \quad C_3 = \begin{bmatrix} 3 \\ 3 \\ 0 \end{bmatrix} \quad C_4 = \begin{bmatrix} 4 \\ 2 \\ 6 \end{bmatrix}$$

and naturally we define the column space of A to be just the span of the columns:

DEFINITION

The **column space** of an m x n matrix A is the span of the columns of A and is a subspace of $\mathfrak{R}^m$. We denote it by col(A).

But if we want to know just what columns are needed to get a basis for the column space of A, we can take the transpose of A if we wish to do row operations. Then doing

141

row operations on the transpose of A will *not destroy the column structure* of A.
In general, doing row operations changes the *column* structure of A. Here we have

$$A^t = \begin{bmatrix} 1 & 0 & 1 \\ 2 & 1 & 3 \\ 3 & -3 & 0 \\ 4 & 2 & 6 \end{bmatrix}$$

Certainly we will be able to eliminate at least one row, since 4 vectors cannot be independent in 3-space. In fact:

$$A^t = \begin{bmatrix} 1 & 0 & 1 \\ 2 & 1 & 3 \\ 3 & -3 & 0 \\ 4 & 2 & 6 \end{bmatrix} \sim \begin{bmatrix} 1 & 0 & 1 \\ 0 & 1 & 1 \\ 0 & -3 & -3 \\ 0 & 2 & 2 \end{bmatrix} \sim \begin{bmatrix} 1 & 0 & 1 \\ 0 & 1 & 1 \\ 0 & 0 & 0 \\ 0 & 0 & 0 \end{bmatrix}$$

so that the column space is spanned by two vectors $\left\{ \begin{bmatrix} 1 \\ 0 \\ 1 \end{bmatrix}, \begin{bmatrix} 0 \\ 1 \\ 1 \end{bmatrix} \right\}$ and you should be able to show

that the last two columns of A are combinations of the *first* two columns.

Alternatively, it can be proven that once the *pivot columns* of A are identified (through row reduction), the **original columns** in which the *pivot elements occur* form a basis for col(A). Here

we could have used the two original columns $C_1 = \begin{bmatrix} 1 \\ 0 \\ 1 \end{bmatrix}, C_2 = \begin{bmatrix} 2 \\ 1 \\ 3 \end{bmatrix}$.

Thus the column space of A is also 2-dimensional. Is this an accident or not? Well, the rank of A gives not only the dimension of the row(A) but also the dimension of col(A).

THEOREM $\dim(\text{col}(A)) = \dim(\text{row}(A))$
Proof: Let $\dim(\text{row}(A)) = \text{rank}(A)$ and let $k = \dim(\text{null}(A))$; then $r + k = n$ so $k = n - r$.
Case I: Suppose $k = 0$; this is left as an exercise.
Case II: Suppose $k > 0$; then $k = n - r$ so there are r pivot columns in the echelon form of A; let U be the echelon form of A and suppose the first r columns of A are the pivot columns. Then A is of the form

$$\begin{bmatrix} C_1 & C_2 & \cdots C_r & C_{r+1} & \cdots & C_{r+k} \end{bmatrix}$$

If we let

$$X = (\, x_1, x_2, ..., x_n \,) \in null(\,A\,)$$

then there are infinitely many solutions of

$$x_1 C_1 + \dots + x_n C_n = \theta$$

By letting $x_{r+1} = 1$ and setting the rest of the free variables $x_{r+2}, x_{r+3}, \dots, x_{r+k}$ equal to zero, we see that we can solve for C_{r+1} in terms of the first r columns; this argument can be repeated for all the non-pivot columns so that $S = \{ C_1, C_2, \dots, C_r \}$ is a spanning set for col(A).

Claim: S is independent. Suppose

$$x_1 C_1 + x_2 C_2 + \dots + x_r C_r = \theta$$

since the reduced echelon form of this system is $I_r \begin{bmatrix} x_1 \\ x_2 \\ \vdots \\ x_r \end{bmatrix} = \begin{bmatrix} 0 \\ 0 \\ \vdots \\ 0 \end{bmatrix}$, it is clear that the only solution is

the trivial one ie.

$$x_1 = 0, x_2 = 0, \dots, x_r = 0$$

so S is independent and thus S is a basis for col(A). Hence dim(col(A)) = r = dim(row(A)).

QED

Be careful, **this does not mean that row(A) = col(A)!** In general, the row(A) is NOT the same as the col(A).

Col(A) and consistent systems

Now the col(A) is also useful since it can tell us for what right-hand sides B the corresponding system AX = B will be consistent. For example, suppose we look at a system with an arbitrary right-hand side B whose coefficient matrix is A; A is 3 x 4 so m = 3 and n = 4:

$$\begin{bmatrix} 1 & 2 & 3 & 4 \\ 0 & 1 & 3 & 2 \\ 1 & 3 & 0 & 6 \end{bmatrix} \begin{bmatrix} x_1 \\ x_2 \\ x_3 \\ x_4 \end{bmatrix} = \begin{bmatrix} a \\ b \\ c \end{bmatrix}$$

Let's write out the system:

$$x_1 + 2 x_2 + 3 x_3 + 4 x_4 = a$$

$$x_2 - 3 x_3 + 2 x_4 = b$$

$$x_1 + 3 x_2 + \quad + 6 x_4 = c$$

Looking at this column-wise

$$x_1 \begin{bmatrix} 1 \\ 0 \\ 1 \end{bmatrix} + x_2 \begin{bmatrix} 2 \\ 1 \\ 3 \end{bmatrix} + x_3 \begin{bmatrix} 3 \\ -3 \\ 0 \end{bmatrix} + x_4 \begin{bmatrix} 4 \\ 2 \\ 6 \end{bmatrix} = \begin{bmatrix} a \\ b \\ c \end{bmatrix}$$

it is clear that the *system will have a solution* only if the right hand side B is a linear combination of the columns of A. But the span of the columns of A is the *column space* of A. So we have:

THEOREM: The system AX = B is consistent iff $B \in col(A)$.

We can verify that the theorem is true by finding all $B = \begin{bmatrix} a \\ b \\ c \end{bmatrix}$ such that the system AX = B is

consistent. If we look at the augmented matrix:

$$\begin{bmatrix} 1 & 2 & 3 & 4 & a \\ 0 & 1 & -3 & 2 & b \\ 1 & 3 & 0 & 6 & c \end{bmatrix} \sim \begin{bmatrix} 1 & 2 & 3 & 4 & a \\ 0 & 1 & -3 & 2 & b \\ 0 & 1 & -3 & 2 & c-a \end{bmatrix} \sim \begin{bmatrix} 1 & 2 & 3 & 4 & a \\ 0 & 1 & -3 & 2 & b \\ 0 & 0 & 0 & 0 & c-a-b \end{bmatrix}$$

We can see that the system is inconsistent unless c - a - b = 0 or c = a + b. But this means that B is of the form:

$$B = \begin{bmatrix} a \\ b \\ c \end{bmatrix} = \begin{bmatrix} a \\ b \\ a+b \end{bmatrix} = \begin{bmatrix} a \\ 0 \\ a \end{bmatrix} + \begin{bmatrix} 0 \\ b \\ b \end{bmatrix} = a \begin{bmatrix} 1 \\ 0 \\ 1 \end{bmatrix} + b \begin{bmatrix} 0 \\ 1 \\ 1 \end{bmatrix}$$

which shows again (using just chapter one stuff) that the system has a solution iff B is a combination of (1,0,1) and (0,1,1) (ie. B lies in col(A)).

Here B is a vector in 3-space but the columns are spanned by just two vectors, so col(A) is a plane through the origin. Most vectors in 3-space do not lie on this plane. If a B is picked at random using the given coefficient matrix A, the corresponding system AX = B would probably be inconsistent. Knowing what the column space looks like is certainly an advantage. Suppose we choose a B like

$B = \begin{bmatrix} 3 \\ 1 \\ 4 \end{bmatrix}$; first, we know the system has a solution since for this B, c = 4 = 3 + 1 = a + b , thus B is

in col(A).

$$A = \begin{bmatrix} 1 & 2 & 3 & 4 & 3 \\ 0 & 1 & -3 & 2 & 1 \\ 1 & 3 & 0 & 6 & 4 \end{bmatrix} \sim \begin{bmatrix} 1 & 2 & 3 & 4 & 3 \\ 0 & 1 & -3 & 2 & 1 \\ 0 & 1 & -3 & 2 & 1 \end{bmatrix} \sim \begin{bmatrix} 1 & 2 & 3 & 4 & 3 \\ 0 & 1 & -3 & 2 & 1 \\ 0 & 0 & 0 & 0 & 0 \end{bmatrix}$$

This means we can solve for X:

$$x_2 = 1 + 3x_3 - 2x_4 \quad x_3 = s \quad x_4 = t \quad \Rightarrow x_2 = 3s - 2t + 1$$

$$x_1 = 3 - 2 x_2 - 3 x_3 - 4 x_4 = x_1 = -9 x_3 + 1 = -9 s + 1$$

$$X = \begin{bmatrix} -9s+1 \\ 3s-2t+1 \\ s \\ t \end{bmatrix} = \begin{bmatrix} -9s \\ 3s \\ s \\ 0 \end{bmatrix} + \begin{bmatrix} 0 \\ -2t \\ 0 \\ t \end{bmatrix} + \begin{bmatrix} 1 \\ 1 \\ 0 \\ 0 \end{bmatrix} = s \begin{bmatrix} -9 \\ 3 \\ 1 \\ 0 \end{bmatrix} + t \begin{bmatrix} 0 \\ -2 \\ 0 \\ 1 \end{bmatrix} + \begin{bmatrix} 1 \\ 1 \\ 0 \\ 0 \end{bmatrix}$$

From chapter 1, we recognize the form of the solution to be $X = X_h + X_p$, where X_h (containing the parameter) is the solution of the corresponding homogeneous system. We see the solution CANNOT be unique, since there are two parameters in X_h. At this point a definition is in order.

DEFINITION

The **null space** of the matrix A is the set of all solutions of the corresponding homogeneous equation. $\quad null(A) = \{ X : AX = \theta \}$

Thus the null space is contained in $\mathfrak{R}^n$ since, if A is m by n, X must be n by 1, with the product being m by 1. Let's show that null(A) is indeed a subspace: suppose $X_1 \& X_2$ are in null(A); then we have:

$$A(X_1 + X_2) = A X_1 + A X_2 = \theta + \theta = \theta \Rightarrow X_1 + X_2 \in null(A)$$

Also

$$A(k X_1) = k(A X_1) = k(\theta) = \theta \Rightarrow k X_1 \in null(A)$$

so that null(A) is closed under addition and scalar multiplication and is definitely a subspace of $\mathfrak{R}^n$. The size of the null space is the nullity of A.

$$nullity(A) = dim(null(A))$$

Thus X_h is just the **null space** of A (the reader can verify this by multiplying A times X_h; you will get the zero matrix). Thus we have already found a basis for null(A):

$$null(A) = span \left\{ \begin{bmatrix} -9 \\ 3 \\ 1 \\ 0 \end{bmatrix}, \begin{bmatrix} 0 \\ -2 \\ 0 \\ 1 \end{bmatrix} \right\}$$

so that this subspace of $\Re^4$ is two-dimensional and nullity(A) = 2. The null space of A is a plane through the origin in 4-space. In Chapter 5 we will show that row(A) and null(A) are actually orthogonal to each other- coming attractions!

It is important to note that the set of all solutions of the non-homogeneous system AX = B is **NOT** a subspace of $\Re^4$, since if X_1 & X_2 are two solutions of AX = B then we have:

$$A(\ X_1+X_2\ )=A\ X_1+A\ X_2=B+B=2\ B\neq B$$

so the set of all solutions of AX = B is NOT closed under addition.

To summarize, the row structure of the system here is that there are 3 hyperplanes in 4-space trying to intersect. They do in fact intersect but not in a point; they intersect in a plane. There are only two leading ones in the row-echelon form of A (since the rank of A is 2) and this means that the nullity of A is also 2 (since there are 4 columns and rank(A) + nullity(A) = # of columns = 4). This means two parameters in the solution of AX = B which will produce a plane (not through the origin though, since the system AX = B is not homogeneous).

And of course a theorem from chapter one is:

$$\text{rank(A) + nullity(A) = n (\# of columns of A)}$$

This means that the rank of A constrains the size of the row space and null space; if one is large the other must be small (and conversely).

EXAMPLE 1 Analyze the matrix

$$A=\begin{bmatrix} 1 & 0 & 1 & 2 & 3 \\ 0 & 1 & 2 & -2 & 2 \\ -1 & -1 & 0 & 0 & 1 \\ 0 & 0 & 3 & 0 & 6 \end{bmatrix}$$

Find a basis for row(A), col(A) and null(A); write the dependent columns as a combination of a basis for col(A).

$$\begin{bmatrix} 1 & 0 & 1 & 2 & 3 \\ 0 & 1 & 2 & -2 & 2 \\ -1 & -1 & 0 & 0 & 1 \\ 0 & 0 & 3 & 0 & 6 \end{bmatrix} \sim \begin{bmatrix} 1 & 0 & 1 & 2 & 3 \\ 0 & 1 & 2 & -2 & 2 \\ 0 & 0 & 1 & 0 & 2 \\ 0 & 0 & 0 & 0 & 0 \end{bmatrix}$$

The echelon form of A tells us that row(A) is 3-dimensional and has basis {(1,0,1,2,3),(0,1,2,-2,2),(0,0,1,0,2)}. The rank of A must be 3. Since the last row was eliminated, it must have been a combination of the first three rows so the rows of A were dependent. Now for col(A); we work with the transpose of A:

146

$$A^t = \begin{bmatrix} 1 & 0 & -1 & 0 \\ 0 & 1 & -1 & 0 \\ 1 & 2 & 0 & 3 \\ 2 & -2 & 0 & 0 \\ 3 & 2 & 1 & 6 \end{bmatrix} \sim \begin{bmatrix} 1 & 0 & -1 & 0 \\ 0 & 1 & -1 & 0 \\ 0 & 0 & 1 & 1 \\ 0 & 0 & 0 & 0 \\ 0 & 0 & 0 & 0 \end{bmatrix}$$

Thus a basis for col(A) is {(1,0,-1,0),(0,1,-1,0), (0,0,1,1)} and dim((col(A)) = 3. Hence col(A) is a hyperplane through the origin in 4-space. Alternatively if you let B = (a,b,c,d) and adjoin the corresponding column to the coefficient matrix A, you'd find that AX = B has a solution only if d - a - b - c = 0 or d = a + b + c; hence B is of the form
B = (a,b,c,a + b + c) = a(1,0,0,1) + b(0,1,0,1) + c(0,0,1,1)
Another option for col(A) is to use the pivot columns of A (not its echelon form!!!) for a basis for col(A). Since rank(A) = 3, then we also know that the columns of A are *dependent*.
The first three columns of A are pivot columns so each of the last two columns must be a linear combination of the first three.
Now for null(A);

$$\begin{bmatrix} 1 & 0 & 1 & 2 & 3 & 0 \\ 0 & 1 & 2 & -2 & 2 & 0 \\ -1 & -1 & 0 & 0 & 1 & 0 \\ 0 & 0 & 3 & 0 & 6 & 0 \end{bmatrix} \sim \begin{bmatrix} 1 & 0 & 1 & 2 & 3 & 0 \\ 0 & 1 & 2 & -2 & 2 & 0 \\ 0 & 0 & 1 & 0 & 2 & 0 \\ 0 & 0 & 0 & 0 & 0 & 0 \end{bmatrix}$$

$$x_3 + 2 x_5 = 0 \Rightarrow x_3 = -2t$$

$$x_2 + 2 x_3 - 2 x_4 + 2 x_5 = 0 \Rightarrow x_2 = -2(-2t) + 2s - 2t = 2s + 2t$$

$$x_1 + x_3 + 2 x_4 + 3 x_5 = 0 \Rightarrow x_1 = - x_3 - 2 x_4 - 3 x_5 = -(-2t) - 2s - 3t = -2s - t$$

If X is in null(A) then

$$X = \begin{bmatrix} x_1 \\ x_2 \\ x_3 \\ x_4 \\ x_5 \end{bmatrix} = \begin{bmatrix} -2s-t \\ 2s+2t \\ -2t \\ s \\ t \end{bmatrix} = s \begin{bmatrix} -2 \\ 2 \\ 0 \\ 1 \\ 0 \end{bmatrix} + t \begin{bmatrix} 1 \\ 2 \\ -2 \\ 0 \\ 1 \end{bmatrix}$$

Hence we have a basis for null(A); it must be 2-dimensional. Thus our basic theorem is verified here since rank(A) + nullity(A) = 3 + 2 = 5 (# of columns of A). Geometrically null(A) is a plane through the origin in 5-space.

Using the null space of A

The null space of A is useful for answering several questions about the matrix A and the solutions of the corresponding system AX = B. By direct observation from ex.1, if null(A) is NOT trivial, then there must be at least one parameter in X_h, and the solution of AX = B CANNOT be unique. But since nullity(A) = n - r, a unique solution can only occur if n - r = 0 or n = r.

Thus if $B \in col(A)$, rank(A) < n implies a *consistent and dependent* system (ie. no uniqueness). From EX1, rank(A) = 3 and since n = 5, if AX = B has a solution, it cannot be unique. Of course, AX = B could be inconsistent, since $col(A) \neq \Re^4$ and col(A) is a hyperplane in 4-space through the origin. These ideas are summarized below.

Suppose B is in col(A) with rank(A) = r and n = number of columns of A:

nullity	null(A)	solution X
n – rank(A) = 0	trivial	unique
n – rank(A) > 0	non-trivial	not unique

The column structure of A from EX1 can be further analyzed also using null(A). We can show that columns 4 and 5 of the original matrix A are *linear combinations* of columns 1, 2 and 3 (the pivot columns). If t = 0 and s = 1, then X = (-2,2,0,1,0) is a solution of $AX = \theta$; this means that

$$-2C_1 + 2C_2 + C_4 = \theta$$

where C_1 is the first column of A, C_2 the second and so forth. Therefore $C_4 = 2C_1 - 2C_2$. Likewise, if s = 0 and t = 1, we have $-1C_1 + 2C_2 - 2C_3 + C_5 = \theta$. Hence we can solve for C_5: $C_5 = C_1 - 2C_2 + 2C_3$. This proves that the last two columns of A were dependent on the first three (since they can be written as linear combinations of $C_1, C_2, \& C_3$.

SUMMARY

Subspace	contained in	dimension
row space	$\Re^n$	rank(A) = r
column space	$\Re^m$	rank(A) = r
null space	$\Re^n$	n - r

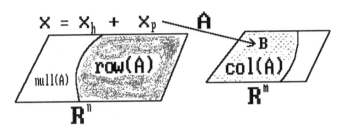

Square Matrices

Square matrices   are special in some respects; only a square matrix can have an inverse (and not all do!), only square matrices have determinants, and only square matrices have eigenvalues(a future topic). What about row space, column space, etc.if A is square? Suppose

$$A = \begin{bmatrix} 1 & 2 & 3 \\ 0 & 1 & -3 \\ 0 & 0 & 6 \end{bmatrix}$$

It is clear than rank(A) = 3 so nullity(A) = 0. Thus AX = B is consistent for all B and not only that, since null(A) is trivial(only the zero vector) the solution must be unique. Without doing really any work, we know all of the important facts about A and the system AX = B.

Since rank(A) = 3, the span of the rows must be all of 3-space; similarly the span of the columns must be all of three space so that means that row(A) = col(A). The rows (columns) of A are independent. This means that A is invertible (recall how the inverse of A is found using Gauss-Jordan elimination). So the determinant of A cannot be zero. We learn a lot just by knowing that the rank of A is 3.

EXAMPLE 2 Given

$$A = \begin{bmatrix} 3 & 6 & -3 \\ -1 & -2 & 1 \\ 2 & 4 & -2 \end{bmatrix}$$

using rank(A), determine if AX = B is consistent for all B.

Clearly the rows of A form a dependent set; all are multiples of (-1,-2,1) so rank(A) = 1. Since rank(A) + nullity(A) = 3, nullity(A) = 2 and null(A) is a plane through (0,0,0).

dim(col(A)) = rank(A) = 1 implies that col(A) is NOT all of 3-space so AX = B will be inconsistent unless B lies on the line X = t(3,-1,2). AX = B cannot have a unique solution, since null(A) is not trivial. The rank of A is the key to deducing all of these conclusions about AX = B.

FUNDAMENTAL THEOREM OF LINEAR ALGEBRA

Given a square matrix of size n by n; the following are equivalent:
1) A is *invertible*
2) $A \sim I_n$
3) rank(A) = n and nullity(A) = 0
4) rows(columns) of A are independent
5) det $(A) \neq 0$
6) *null $(A) = \{ \theta \}$*
7) row(A) = col(A) = $\Re^n$
8) AX = B has a unique solution

EXAMPLE 3 Discuss the matrix A in light of the fundamental theorem.

$$A = \begin{bmatrix} 2 & 2 & -4 & 1 \\ -1 & -2 & 1 & 2 \\ 1 & 0 & -3 & 3 \\ 2 & 0 & -6 & 6 \end{bmatrix}$$

Notice that the 4th row is the sum of the first three. Thus the rows are NOT independent so rank(A) < 4. Hence A is singular and is not row-equivalent to the identity matrix. The nullity(A) > 0 and the columns are dependent. The det(A) = 0 and both row(A) and col(A) are proper subspaces of 4-space. AX = B may be inconsistent or have infinitely-many solutions; thus AX = B cannot have a unique solution..

EXERCISES 3.5

Part A. For each matrix A given, find a)basis for row(A) b) basis for col(A) c) rank(A)

1. $A = \begin{bmatrix} 1 & 0 & 1 & 0 \\ 2 & 2 & 2 & 0 \\ 3 & 2 & 3 & 1 \end{bmatrix}$ ans. {(1,0,1,0),(0,1,0,0),(0,0,0,1)},{(1,1,3),(0,1,2),(0,0,1)}, rank(A) = 3

2. $A = \begin{bmatrix} 1 & 0 & 1 & 0 \\ 2 & 2 & 2 & 0 \end{bmatrix}$

3. $A = \begin{bmatrix} 1 & 0 & 1 & 0 \\ 2 & 2 & 2 & 0 \\ 3 & 2 & 3 & 1 \\ 6 & 4 & 6 & 1 \end{bmatrix}$ ans. {(1,0,1,0),(0,1,0,0),(0,0,0,1), {(1,2,3,6),(0,1,1,2),(0,0,1,1)},

rank(A) = 3

4. $A = \begin{bmatrix} 1 & 0 & 1 & 0 & 1 \\ 2 & 2 & 2 & 0 & 4 \\ 3 & 2 & 3 & 1 & 0 \end{bmatrix}$

5. $A = \begin{bmatrix} 1 & 0 & 1 \\ 2 & 2 & 2 \\ 3 & 2 & 3 \end{bmatrix}$ ans. $\{(1,0,1),(0,1,0)\}$), $\{(1,2,3),(0,1,1)\}$, rank(A) = 2

Part B. For each matrix in part A, find a basis for null(A) and find the nullity of A. Verify the "rank + nullity" theorem for each matrix A.

6.

7. ans. B = $\{(-1,0,1,0),(0,0,1,0)\}$, nullity(A) = 2, 2 + 2 = 4

8.

9. ans. B = $\{(-1,0,1,0,0),(-1,-1,0,5,1)\}$, nullity(A) = 2, 3 + 2 = 5

10.

$$x_1 + 2 x_2 + \quad + x_4 = 4$$

Part C. Given the system $-x_1 + 2 x_2 + 4 x_3 - 5 x_4 = 2$;

$$-x_1 + 2 x_2 - x_3 + 10 x_4 = 2$$

11. Write the system in the form AX = B. ans. $\begin{bmatrix} 1 & 2 & 0 & 1 \\ -1 & 2 & 4 & -5 \\ -1 & 2 & -1 & 10 \end{bmatrix} \begin{bmatrix} x_1 \\ x_2 \\ x_3 \\ x_4 \end{bmatrix} = \begin{bmatrix} 4 \\ 2 \\ 2 \end{bmatrix}$

12. Find the rank and nullity of A.

13. Find a basis for row(A) and its dimension. ans. $\{(1,2,0,1),(0,1,1,-1),(0,0,1,-3)\}$, 3

14. Find a basis for col(A) and its dimension; show dim(row(A)) = dim(col(A)).

15. Find null(A) and a basis for this subspace; give its dimension. Show dim(null(A)) = nullity(A) from # 12. ans. null(A) = span($\{(3,-2,3,1)\}$), nullity(A) = 1

16. Determine if the given B is in col(A); if so, find the solution X and write X in the form X = X$_h$ + X$_p$.

17. Prove that AX = B has a solution for *any* B in $\Re^3$. ans. rank(A) = 3

18. Which columns of A form a basis for col(A)?

19. Show that the remaining column(s) can be written as linear combination(s) of the columns from #18 using the null space of A. Hint: look back at "Using the null space" ans. $C_4 = -3 C_1 + 2 C_2 - 3 C_3$

20. Show that the only vector both in row(A) and null(A) is the zero vector. Hint: let X = (x,y,z,v) be in row(A) and in null(A); write out what this means in terms of linear combinations.

Part D. Given $A = \begin{bmatrix} 1 & 2 & 0 \\ 3 & 3 & 0 \\ 2 & 2 & 2 \end{bmatrix}$ which forms the linear system AX = B.

21. Find rank(A) and nullity(A). Is A invertible? Why? ans. 3,0,YES, rank(A) = 3
22. Determine for what B AX = B is consistent. Can AX = B have infinitely-many solutions? Why or why not?

Part E. Given $A = \begin{bmatrix} 1 & 2 & 1 \\ 3 & 3 & 0 \\ 4 & 5 & 1 \end{bmatrix}$ which forms the linear system AX = B.

23. Find rank(A) and nullity(A). Is A invertible? Why? ans. 2,1,NO, rank(A) < 3
24. Find a basis for row(A).
25. Find a basis for col(A). ans. $\{ \begin{bmatrix} 1 \\ 3 \\ 4 \end{bmatrix}, \begin{bmatrix} 0 \\ 1 \\ 1 \end{bmatrix} \}$

26. Determine for what B AX = B is consistent. Can AX = B have infinitely-many solutions? Why or why not?
27. Find an equation for row(A). ans. x - y + z = 0
28. Find an equation for col(A). Remember both are subspaces of $\Re^3$.
29. Find a basis for null(A). ans. (1,-1,1)
30. Show that only (0,0,0) is common to both row(A) and null(A).

Part F. Given the coefficient matrix $A = \begin{bmatrix} 1 & 0 & 2 & 2 \\ 0 & 1 & 4 & 8 \\ 1 & 1 & 6 & 10 \end{bmatrix}$ of a system AX = B;

31. Find a basis for col(A) using A^t. ans. {(1,0,1),(0,1,1)}
32. Find an equation for col(A); describe geometrically what col(A) is (eg. line, plane, etc).
33. Find a basis for row(A). ans. {(1,0,2,2),(0,1,4,8)}
34. Which columns of A depend on the pivot columns?
35. Find the row-echelon form of A; show that the **pivot columns of** A(NOT its echelon form!) also form a basis for col(A). ans. $\begin{bmatrix} 1 & 0 & 2 & 2 \\ 0 & 1 & 4 & 8 \\ 0 & 0 & 0 & 0 \end{bmatrix}$, first TWO columns of A are pivot columns, span{col's} = {(x,y,z): x + y - z = 0}
36. Find the form of any B for which AX = B is consistent. Can the system have a unique solution? Why or why not?
37. Find rank(A) and nullity(A). ans. 2,2
38. Find a basis for null(A).

39. Solve AX = B if $B = \begin{bmatrix} 2 \\ 3 \\ 5 \end{bmatrix}$; write the answer in the form X = Xh + Xp.

ans. $X = \begin{bmatrix} -2s - 2t \\ -4s - 8t \\ s \\ t \end{bmatrix} + \begin{bmatrix} 2 \\ 3 \\ 0 \\ 0 \end{bmatrix}$

40. Find the null space of A^t ; show that any vector Y = (a,b,c) can be written as the sum of two vectors, one from col(A) and one from the null space of A^t .

41. Show that $dim(\ col(A\ )\ ) + dim(\ null(\ A^t\ )\ ) = \#\ of\ ROWS\ of\ A$.

42. Show that the basis for row(A) and the basis for null(A) together form a basis for $\Re^4$.
Hint: Count and consider independence.

Part G. Given the matrix $A = \begin{bmatrix} 1 & 0 & 2 & -1 & 0 \\ 0 & 1 & 0 & 3 & 4 \\ 0 & 0 & 1 & 0 & 1 \end{bmatrix}$;

43. Find a basis for row(A) and rank(A). ans. original rows of A, 3
44. Find a basis for null(A).
45. Show that every vector X = (a,b,c,d,e) can be written (uniquely) as a linear combination of the vectors in a basis for the row(A) and null(A). NB. There should be a total of FIVE vectors there!
46. Find a basis for col(A) consisting of *original columns* of A. Hint: no work is required, think pivot columns.
47. Find the "left null space of A" (the null space of A^t) NB. The name "left null space" comes from the fact that if Y is in the left null space, then $A^t Y = \theta \Rightarrow Y^t A = \theta$. ans.{(0,0,0)}
48. If m = # of rows of A, show that rank(A) + $dim(\ null(\ A^t\ )\ ) = m$.
49. Is AX = B consistent for all B? Why or why not? ans. YES, rank(A) = 3
50. Use null(A) to write column 4 as a linear combination of the first 3 columns of A.
51. Use null(A) to write column 5 as a linear combination of the first 3 columns of A.
ans. C5 = -2C1 + 4C2 + C3

Part H. Use the given vectors to form the rows of matrix A; then put A in echelon form to determine whether the given set of vectors in independent or dependent. If dependent, determine which vector(s) can be discarded from S without affecting span(S).
52. S = {(1,0,1,2),(2,3,1,1),(3,3,2,3)}
53. S = {(1,0,1,1,2),(0,1,-1,1,3),(2,1,1,3,7)} ans. dependent, last vector

Part I. Given that col(A) = $span\{ \begin{bmatrix} 1 \\ 0 \\ 2 \\ 4 \end{bmatrix}, \begin{bmatrix} 0 \\ 1 \\ 2 \\ -2 \end{bmatrix} \}$ and null(A) = $span\{ \begin{bmatrix} -1 \\ -1 \\ 1 \end{bmatrix} \}$;

54. What is the rank of A and nullity of A?

55. What size matrix must A be? ans. 4 by 3

56. Find a matrix A such that the column space of A and the null space of A are as given above. Hint: you should know what the first two columns of A look like; let the third column be arbitrary

of the form $\begin{bmatrix} a \\ b \\ c \\ d \end{bmatrix}$ and use the fact that you know a basis for null(A).

Part J. Given the matrix

$$A = \begin{bmatrix} 1 & 2 & 3 \\ 0 & 2 & 0 \\ 0 & 0 & 2 \end{bmatrix} \quad B = A - \lambda \begin{bmatrix} 1 & 0 & 0 \\ 0 & 1 & 0 \\ 0 & 0 & 1 \end{bmatrix} = \begin{bmatrix} 1-\lambda & 2 & 3 \\ 0 & 2-\lambda & 0 \\ 0 & 0 & 2-\lambda \end{bmatrix}$$

57. Find all values for which B is singular. ans. $\lambda = 1, 2$

58. For each value of λ found in # 57, find the null space of the matrix B. Find a basis for the null space of B for each lambda and give its dimension.

Part K. Given twice-differentiable functions f,g, & h; let W =

$$W(x) = \begin{bmatrix} f & g & h \\ f' & g' & h' \\ f'' & g'' & h'' \end{bmatrix}$$

be the Wrońskian matrix of the functions f,g & h.

59. Find the matrix W for $f = 1, g = x, h = x^2$. ans. $W(x) = \begin{bmatrix} 1 & x & x^2 \\ 0 & 1 & 2x \\ 0 & 0 & 2 \end{bmatrix}$

60. Using the answer from #59, show that rank(W) = 3, det(W)$\neq 0$, and S = {f,g,h} is an independent set.

61. Repeat #59 for $S = \{ e^x, \sin(x), \cos(x) \}$. ans. $W(x) = \begin{bmatrix} e^x & \sin(x) & \cos(x) \\ e^x & \cos(x) & -\sin(x) \\ e^x & -\sin(x) & -\cos(x) \end{bmatrix}$

62. Using the answer from #61, show rank(W) = 3, det(W) $\neq 0$, and S is independent.

63. Let $S = \{ 1 + x^2, 1 - x^2, x^2 \}$; find the matrix W and show rank(W) < 3 for all x.

ans. $W(x) = \begin{bmatrix} 1+x^2 & 1-x^2 & x^2 \\ 2x & -2x & 2x \\ 2 & -2 & 2 \end{bmatrix}$, rank(W) = 2

64. Using the matrix W from #63, show det(W) = 0 and that S is a dependent set.

3.6 Coordinates

In this section we wish to explore the notion of coordinate representation. Suppose we pick a particular vector in 3-space, say X = (1,2,3). What do we mean by (1,2,3)? It sounds like a silly question- the physicist might answer that

$$X = (1,2,3) = 1\vec{i} + 2\vec{j} + 3\vec{k}$$

In other words we must have a **basis** in mind when we say that X = (1,2,3) if everyone is to be thinking of the same vector. In fact, the basis that we use is an underlined set, in other words, the standard basis for 3-space is $B = \{\vec{i}, \vec{j}, \vec{k}\}$ and we don't write the set in a different order. If we choose X to be (1,2,3,4) (ie. a vector in 4-space) we mean that

$$X = 1(1,0,0,0) + 2(0,1,0,0) + 3(0,0,1,0) + 4(0,0,0,1)$$

where we are writing X as a *linear combination* of the basis vectors in $\Re^4$. It is important that this representation is **unique**- that is, there is no other vector that can be written in this fashion.
Suppose a vector X in V with basis $B = \{X_1, X_2, ..., X_n\}$ can be written as a linear combination in two ways:

$$X = k_1 X_1 + k_2 X_2 + ... + k_n X_n \qquad X = j_1 X_1 + j_2 X_2 + ... + j_n X_n$$

Then by subtracting we have

$$(k_1 - j_1)X_1 + (k_2 - j_2)X_2 + ... + (k_n - j_n)X_n = \theta$$

$$(k_1 X_1 + k_2 X_2 + ... + k_n X_n) - (j_1 X_1 + j_2 X_2 + ... + j_n X_n) = \theta$$

But since the set B must be independent, the only way this combination can be zero is if all the scalars are zero:

$$k_1 - j_1 = 0 \Rightarrow k_1 = j_1, \quad k_2 - j_2 = 0 \Rightarrow k_2 = j_2, \quad ... , k_n - j_n = 0 \Rightarrow k_n = j_n$$

In order to emphasize the particular basis that we are using, we can write this vector X as $(X)_B = (1,2,3,4)$ in order to keep in mind the fact that we are using the basis

$$B = \{(1,0,0,0), (0,1,0,0), (0,0,1,0), (0,0,0,1)\}$$

Of course we like to get matrices into this discussion so we can also use the **coordinate matrix** representation for X:

$$[X]_B = \begin{bmatrix} 1 \\ 2 \\ 3 \\ 4 \end{bmatrix}$$

Actually we will find it more useful to use the coordinate matrix in our subsequent work. In the future, if *no subscript* is used it is understood that the coordinate matrix is given with respect to the *standard basis* in V!

EXAMPLE 1 Find the coordinate matrix for the vector $p(x)= 2 + 3 x$ if the basis B is given by B = {1 + 3x, 4 - 5x}. We need to write p as a linear combination of the two vectors given:

$$p(x)= 2 + 3 x = k_1(1 + 3 x)+ k_2(4 - 5 x)$$

This means that we need to find the two scalars k_1 & k_2. Combining like terms implies:

$$p(x)= 2 + 3 x = (k_1 + 4 k_2)+(3 k_1 - 5 k_2)x$$

The linear system is thus

$$k_1 + 4 k_2 = 2$$

$$3 k_1 - 5 k_2 = 3$$

which can be represented by its augmented matrix or written in the form AK = B:

$$\begin{bmatrix} 1 & 4 & 2 \\ 3 & -5 & 3 \end{bmatrix} \text{ or } A = \begin{bmatrix} 1 & 4 \\ 3 & -5 \end{bmatrix} \begin{bmatrix} k_1 \\ k_2 \end{bmatrix} = \begin{bmatrix} 2 \\ 3 \end{bmatrix}$$

Choosing the tried and true Gaussian elimination method, we have:

$$\begin{bmatrix} 1 & 4 & 2 \\ 3 & -5 & 3 \end{bmatrix} \underset{-3\,row_1 + row_2}{\sim} \begin{bmatrix} 1 & 4 & 2 \\ 0 & -17 & -3 \end{bmatrix} \underset{\frac{-1}{17}row_2}{\sim} \begin{bmatrix} 1 & 4 & 2 \\ 0 & 1 & \frac{3}{17} \end{bmatrix}$$

so $k_2 = \dfrac{3}{17}$ which means

$$k_1 + 4(\frac{3}{17})= 2 \Rightarrow k_1 = \frac{34 - 12}{17} = \frac{22}{17}$$

so finally

$$[\;p\;]_B=\begin{bmatrix}\dfrac{22}{17}\\[2mm]\dfrac{3}{17}\end{bmatrix}$$

This can be checked directly:

$$\frac{22}{17}(\;1+3\,x\;)+\frac{3}{17}(\;4-5\,x\;)=\frac{22+12}{17}+\frac{66-15}{17}\,x=2+3\,x=p(\;x\;)$$

So we found the coordinate matrix of p with respect to the basis B given.

EXERCISES 3.6

Part A. Find the coordinates of the vector X wrt. the basis B given:
1. X = (2,2,4), B = {(1,1,0),(0,0,2),(1,1,3)}ans. $(\;X\;)_B=(\;2,2,0\;)$

2. X = (3,3,4,3), B = {(1,0,1,0),(1,0,-1,0),(0,1,0,1),(0,1,0,-1)}

3. X = (1,2,3), B = {(1,0,1),(1,0,-1),(0,1,0)}ans. $(\;X\;)_B=(\;2,-1,2\;)$

4. X = (2,4,2,4), B = {(1,0,0,0),(0,1,1,1),(0,0,1,1),(0,0,0,1)}

5. X = (1,2,3,4), B = {(1,0,0,0),(1,1,0,0),(1,1,1,0),(1,1,1,1)} ans. $(\;X\;)_B=(\;-1,-1,-1,4\;)$

Part B. Find the coordinates of the polynomial p with respect to the basis B given:
6. $p(\;x\;)=2+x^2,\quad B=\{\;3,2-x,x-x^2\;\}$

7. $p(\;x\;)=1+2\,x+3\,x^2,\quad B=\{\;1-x,1+x,4\,x^2\;\}$ ans. $(\;p\;)_B=(\;-\frac{1}{2},\frac{3}{2},\frac{3}{4}\;)$

8. $p(\;x\;)=-2+2\,x+3\,x^2,\quad B=\{\;1,x,x^2-\frac{1}{3}\;\}$

9. $p(\;x\;)=-x+3\,x^2+5\,x^3,\quad B=\{\;1,x,x^2-\frac{1}{3},x^3-\frac{3}{5}\,x\;\}$ ans. $(\;p\;)_B=(\;1,2,3,5\;)$

10. $p(\;x\;)=1+2\,x+3\,x^2,\quad B=\{\;1,x,2\,x^2-1\;\}$

11. $p(\;x\;)=2\,x+4\,x^2,\quad B=\{\;1,2\,x,4\,x^2-1\;\}$ ans. $(\;p\;)_B=(\;1,1,1\;)$

12. $p(\;x\;)=1+2\,x+3\,x^2,\quad B=\{\;1,1-x,2-4\,x+x^2\;\}$

158

13. $p(x) = 1 + x + x^2 + x^3$, $B = \{ 1, x, 2x^2 - 1, 4x^3 - 3x \}$ ans. $(p)_B = (\dfrac{3}{2}, \dfrac{5}{2}, \dfrac{1}{2}, \dfrac{1}{4})$

Part C. Find the coordinates of the matrix A with respect to the given basis B:

14. $B = \{ \begin{bmatrix} 2 & 0 \\ 0 & 0 \end{bmatrix}, \begin{bmatrix} 0 & -1 \\ 0 & 0 \end{bmatrix}, \begin{bmatrix} 0 & 0 \\ 3 & 0 \end{bmatrix}, \begin{bmatrix} 0 & 0 \\ 0 & -4 \end{bmatrix} \}$ $A = \begin{bmatrix} 1 & 2 \\ 3 & 4 \end{bmatrix}$

15. $B = \{ \begin{bmatrix} 1 & 0 \\ 0 & 0 \end{bmatrix}, \begin{bmatrix} 0 & 0 \\ 0 & 1 \end{bmatrix}, \begin{bmatrix} 0 & 1 \\ 1 & 0 \end{bmatrix} \}$ $A = \begin{bmatrix} 3 & 2 \\ 2 & 5 \end{bmatrix}$

ans. $(A)_B = (3, 4, -5)$

16. $B = \{ \begin{bmatrix} 0 & 1 & 0 \\ -1 & 0 & 0 \\ 0 & 0 & 0 \end{bmatrix}, \begin{bmatrix} 0 & 0 & 1 \\ 0 & 0 & 0 \\ -1 & 0 & 0 \end{bmatrix}, \begin{bmatrix} 0 & 0 & 0 \\ 0 & 0 & 1 \\ 0 & -1 & 0 \end{bmatrix} \}$

$A = \begin{bmatrix} 0 & 3 & 4 \\ -3 & 0 & -5 \\ -4 & 5 & 0 \end{bmatrix}$

17. $B = \{ \begin{bmatrix} 1 & 0 & 0 \\ 0 & 0 & 0 \\ 0 & 0 & -1 \end{bmatrix}, \begin{bmatrix} 0 & 0 & 0 \\ 0 & 1 & 0 \\ 0 & 0 & -1 \end{bmatrix} \}$ $A = \begin{bmatrix} 2 & 0 & 0 \\ 0 & 4 & 0 \\ 0 & 0 & -6 \end{bmatrix}$

ans. $(A)_B = (2, 4)$

Part D. Given the basis $B = \{1, \cos(x), \sin(x)\}$ where $W = \text{span}(B)$ find the coordinate matrix for:

18. $f = 3 - 4 \sin(x)$

19. $f = -2 + 3 \cos(x) - 5 \sin(x)$ ans. $[f]_B = \begin{bmatrix} -2 \\ 3 \\ -5 \end{bmatrix}$

20. $f = 2 \cos(x) + 5 \sin(x) - 3(\cos^2(x) + \sin^2(x))$

Part E. Find the coordinate matrix for each matrix A using the standard basis in $M_{2,2}$.

21. $A = \begin{bmatrix} 1 & 3 \\ 4 & 7 \end{bmatrix}$ ans. $[A]_B = \begin{bmatrix} 1 \\ 3 \\ 4 \\ 7 \end{bmatrix}$

22. $A = \begin{bmatrix} -3 & 5 \\ 7 & 11 \end{bmatrix}$

159

23. $A = 2 \begin{bmatrix} 3 & -4 \\ 5 & -2 \end{bmatrix}$ ans. $[A]_B = \begin{bmatrix} 6 \\ -8 \\ 10 \\ -4 \end{bmatrix}$

Part F. Given B = $\{1, x, e^{-x}, e^x\}$ find the coordinate matrix for each vector, where W = span(B).

24. $f = 2x - e^x$

25. f = cosh(x) ans.

$$(f)_B = \begin{bmatrix} 0 \\ 0 \\ \frac{1}{2} \\ \frac{1}{2} \end{bmatrix}$$

26. f = sinh(x)

27. f = 2 + 3x + 2 cosh(x) + 4sinh(x) ans. $(f)_B = \begin{bmatrix} 2 \\ 3 \\ -1 \\ 3 \end{bmatrix}$

Part G. Given the basis B = $\{1, \cos(x), \sin(x), \cos(2x), \sin(2x)\}$; if W = span(B), find the coordinates of the function f with respect to the basis B if f is in W.

28. f = 3 - 3cos(x) + 4 sin(2x)

29. $f = 3 + 4\cos(x) + 2\cos^2(x)$ ans. (4,4,0,1,0)

30. $f = 5 - 4\sin(x) + 6\sin^2(x)$

31. $f = \cos^2(x) - \sin^2(x)$ ans. (0,0,0,1,0)

32. $f = 3\sin(x) + 4\cos^2(x) - 4\sin^2(x) + 6\cos(x)\sin(x)$

160

3.7 Chapter Three Outline and Review

3.1. Vector Spaces
"Model" vector space is n-space; know the properties of vector addition and scalar multiplication in n-space. A vector space is a set V of vectors with two operations:
1) vector addition and 2) scalar multiplication
such that the 5 vector space axioms hold.
Alternative: (V,+) forms a **commutative group**; V is closed under scalar multiplication and:
   a) $k(X + Y) = kX + kY$
   b) $(k + j)X = kX + jX$
   c) $(kj)X = k(jX)$
   d) $1X = X$
Elements in a vector space behave like vectors in n-dimensional space under addition and scalar multiplication. Know the 5 axioms and notation for classic vector spaces.

3.2 Subspaces
If $W \subseteq V$ and W is itself a vector space then W is a subspace of V.
THEOREM If W is a subset of V which is closed under addition and scalar multiplication, then W is subspace of V. This means if X,Y are in W then so are 1)X + Y and 2) kX
Alternative: kX + jY is in W for all X,Y in W and scalars k,j implies W is a subspace of V
In n-space a line, plane or hyperplane through the origin are subspaces

3.3 Linear Independence and Spanning
Linear combination: given

$$S = \{ X_1, X_2, \dots, X_n \} \subseteq V$$

if there exists scalars k_1, k_2, etc. such that

$$X = k_1 X_1 + k_2 X_2 + \dots + k_n X_n$$

then X is a **linear combination** of the vectors in S
Spanning set: if

$$W = \{ k_1 X_1 + k_2 X_2 + \dots + k_n X_n : k_i \in \Re \}$$

then W is the collection of *all linear combinations* of vectors in S; we say S **spans** W and we write
W = span(S)
span(S) is always a subspace of V
Independence equation:if

$$k_1 X_1 + k_2 X_2 + \dots + k_n X_n = \theta \implies k_1 = 0, k_2 = 0, \dots k_n = 0$$

then the set S = $\{ X_1, X_2, \dots, X_n \}$ is **independent**; if non-trivial solutions exist (ie. even one scalar k_i is NOT zero, then the set is *dependent*.
NB. Any set S containing the zero vector is automatically dependent.

THEOREM A set S is dependent iff at least one vector in S is a linear combination of the other vectors in S (so in an independent set, no vector is a combination of the others).

3.4 Basis and Dimension

Basis- if

$$B = \{\ X_1, X_2, ..., X_n\ \} \subseteq V$$

and B is independent and spans V, then B is a **basis** for V. A basis is thus a
1) minimal spanning set and a 2) maximal independent set in V.
If B has finitely-many vectors, then V is *finite-dimensional* and its **dimension** is n; ie.
$\dim(V) = n$. If no such n exists, then V is infinite-dimensional (eg. C[a,b] - set of all continuous functions defined on [a,b]).
THEOREM If $\dim(V) = n$, then
1) If S has n vectors and is <u>independent</u>, then S *spans* V
2) If S has n vectors and *spans* V, then S is <u>independent</u>
3) Any collection S with $>$ n vectors is dependent; any collection S with $<$ n vectors cannot span V.
Know the standard bases for all classic vector spaces $\Re^n, M_{m,n}, P_n$

3.5 Row and Column Spaces

Given any m by n matrix A; consider the rows as vectors in $\Re^n$ and the columns as vectors in $\Re^m$; then
1) **row space** = span of rows and 2) *column space* = span of columns
where we write 1) row(A) and 2) col(A) respectively
Null space- all solutions of $AX = \theta$; we write null(A).
$\text{Rank}(A) = \dim(\text{row}(A)) = \dim(\text{col}(A))$ and $\dim(\text{null}(A)) = \text{nullity}(A)$ where
THEOREM $\text{rank}(A) + \text{nullity}(A) = n$ (no. of columns of A)
Finding bases for row(A) and col(A):
1) for row(A) just put A is echelon form- nonzero rows form a basis for row(A)
2) for col(A) put A^t in echelon form and use its ROWS or (use pivot columns of A-not its echelon form!)
THEOREM AX = B has a solution iff B is in col(A).
AX = B has a unique solution if B is in col(A) and null(A) is trivial. Nullity(A) $>$ 0 means no unique solution .
Know the fundamental theorem! If A is invertible, then...

3.6 Coordinates

Given $B = \{\ X_1, X_2, ..., X_n\ \} \subseteq V$ where B is a basis for V; if

$$X = k_1 X_1 + k_2 X_2 + ... + k_n X_n$$

then the **coordinates** of X are the (unique) scalars $k_1, k_2, ..., k_n$ and we write

$$(X)_B = (k_1, k_2, \ldots, k_n)$$

The *coordinate matrix* of X is the column matrix $[X]_B = \begin{bmatrix} k_1 \\ k_2 \\ \ldots \\ k_n \end{bmatrix}$. If B is the standard basis for V,

we may omit the subscript.

Keep in mind a basis is an *ordered* set!

Review Problems

Part A: Given the set W contained in the vector space V; determine if W is a subspace of V. If it is, give its dimension if it is finite-dimensional.

1. $W = \left\{ \begin{bmatrix} 0 & a \\ b & 0 \end{bmatrix} : a, b \in \Re \right\} \subseteq M_{2,2}$ ans. yes, 2

2. $W = \left\{ \begin{bmatrix} 1 & a \\ b & 1 \end{bmatrix} : a, b \in \Re \right\} \subseteq M_{2,2}$ ans. no

3. $W = \{ a + b x^2 : a, b \in \Re \} \subseteq P_2$ ans. yes, 2

4. $W = \{ a + x^2 : a \in \Re \} \subseteq P_2$ ans. no

5. W = {f: f is differentiable} where $W \subseteq C [a, b]$ ans. yes

6. $W = \{ f : f' + 2 f = 0 \} \subseteq C [a, b]$ NB. solution of y' + 2y = 0 is $y = C e^{-2x}$ ans. yes, 1

7. W = span({(1,2,0,1),(0,1,1,1)}) ans. yes, 2

8. W = span({(1,2,0,1),(0,1,1,1),(0,1,0,1)}) ans. yes, 3

9. W = {(x,y,z): x - y + z = 0} ans. yes, 2

10. W = {(x,y,z): x + y - z = 1} ans. no

11. $W = \left\{ \begin{bmatrix} a & b \\ c & -a \end{bmatrix} : a, b, c \in \Re \right\} \subseteq M_{2,2}$ ans. yes, 3

12. $W = \left\{ \begin{bmatrix} a & b & c \\ c & a & b \\ b & c & a \end{bmatrix} : a, b, c \in \Re \right\} \subseteq M_{2,2}$ ans. yes, 3

Part B: Write the vector X as a linear combination of the vectors in S.

13. $X = (2,1,2,1)$, $S = \{(1,0,0,1), (0,1,1,0,), (1,0,1,0)\}$ ans. $X = 1(1,0,0,1) + 1(0,1,1,0,) + 1(1,0,1,0)$

14. $X = (6,6,7,9)$, $S = \{(1,1,0,0), (0,0,1,1), (1,1,1,3)\}$ ans. $X = 5(1,1,0,0) + 6(0,0,1,1) + 1(1,1,1,3)$

15. $p = 3 + 2x + 3x^2$, $S = \{ 1 + x, 1 - x, x + x^2 \}$ ans. $p = 1(1 + x) + 2(1 - x) + 3x^2$

16. $p = 3 + 3x - x^2$, $S = \{ 1 + x^2, 1 - x^2, x \}$ ans. $p = 1(1 + x^2) + 2(1 - x^2) + 3x$

17. $X = \begin{bmatrix} 4 & 5 \\ 2 & 4 \end{bmatrix}$ $S = \left\{ \begin{bmatrix} 1 & 0 \\ 0 & 1 \end{bmatrix}, \begin{bmatrix} 0 & 1 \\ 1 & 0 \end{bmatrix}, \begin{bmatrix} 1 & 1 \\ 0 & 1 \end{bmatrix} \right\}$

ans. $X = \begin{bmatrix} 4 & 5 \\ 2 & 4 \end{bmatrix} = 1 \begin{bmatrix} 1 & 0 \\ 0 & 1 \end{bmatrix} + 2 \begin{bmatrix} 0 & 1 \\ 1 & 0 \end{bmatrix} + 3 \begin{bmatrix} 1 & 1 \\ 0 & 1 \end{bmatrix}$

18. $X = \begin{bmatrix} 2 & 9 \\ 4 & 2 \end{bmatrix}$ use S from #17

ans. $X = \begin{bmatrix} 2 & 9 \\ 4 & 2 \end{bmatrix} = -3 \begin{bmatrix} 1 & 0 \\ 0 & 1 \end{bmatrix} + 4 \begin{bmatrix} 0 & 1 \\ 1 & 0 \end{bmatrix} + 5 \begin{bmatrix} 1 & 1 \\ 0 & 1 \end{bmatrix}$

Part C: Determine if S spans W; if so state whether S is independent or not. Give the dimension of W also.

19. $S = \{ 1 - x, 1 + x, 2x^2, 2 + 2x + 4x^2 \}$, $W = P_2$

ans. YES, dependent($2 + 2x + 4x^2$ is a linear combination of the first 3),3

20. $S = \{ 1 - x, 1 + x \}$, $W = P_1$ ans. YES, independent, 2

21. $S = \{(1,0,0,1),(0,1,1,0),(2,3,3,2)\}$, $W = \Re^4$ ans. NO, dependent (3rd vector is a linear combination of the first 2), 2

22. $S = \{(1,0,0,1),(0,1,1,0),(0,1,1,1),(1,1,1,1)\}$, $W = \Re^4$ ans. NO, dependent (4th vector is a linear combination of the first 3), 3

23. $S = \{(1,0,0,1),0,1,0,-2),(0,0,1,3)\}$, W is the hyperplane $x - 2y + 3z - w = 0$ ans. YES, independent, 3

24. $S = \{(1,0,1),(0,1,-2),(2,3,-4)\}$, W is the plane $x - 2y - z = 0$ ans. YES, dependent, (3rd vector is a linear combination of the first 2), 2

164

25. $S = \left\{ \begin{bmatrix} 1 & 0 \\ 0 & -1 \end{bmatrix}, \begin{bmatrix} 0 & 1 \\ 0 & 0 \end{bmatrix}, \begin{bmatrix} 0 & 0 \\ 1 & 0 \end{bmatrix} \right\}$, $W = \left\{ \begin{bmatrix} a & b \\ c & -a \end{bmatrix} : a, b, c \in \Re \right\}$ ans. YES, independent, 3

Part D: For each set S, describe W = span(S) (ie. give the form of the vectors in W) and give dim(W). Also determine if S is independent or not.

26. $S = \{(1,0,1),(0,1,2)\}$ ans. $W = \{(x,y,z): x + 2y - z = 0\}$, 2, independent

27. $S = \{ 1 + x, 1 - x \}$ ans. $W = P_1$, 2, independent

28. $S = \{(1,0,1,0),(0,1,0,1),(1,1,1,2)\}$ ans. $W = \{(x,y,z,w): x - z = 0\}$, 3, independent

29. $S = \{(1,1,2,0),(0,1,0,3),(1,2,2,3)\}$ ans. $W = \{(x,y,z,w): 2x - z = 0, 3x - 3y + w = 0\}$, 2, dependent

30. $S = \{ 1 - x, 1 + x, 3 x^2 \}$ ans. $W = P_2$ 134, 3, independent

31. $S = \left\{ \begin{bmatrix} 1 & 0 \\ 0 & 1 \end{bmatrix}, \begin{bmatrix} 0 & 1 \\ 1 & 0 \end{bmatrix}, \begin{bmatrix} 1 & 1 \\ 0 & 1 \end{bmatrix} \right\}$

ans. $W = \left\{ \begin{bmatrix} a & b \\ c & a \end{bmatrix} : a, b, c \in \Re \right\}$, 3, independent

32. $S = \left\{ \begin{bmatrix} 1 & 0 \\ 0 & -1 \end{bmatrix}, \begin{bmatrix} 0 & 1 \\ 0 & 0 \end{bmatrix}, \begin{bmatrix} 0 & 0 \\ 1 & 0 \end{bmatrix}, \begin{bmatrix} 1 & 1 \\ 1 & -1 \end{bmatrix} \right\}$

ans. $W = \left\{ \begin{bmatrix} a & b \\ c & -a \end{bmatrix} : a, b, c \in \Re \right\}$, 3, dependent

Part E: Given W is a subspace of n-space; find a basis and thus the dimension of each subspace.
33. $W = \{ (x, y, z, w): x - 2 y + 3 z - w = 0 \} \subseteq \Re^4$ Hint: write w in terms of x,y and z and just separate. ans. $\{(2,1,0,0),(-3,0,1,0),(1,0,0,1)\}$, dim(W) = 3
34. $W = \{ (x, y, z, v, w): x - 2 y + 3 z - 4 v - w = 0 \} \subseteq \Re^5$ ans. B = $\{(1,0,0,0,1), (0,1,0,0,-2),$ $(0,0,1,0,3),((0,0,0,1,-4)\}$, dim(W) = 4

Part F: Given the matrix $A = \begin{bmatrix} 1 & 0 & 1 & 0 & 3 \\ 0 & 1 & 2 & -2 & 2 \\ -1 & -1 & 0 & 0 & 1 \\ 0 & 0 & 3 & -2 & 7 \end{bmatrix}$;

35. Find a basis for row(A). ans. $\{(1,0,1,0,3),(0,1,2,-2,2),(0,0,1,-2/3,2),(0,0,0,0,1)\}$
36. Find a basis for col(A). What is col(A) here?
ans. $\{(1,0,-1,0),(0,1,-1,0),(0,0,1,1),(0,0,1,7/6)\}$, $col(A)= \Re^4$
37. What is rank(A)? ans. 4
38. Find null(A) and give its dimension. null(A) = span($\{(-2,2,2,3,0)\}$), ans. 1

165

39. Verify the "rank + nullity" theorem here. $4 + 1 = 5$

40. For what B does AX = B have a solution? ans. for all B

41. Describe row(A) & null(A) geometrically. ans. row(A) = hyperplane, col(A) = 4-space

42. Write the dependent column(s) of A as combinations of the independent columns in A.

ans. $C_4 = \dfrac{2}{3} C_1 - \dfrac{2}{3} C_2 - \dfrac{2}{3} C_3$

43. Can AX = B have a unique solution? Why or why not? ans. no; null(A) not trivial

Part G: Let $C = A^t$; repeat #35 - #43 using C.

44. ans. {(1,0,-1,0),(0,1,-1,0),(0,0,1,1),(0,0,1,7/6)}

45. ans. {(1,0,1,0,3),(0,1,2,-2,2),(0,0,1,-2/3,2),(0,0,0,0,1)}

46. ans. 4

47. ans. null(C) = {(0,0,0,0,0)}, 0

48. ans. $4 + 0 = 4$

49. ans. B = (a,b,c,d,e) implies CX = B has a solution if 2a-2b-2c-3d = 0

50. ans. row(C) = $\mathfrak{R}^4$, col(C) = hyperplane through origin

51. ans. columns of C are independent

52. ans. YES; null(C) is trivial

Part H: Given $A = \begin{bmatrix} 1 & 0 & 1 & 0 \\ 0 & 1 & 2 & -2 \\ -1 & -1 & 0 & 0 \\ 0 & 0 & 3 & 4 \end{bmatrix}$;

53. Find a basis for row(A) and col(A). ans. any set of original rows/cols

54. What is rank(A) and nullity(A)? ans. 4,0

55. Are the rows/columns of A independent? Why/why not? ans. YES, rank(A) = 4

56. Using the fundamental theorem, what do you know about det(A), A^{-1} 135, $AX = \theta$, and

AX = B for any $B \in \mathfrak{R}^4$? ans. det(A) is not zero, A is invertible, only solution is $X = \theta$, a unique solution exists for all B

Part I: Given $A = \begin{bmatrix} 1 & 0 & 2 & 0 & 1 \\ 0 & 1 & 0 & 1 & -1 \\ 0 & 0 & 1 & 1 & 1 \end{bmatrix}$;

57. Let X = (x,y,z,v,w) be in row(A); this means there exist 3 scalars k_1, k_2, & k_3 such that k_1 row1 + k_2 row2 + k_3 row3 = X. Set up the corresponding augmented matrix and use Gaussian elimination; show that a homogeneous system of TWO equations in x,y,z,v & w results.
ans. x+y-z+w=0,2x-y-z-v=0

58. Find a basis for null(A). ans. {(2,-1,-1,1,0),(1,1,-1,0,1)}

59. Let X = (x,y,z,v,w) be in null(A); mimic what you did for #57 and show you get a homogeneous system of THREE equations in x,y,z,v & w.
ans. x-2v-w=0,y+v-w=0,z+v+w=0

60. Put the two systems (from #57 & #59) together to get a 5 by 5 linear system. Solve the system and show that the only solution is the trivial solution. You just proved that *the only vector in row(A) and null(A) is the zero vector.*

Part J: Find the coordinates of the given vector with respect to the given basis vectors.

61. $p = 4 - x^2$, $B = \{ 1 - x, 1 + x, - x^2 \}$ ans. (2,2,1)

62. $p = 7 + 10 x + x^2$, $B = \{ 1 - x^2, 1 + x^2, 2 x \}$ ans. (3,4,5)

63. $f = 2 + \cos(x) + 3 \sin(x)$, $B = \{ 1 - \cos(x), 1 + 2 \cos(x), \sin(x) \}$ ans. (1,1,3)

64. $X = \begin{bmatrix} 4 & 5 \\ 2 & 4 \end{bmatrix}$ $S = \{ \begin{bmatrix} 1 & 0 \\ 0 & 1 \end{bmatrix}, \begin{bmatrix} 0 & 1 \\ 1 & 0 \end{bmatrix}, \begin{bmatrix} 1 & 1 \\ 0 & 1 \end{bmatrix} \}$ ans.(1,2,3)

65. X = (3,4,3), B = {1,2,0),(1,1,0),(0,0,1)}
ans. $(X)_B = (1, 2, 3)$

Part K: Given an m x n matrix A with rank(A) = r and nullity(A) = k;
66. What is k in terms of n and r? ans. k = n - r
67. Let nullity(A') = k'. Show k' = m - n + k.
68. Show that if A is square, then k' = k; otherwise $k' \neq k$.

69. Verify these results using $A = \begin{bmatrix} 1 & 0 & 2 & 0 \\ 0 & 0 & 2 & 4 \\ 3 & 0 & 8 & 6 \end{bmatrix}$; ie. find r, k and k' (you know m and n by

observation). ans. r = 3, k = 1, k' = 0

Part L: An *algebra* is a set V which is both a *ring* and a *vector space*. So an algebra V satisfies the vector space axioms and
1) V is closed under *vector multiplication,*
2) X(Y + Z) = XY + XZ and (X + Y)Z = XZ + YZ and
3) for any scalar k, it must be that k(XY) = (kX)Y = X(kY) for any vectors X and Y in V.
70. Show the for any n, the set M_n of square n by n matrices forms an algebra.
71. Show that P_2 clearly cannot form an algebra. ans. (deg2)times(deg2) = deg4
72. Show that C[a,b] (vector space of functions continuous on [a,b]) forms an algebra using (fg)(x) = f(x)g(x) and (kf)(x) = kf(x). Hint: what do you know from calculus about the product of two continuous functions?
73. Consider the set of complex numbers $\mathcal{C} = span(\{ 1, i \}) = \{ a + bi : a, b \in \mathfrak{R} \}$; show that this set is an algebra since it is a vector space over $\mathfrak{R}$ (ie. the scalars are real numbers) but vectors can also be *multiplied*, since the product of two complex numbers is a linear combination of 1 and i. NB. $i^2 \equiv -1$

CHAPTER 4

TRANSFORMATIONS

CHAPTER OVERVIEW

A **transformation** is just another name for a function T. In this chapter we are interested in transformations from one vector space into another vector space. The only transformations that we are interested in are the **linear** transformations; for a linear transformation, the image of a sum is the sum of the images and the image of a scalar multiple is the scalar multiple of the image. In section two we show that any matrix represents a linear transformation and conversely any linear transformation can be represented by a matrix. Two subspaces associated with any linear transformation are its **kernel** and its **range**; these turn out to be old friends (the *null space* and the *column space*). Then we are interested in determining when a transformation is **one-to-one** or **onto**; it turns out that the matrix which represents the transformation (its standard matrix) contains the answer. The rank of this matrix determines the size of the range of T and also the size of the kernel. A maximal rank implies ontoness; a minimal kernel (ie. the trivial space) implies a one-to-one transformation.

If a transformation deals only with one vector space, then this transformation is called an **operator**; these are studied in Section 4. Finally we will learn that choosing a *different basis* for a vector space V affects the matrix which represents it; we will figure out how to get the "new" matrix for a transformation by using the standard matrix. Another important idea from the last section is that two matrices which represent the same linear operator are **similar** matrices ie.

$$A' = P^{-1} A P$$

4.1 Linear Transformations

Transformations

Consider the function T from the vector space V into the vector space W:

$$T \mid V \rightarrow W$$

such that $T(X) = Y$ where the domain of T is V and the range of T is contained in W^{1}. We say that Y is the *image* of X under the transformation T and X is the pre-image of Y (images in the range of T and pre-images in the domain of T).

Here T is a **transformation** (or function) of V into W. It is important that V and W are vector spaces.

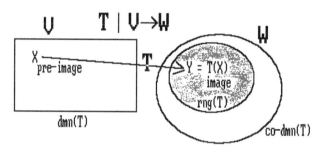

EXAMPLE 1 consider the transformation T(p) = xp where $T \mid P_{1} \rightarrow P_{2}$. Clearly T is a function (transformation or map) of P_{1} into P_{2}.

For example, $T(2 + 3x) = x(2 + 3x) = 2x + 3x^{2}$

so that with our new terminology, the image of p = 2 + 3x is $q = 2x + 3x^{2}$.

The vector p = 2 + 3x is in the dmn(T) and the vector $q = 2x + 3x^{2}$ is in the rng(T).

Also we can say that the pre-image of $q = 2x + 3x^{2}$ is p = 2 + 3x.

This terminology boils down to the simple statement that T(p) = q.

EXAMPLE 2 Suppose we consider the transformation T(x,y) = (x + y, xy); here $T \mid \Re^{2} \rightarrow \Re^{2}$. If we let X = (2,3) then T(2,3) = (5,6) = Y and T(X) = Y. So Y = (5,6) is the **image** of X = (2,3) or we can say that X = (2,3) is the *pre-image* of Y = (5,6). So (2,3) is in dmn(T) and (5,6) is in rng(T).

---

[1] W is sometimes called the *co-domain* of T

In fact we know that

$$dmn(T) = \Re^2 \quad rng(T) = \{ (u,v) : u \in \Re, v \geq 0 \}$$

EXAMPLE 3 If you like calculus, consider the derivative transformation $T | P_2 \rightarrow P_1$ where $T(p) = \dfrac{dp}{dx}$. Clearly the derivative of a degree two polynomial is a degree one polynomial; for example

$$p = x^2 + 4x + 6 = \Rightarrow T(p) = T(x^2 + 4x + 6) = \frac{dp}{dx} = \frac{d}{dx}(x^2 + 4x + 6) = 2x + 4$$

Thus the image of $p = x^2 + 4x + 6$ is q = 2x + 4 or the pre-image of 2x + 4 under this transformation is $x^2 + 4x + 6$. In any case T(p) = q.

EXAMPLE 4 Consider the transformation T(A) = det (A) where $T | M_{2,2} \rightarrow \Re$. In other words, the determinant function takes a 2 by 2 matrix and sends it to a unique real number (its determinant). Here $T | M_{2,2} \rightarrow \Re$.

Linear transformations

DEFINITION	A transformation $T | V \rightarrow W$ is **linear** iff

$$T(X_1 + X_2) = T(X_1) + T(X_2)$$

$$T(kX) = kT(X)$$

for all X in V and $k \in \Re$. If a transformation T is linear, the **image of a sum** is the *sum of the images* and the **image of a scalar multiple** is the *scalar multiple of the image*[2]. This is the most important type of transformation from our point of view!

EXAMPLE 5 Consider T(x,y,z) = (x + y, 3z). Show that this transformation is linear.
We need to pick two arbitrary vectors in $\Re^3$; let's let $X_1 = (x_1, y_1, z_1)$ & $X_2 = (x_2, y_2, z_2)$. If we add first, we get:

$$T(X_1 + X_2) = T(x_1 + x_2, y_1 + y_2, z_1 + z_2) = (x_1 + x_2 + y_1 + y_2, 3(z_1 + z_2)) =$$

$$T(X_1 + X_2) = ((x_1 + y_1) + (x_2 + y_2), 3z_1 + 3z_2) =$$

---

[2] These properties are called additivity and homogeneity, respectively.

171

$$(\, x_1 + y_1, 3 \, z_1 \,) + (\, x_2 + y_2, 3 \, z_2 \,) = T(\, X_1 \,) + T(\, X_2 \,)$$

so additivity holds. Now we check homogeneity:

$$T(\, k \, X_1 \,) = T(\, k(\, x_1, y_1, z_1 \,) \,) = T(\, k \, x_1, k \, y_1, k \, z_1 \,) = (\, k \, x_1 + k \, y_1, 3 \, k \, z_1 \,) =$$
$$T(\, k \, X_1 \,) = k(\, x_1 + y_1, 3 \, z_1 \,) = k \, T(\, x_1, y_1, z_1 \,)$$

This holds also so this transformation is linear. Just by looking at the given transformation, we see it is a transformation from 3-space into 2-space and we write $T \mid \mathfrak{R}^3 \to \mathfrak{R}^2$.

EXAMPLE 6 Consider the transformation

$$T \mid P_1 \to P_2 \; where \, T(\, p(\, x \,) \,) = \int_0^x p(\, t \,) dt$$

Two basic properties of integration[3] immediately give away the fact that this transformation is linear- namely that $\int (\, f + g \,) d \, x = \int f \, d \, x + \int g \, d \, x$ and $\int (\, k \, f \,) d \, x = k \int f \, d \, x$. And it makes sense that the image of any degree one polynomial (in P_1) must be in P_2 since integrating a polynomial *increases* its degree by one.

EXAMPLE 7 Using the transformation in EX5, find all pre-images of Y = (7,15).
We want the image to be Y = (7,15) so set T(x,y,z) = (x + y,3z) = (7,15). This means that

$$x + y \quad = 7$$

$$3 \, z = 15$$

Right away, we write down the augmented matrix; we then put it in echelon form:

$$\begin{bmatrix} 1 & 1 & 0 & 7 \\ 0 & 0 & 3 & 15 \end{bmatrix} \sim \begin{bmatrix} 1 & 1 & 0 & 7 \\ 0 & 0 & 1 & 5 \end{bmatrix}$$

Clearly this system is consistent and dependent and we have z = 5 and x + y = 7 so x = 7 - y = 7 - t. Thus the set of all pre-images is of the form X = (7 - t, t, 5). This could be checked easily by just finding the image of X; it must be Y = (7,15)!

EXAMPLE 8: Suppose $T \mid \mathfrak{R}^3 \to \mathfrak{R}^2$ where T(1,0,0) = (1,0), T(0,1,0) = (2,0), and T(0,0,1) = (0,3). Find T(3,4,5) if T is linear.
Let's write X = (3,4,5) as a linear combination of the basis vectors:

$$X = (\, 3,4,5 \,) = 3(\, 1,0,0 \,) + 4(\, 0,1,0 \,) + 5(\, 0,0,1 \,)$$

---

[3] Thus both differentiation and integration are really linear transformations!

Using the linearity of T we have:

$$T(3,4,5)=T(3(1,0,0)+4(0,1,0)+5(0,0,1))=T(3((1,0,0))+T(4(0,1,0))+T(5(0,0,1))=$$

$$3T((1,0,0))+4T((0,1,0))+5T((0,0,1))=3(1,0)+4(2,0)+5(0,3)=(8,15)$$

This example brings out the important fact that if we know the effect of a linear transformation on a *basis* for V, then we can map *any vector* in V under T since we can write every vector X in V as a linear combination of the basis vectors.

Properties of Linear Transformations

Now suppose we are given $T \mid V \rightarrow W$ such that T is linear. If T is linear, then we can prove that:

THEOREM: 1) $T(\theta_v)=\theta_w$ and 2) $T(X_1-X_2)=T(X_1)-T(X_2)$

These two results say that 1) the image of the zero vector is the zero vector and
2) the image of a difference of two vectors is the difference of the images.
The proof of 2) is left as an exercise.
Proof(1): Suppose T is linear but $T(\theta_v)\neq\theta_w$. Let $T(\theta_v)=Y$ where $Y\neq\theta_w$.
Then

$$T(\theta_v-\theta_v)=T(\theta_v)-T(\theta_v)=Y-Y=\theta_w$$

But $T(\theta_v-\theta_v)=T(\theta_v)$ so therefore $T(\theta_v)=\theta_w$ (which is a contradiction).
Therefore $T(\theta_v)=\theta_w$.

EXAMPLE 9: Consider the transformation T(x) = 3x + 4; show that this transformation is not linear. Here $T \mid \Re \rightarrow \Re$; using the above theorem, if T(0) is NOT 0, then T cannot be linear. But T(0) = 0 + 4 = 4. Hence this transformation is not linear.

Transformations involving matrices

Since a matrix is the "star" of this book, we might hope that somehow matrices would give rise to linear transformations. Think about all the possible operations that you know of concerning matrices: a) row operations b) multiplication c) addition d) scalar multiplication e) determinants f) transposition g) inversion ... etc. Let's check out some of these operations. Suppose we use the vector space of all 2 x 2 matrices for simplicity.

Consider the transformation T where

$$T(M)=M+\begin{bmatrix} 1 & 2 \\ 3 & 4 \end{bmatrix}$$

for any 2 x 2 matrix M. Check out whether T(M + N) = T(M) + T(N):

$$T(M+N)=M+N+\begin{bmatrix} 1 & 2 \\ 3 & 4 \end{bmatrix}$$

$$T(M)+T(N)=M+\begin{bmatrix} 1 & 2 \\ 3 & 4 \end{bmatrix}+N+\begin{bmatrix} 1 & 2 \\ 3 & 4 \end{bmatrix}=M+N+\begin{bmatrix} 2 & 4 \\ 6 & 8 \end{bmatrix}$$

Thus this rather simplistic transformation is NOT linear. What about a matrix multiplication transformation? Consider T(M) = AM where A is a particular 2 x 2 matrix.

T(M + N) = A(M + N) = AM + AN = T(M) + T(N)

T(kM) = A(kM) = k(AM) = kT(M)

So *multiplication by a constant matrix is a linear transformation.* Now suppose we let $T(M)=M^2$; is this transformation linear? We expect that it is not.

$$T(M+N)=(M+N)^2=(M+N)(M+N)=M^2+MN+NM+N^2$$

But

$$T(M)+T(N)=M^2+N^2$$

which is clearly not equal to the above unless MN = -NM. In general this is NOT true so squaring a matrix is NOT a linear transformation.

How about the determinant function? This would be a transformation from the vector space of square matrices into the vector space of real numbers. But unfortunately

$$\det(M+N)\neq\det(M)+\det(N)$$

in general so the determinant map is NOT a linear map. How about finding the inverse of a matrix? Now the inverse of a 2 x 2 is a 2 x 2 so here matrix inversion would be a transformation from $M_{2,2}\rightarrow M_{2,2}$; however,

$$T(M+N)=(M+N)^{-1}\neq M^{-1}+N^{-1}$$

so this is NOT a linear transformation. In general, determinants and matrix inversion deal well with products of two matrices but not with sums of two matrices. Last one- what about scalar multiplication? Let T(M) = k M where k is any scalar (ie. real no.). Then we have:

$$T(M+N)=k(M+N)=kM+kN=T(M)+T(N)$$

And we see also that

$$T(\,cM\,)=k(\,cM\,)=k\,c\,M=c\,k\,M=c(\,kM\,)=cT(\,M\,)$$

so multiplication by a scalar is actually a linear transformation!

Linear Operators

Pick any 2 by 2 matrix, call it A and define a transformation from $\Re^2$ into $\Re^2$

$$[\,T(\,x,y\,)\,]=\begin{bmatrix}1 & 2 \\ 3 & 4\end{bmatrix}\begin{bmatrix}x \\ y\end{bmatrix}=\begin{bmatrix}x+2\,y \\ 3\,x+4\,y\end{bmatrix}=[Y] \quad or\,[\,T(\,X\,)\,]=A\,[\,X\,]=[Y]$$

This is a transformation from 2-space into 2-space; if a transformation "stays inside" one vector space, we call it an **operator** on that vector space. So here T is an operator on $\Re^2$. Is this operator linear?

$$[\,T(\,(\,x_1,y_1\,)+(\,x_2,y_2\,)\,)\,]=\begin{bmatrix}1 & 2 \\ 3 & 4\end{bmatrix}(\begin{bmatrix}x_1 \\ y_1\end{bmatrix}+\begin{bmatrix}x_2 \\ y_2\end{bmatrix})=$$

$$\begin{bmatrix}1 & 2 \\ 3 & 4\end{bmatrix}\begin{bmatrix}x_1 \\ y_1\end{bmatrix}+\begin{bmatrix}1 & 2 \\ 3 & 4\end{bmatrix}\begin{bmatrix}x_2 \\ y_2\end{bmatrix}=[\,T(\,x_1,y_1\,)\,]+[\,T(\,x_2,y_2\,)\,]$$

For scalar multiplication:

$$[\,T(\,kX\,)\,]=A[\,kX\,]=k(\,A[\,X\,]\,)=k\,[\,T(\,X\,)\,]$$

Both linearity properties hold so T is a linear operator on 2-space.
This type of transformation from 2-space into 2-space represented simply by
[T(X)] = A [X] is *always* a linear transformation.
Thus **every square matrix represents a *linear operator* on n-space.**

EXERCISES 4.1

Part A. For each transformation T, find the image of the given vector. State whether the transformation is linear or not.
1. T(x,y) = x + y   X = (1,3)
ans. 4, linear
2. T(x,y) = (x + y, y)   X = (2,5)

3. T(x,y) = (x,0)   X = (4,7)
ans. (4,0), linear
4. $T(\,x,y\,)=(\,x\,y\,,x^2+y^2\,)$ X = (1,2)

5. T(f) = df/dx, f = sin(x)
ans. cos(x), linear

6. T(M) = det(M), $M = \begin{bmatrix} 1 & 2 \\ 2 & 5 \end{bmatrix}$

7. $T(x, y) = \sqrt{x^2 + y^2}$, X = (3,4)

ans. 5, non-linear

8. T(x,y) = sin(x + y), $X = (\frac{\pi}{4}, \frac{\pi}{2})$

9. $T(f) = \int_0^1 f \, dx$, $f = e^x$

ans. e - 1, linear

Part B. For each transformation T, find the pre-image(s) of Y.

10. T(x,y) = x + y, Y = (4,4)

11. T(x,y,z) = (x + y, y - 2z), Y = (3,6)
ans. X = (-3-2t,6+2t,t)

12. T(f) = df/dx, $Y = x^2 + 4x$

13. T(x,y,z) = (x - 2y, y + 3z), Y = (-3,11)
ans. X = (19-6t,11-3t,t)

14. T(x,y) = (x + y, 2y, x - y), Y = (3,4,-1)

Part C. For each linear transformation T, find the image of X with the given information:

15. X = (-2,3), T(1,0) = (2,3,4), T(0,1) = (-4,-2,10) ans. (-16,-12,22)
16. X = (-6,18), T(1,2) = (2,4), T(3,6) = (-5,5)
17. X = (1,2,3), T(1,0,0) = (2,0,0), T(0,1,0) = (0,2,0), T(0,0,1) = (0,0,0) ans. (2,4,0)
18. X = (2,2,2,3), T(1,0,1,0) = (2,4,2), T(0,1,0,1) = (5,4,3), T(1,1,1,2) = (-3,-2,-1)
19. X = (4,6,-2,-2), T(1,0,1,0) = (1,2,3), T(0,1,0,1) = (2,2,2), T(1,0,-1,0) = (-3,6,-3),
T(0,1,0,-1) = (10,5,5) ans. (36,44,18)

Part D. Given the transformation $T(x, y, z) = (\frac{x + z}{2}, y, \frac{x + z}{2})$;

20. Show that T is linear transformation from 3-space into 3-space.
21. Find T(X) if X = (t,0,t). ans. (t,0,t)
22. Find T(X) = X = (0,s,0).
23. Using the fact that T is linear, show that if X = (t,s,t), then T(X) = X. Here X is called a *fixed point* of the transformation T (ie. image = pre-image).
24. T projects any vector in 3-space onto the plane W = span{(1,0,1),(0,1,0)}. Pick any vector in 3-space not in W (for eg. take X = (4,6,10)). Show that the image of X lies in W.

Part E. Given the transformation $T(X) = AX$ where $A = \begin{bmatrix} 1 & 2 \\ 0 & 0 \end{bmatrix}$ and $X = \begin{bmatrix} a & b \\ c & d \end{bmatrix}$;

25. Show that T is linear and $T \mid M_{2,2} \rightarrow M_{2,2}$.

26. Find the image of $X = \begin{bmatrix} 1 & 2 \\ 3 & 4 \end{bmatrix}$.

27. Find the pre-image(s) of $Y = \begin{bmatrix} 2 & 4 \\ 0 & 0 \end{bmatrix}$.

ans. $X = \begin{bmatrix} 2-2c & 4-2d \\ c & d \end{bmatrix}$

28. Find two matrices A and B such that $T(A) = \begin{bmatrix} 3 & 6 \\ 0 & 0 \end{bmatrix} = T(B)$.

29. Can you find C such that $T(C) = \begin{bmatrix} 3 & 6 \\ 1 & 1 \end{bmatrix}$? ans. NO

Part F. Use the integral transformation in **EX 6**;
30. Show in detail that T is linear. Let $p_1(x) = a_1 + b_1 x$, $p_2(x) = a_1 + b_2 x$; show $T(p_1 + p_2) = T(p_1) + T(p_2)$ and then show $T(k p_1) = k T(p_1)$.

31. Find T(1) and T(x).ans. $x, \dfrac{x^2}{2}$

32. Since T is linear, using the answers to #31, find T(a + bx).
33. If $q(x) = b x + c x^2$, find its pre-image(s) for any b & c. ans.p(x) = b + 2cx
34. Can two different pre-images have the same image?
35. Does every vector in P_2 have a pre-image? ans. NO; any polynomial with a constant term has no pre-image under T

Part G. Given the transformation $T(\begin{bmatrix} a & b \\ c & d \end{bmatrix}) = \begin{bmatrix} a & c \\ b & d \end{bmatrix}$ where $T \mid M_{2,2} \rightarrow M_{2,2}$;

36. Show that T is linear.
37. Find the image of $X = \begin{bmatrix} 1 & 2 \\ 3 & 4 \end{bmatrix}$. ans. $\begin{bmatrix} 1 & 3 \\ 2 & 4 \end{bmatrix}$

38. Find the pre-image(s) of $Y = \begin{bmatrix} 2 & 4 \\ -6 & -8 \end{bmatrix}$.

39. Determine if there exist two *different* matrices A and B such that T(A) = T(B). ans. NO
40. Show that for any matrix $Y = \begin{bmatrix} x & y \\ z & w \end{bmatrix}$ there exists a matrix X such that T(X) = Y.

41. Show the there exists a matrix A such that T(A) = A. In this case A is called a **fixed point** of the transformation T. What kind of matrix does A have to be??? ans. symmetric
NB. The transformation T here is just $T(A) = A^t$.

Part H. Given T is linear such that $T(1) = 2 + x$ and $T(x) = 2x + x^2$;

42. Find $T(a + bx)$.

43. Find the pre-image(s) of $6 - 7x - 5x^2$. ans. $p(x) = 3 - 5x$

44. Find all $p(x) = a + bx$ such that $T(p(x)) = 0$.

45. If $p(x) = a + bx$ and $q(x) = c + dx$ and $T(p(x)) = T(q(x))$ then what must be true about a,b,c and d? ans. a = c, b = d

46. Can you find the pre-image of $r(x) = d_0 + d_1 x + d_2 x^2$ for any such r(x)? If not, determine a constraint on the coefficients.

Part I. Population models; suppose
$A = \begin{bmatrix} 0.6 & 0.5 \\ -0.1 & 1.2 \end{bmatrix}$ $X_0 = \begin{bmatrix} 100 \\ 20 \end{bmatrix}$ where x = number of rabbits and y = number of foxes.

47. Suppose $T(X) = AX$ where $X = \begin{bmatrix} x \\ y \end{bmatrix}$; let $X_1 = T(X_0)$, $X_2 = T(X_1)$ etc. This means that the

output(image) in step one becomes the pre-image(input) in step 2 and so forth. Find X_1.
ans. x = 70, y = 14

48. Apply A three more times; what is happening to the number of rabbits and foxes??

Part J. Prove: If $T | V \to W$, then

49. $T(X_1 - X_2) = T(X_1) - T(X_2)$

50. $T(k_1 X_1 + k_2 X_2 + k_3 X_3) = k_1 T(X_1) + k_2 T(X_2) + k_3 T(X_3)$

Part K. Some books put the two linearity properties together and say that T is linear iff $T(k X_1 + j X_2) = k T(X_1) + j T(X_2)$. Use this "two in one" definition to prove that the following transformations are linear; if you like this definition, use it!

51. $T | P_2 \to P_1$ $T(p) = \dfrac{dp}{dx}$

52. $T | M_{2,2} \to M_{2,2}$ $T(M) = M'$

53. $T | \Re^2 \to \Re^3$ $T(x, y) = (2x, x + y, 3y)$

Part L. Given $T(f) = \int_0^\infty f e^{-st} dt = F(s)$, where $T | E \to L$

(E - set of bounded, piecewise continuous functions defined on $[0, \infty)$), show:

54. $T(f + g) = T(f) + T(g)$

55. $T(kf) = kT(f)$. Thus T is a linear transformation.

56. If $T(1) = 1/s$ and $T(t) = \dfrac{1}{s^2}$ find $T(3 + 4t)$.

4.2 Matrices and Linear Transformations

Introduction: Matrices as Linear Transformations

Consider the 2 x 3 matrix $A = \begin{bmatrix} 1 & 2 & 3 \\ -2 & 4 & 0 \end{bmatrix}$; let X = (x,y,z) and [X] be the corresponding column matrix and define [T(X)] to be A[X]. Then the product is a 2 x 1 column matrix Y (representing a 2-vector):

$$[T(X)] = A[X] = \begin{bmatrix} 1 & 2 & 3 \\ -2 & 4 & 0 \end{bmatrix} \begin{bmatrix} x \\ y \\ z \end{bmatrix} = \begin{bmatrix} x+2y+3z \\ -2x+4y \end{bmatrix} = [Y]$$

$$(2 \times 3) \text{ times } (3 \times 1) = (2 \times 1)$$

Here X is the pre-image (in dmn(T)) and Y = T(X) is the image (in rng(T)).
Let's prove that this transformation is linear:
1)

$$[T(X_1 + X_2)] = A([X_1] + [X_2]) = A[X_1] + A[X_2] = [T(X_1)] + [T(X_2)]$$

(since matrix multiplication is distributive)
2)

$$[T(kX)] = A[kX] = kA[X] = k[T(X)]$$

(by a property of scalar multiplication of matrices)
This example can be generalized to any size matrix; suppose A is an *m* x **n** matrix. Then if we let [T(X)] = A[X] then [X] must be an **n** x 1 column matrix and [T(X)] = [Y] is an *m* x 1 column matrix. The fact that this transformation is linear is a consequence of the properties of matrix multiplication. We therefore have the following theorem:

THEOREM Every m x n matrix A represents a *linear transformation* from $\Re^n$ into $\Re^m$.

EXAMPLE 1 Using the matrix A given above, find the corresponding linear transformation T(X).

$$[T(x,y,z)] = A[X] = \begin{bmatrix} 1 & 2 & 3 \\ -2 & 4 & 0 \end{bmatrix} \begin{bmatrix} x \\ y \\ z \end{bmatrix} = \begin{bmatrix} x+2y+3z \\ -2x+4y \end{bmatrix} = [Y]$$

This means that T(x,y,z) = (x + 2y + 3z, -2x + 4y).

You should be able to verify that T is indeed a *linear transformation* from $\Re^3$ into $\Re^2$ by using the definition from Section 4.1 in a horizontal fashion (ie. without matrices).
Now we wish to consider the converse of the above statement; can we represent a linear transformation by a matrix? We predict the answer must be YES!

The key to doing this is to see what happens to a basis for the vector spaces involved which make up the domain and co-domain of T. Going back to the previous example, let's map the basis vectors for the domain of the transformation B = {(1,0,0),(0,1,0),(0,0,1)}; the images of course are given in terms of the standard basis for the co-domain, $B' = \{ (1,0),(0,1) \}$.

We need to write the basis vectors in B as 3 by 1 *column* matrices if we are to multiply them by the 2 by 3 matrix A:

$$[T(1,0,0)] = \begin{bmatrix} 1 & 2 & 3 \\ -2 & 4 & 0 \end{bmatrix} \begin{bmatrix} 1 \\ 0 \\ 0 \end{bmatrix} = \begin{bmatrix} 1 \\ -2 \end{bmatrix}$$

$$[T(0,1,0)] = \begin{bmatrix} 1 & 2 & 3 \\ -2 & 4 & 0 \end{bmatrix} \begin{bmatrix} 0 \\ 1 \\ 0 \end{bmatrix} = \begin{bmatrix} 2 \\ 4 \end{bmatrix}$$

$$[T(0,0,1)] = \begin{bmatrix} 1 & 2 & 3 \\ -2 & 4 & 0 \end{bmatrix} \begin{bmatrix} 0 \\ 0 \\ 1 \end{bmatrix} = \begin{bmatrix} 3 \\ 0 \end{bmatrix}$$

We can see how the coordinates of the images (with respect to B') of the basis vectors in B form the **columns** of the matrix A which represents the transformation T. This is the key to matrix representation of a linear transformation:

the images of the basis vectors in V form the COLUMNS of the matrix A

$$A = \begin{bmatrix} 1 & 2 & 3 \\ -2 & 4 & 0 \end{bmatrix} = \begin{bmatrix} [T(1,0,0)] \vdots [T(0,1,0)] \vdots [T(0,0,1)] \end{bmatrix}$$

If the standard bases for V and W are used, then the matrix A is called the **standard matrix** of the transformation T. Alternatively one may write the image in a column matrix format and use properties of matrix multiplication:

$$[T(x, y, z)] = \begin{bmatrix} x + 2 y + 3 z \\ -2 x + 4 y \end{bmatrix} = \begin{bmatrix} x \\ -2 x \end{bmatrix} + \begin{bmatrix} 2 y \\ 4 y \end{bmatrix} + \begin{bmatrix} 3 z \\ 0 \end{bmatrix}$$

$$[T(x, y, z)] = x \begin{bmatrix} 1 \\ -2 \end{bmatrix} + y \begin{bmatrix} 2 \\ 4 \end{bmatrix} + z \begin{bmatrix} 3 \\ 0 \end{bmatrix} = \begin{bmatrix} 1 & 2 & 3 \\ -2 & 4 & 0 \end{bmatrix} \begin{bmatrix} x \\ y \\ z \end{bmatrix} = A[X]$$

Clearly here it is the *column* structure of A which is key to understanding how this linear transformation T (represented by the matrix A) works!

EXAMPLE 2 Find a matrix which represents the transformation T where $T \mid P_2 \rightarrow P_1$ given by T(p) = dp/dx (ie. the derivative transformation). First, we must choose a bases for P_2 and P_1; unless stated otherwise we would choose the standard basis for each vector space:

$$B = \{ 1, x, x^2 \} \subseteq P_2 \qquad B' = \{ 1, x \} \subseteq P_1$$

Transform each basis vector in the domain (ie. P_2) and find the *coordinates* of each image with respect to the basis for the range of T (ie. P_1):

$$T(1) = \frac{d}{dx}(1) = 0 \Rightarrow 0 = 0 \cdot 1 + 0 \cdot x$$

The coordinate matrix of T(1) wrt. B' is:

$$[T(1)]_{B'} = \begin{bmatrix} 0 \\ 0 \end{bmatrix}$$

Now find the image of x:

$$T(x) = \frac{d}{dx}(x) = 1 \Rightarrow 1 = 1 \cdot 1 + 0 \cdot x$$

The coordinate matrix of T(x) wrt. B' is:

$$[T(x)]_{B'} = \begin{bmatrix} 1 \\ 0 \end{bmatrix}$$

Finally we want the image of x^2:

$$T(x^2) = \frac{d}{dx}(x^2) = 2x \Rightarrow 2x = 0 \cdot 1 + 2 \cdot x$$

so the coordinate matrix of T(x^2) wrt. B' is:

$$[T(x^2)]_{B'} = \begin{bmatrix} 0 \\ 2 \end{bmatrix}$$

The matrix which represents the transformation T has its columns formed by the coordinate matrices which represent the image of each basis vector in P_2 :

$$A = \begin{bmatrix} 0 & 1 & 0 \\ 0 & 0 & 2 \end{bmatrix}$$

Now, how do we use this matrix? Suppose we wish to find the image of a particular vector in P_2, say p = $2 + 4 - 5\,x^2$. We must write p as a linear combination of the basis vectors in P_2 and then use the components to form the coordinate matrix of p wrt. the basis B:

$$p = 2 \cdot 1 + 4 \cdot x - 5 \cdot x^2 \Rightarrow [\ p\ ]_B = \begin{bmatrix} 2 \\ 4 \\ -5 \end{bmatrix}$$

Therefore, to find the coordinates of q = T(p) wrt. B', we multiply by A:

$$[\ T(\ p\ )\ ]_{B'} = A[\ p\ ]_B = \begin{bmatrix} 0 & 1 & 0 \\ 0 & 0 & 2 \end{bmatrix} \begin{bmatrix} 2 \\ 4 \\ -5 \end{bmatrix} = \begin{bmatrix} 4 \\ -10 \end{bmatrix} = [\ q\ ]_{B'}$$

This means that the actual image of p is found using these components with the basis vectors in $P_1: T(\ p\ ) = 4 \cdot 1 - 10 \cdot x = 4 - 10\,x = q$. As a check, simply differentiate p:

$$T(\ p\ ) = \frac{d}{dx}(\ 2 + 4 - 5\,x^2\ ) = 4 - 10\,x = q$$

To summarize:

1) Any m x n matrix A represents a linear transformation T from $\Re^n$ into $\Re^m$; we write $T \mid \Re^n \rightarrow \Re^m$. This transformation is accomplished by a simple matrix multiplication:

[T(X)] = A [X] and the fact that T is linear follows from the properties of matrix multiplication.

2) Conversely, any m x n matrix A represents a linear transformation from n-space into m-space. The **columns** of A are the images (in order) of the n basis vectors of $\Re^n$.

Thus there is a one-to-one correspondence between the set of **m by n matrices** and the *set of linear transformations* from n-space into m-space.

EXAMPLE 3 Create a matrix A representing a linear transformation from 2-space into 3-space such that T(1,0) = (2,3,4) and T(0,1) = (-5,5,-5). This is easy to do- the images of each basis vector must be the *columns* of A: $A = \begin{bmatrix} 2 & -5 \\ 3 & 5 \\ 4 & -5 \end{bmatrix}$. If one wants to know what T(X) is, a simple matrix

multiplication accomplishes the task:

$$[T(X)] = A[X] = \begin{bmatrix} 2 & -5 \\ 3 & 5 \\ 4 & -5 \end{bmatrix} \begin{bmatrix} x \\ y \end{bmatrix} = \begin{bmatrix} 2x - 5y \\ 3x + 5y \\ 4x - 5y \end{bmatrix}$$

so T(x,y) = (2x - 5y, 3x + 5y, 4x - 5y). Clearly the transformation goes from 2-space into 3-space. Note that the dimensions of the matrix give the dimensions of the vector spaces involved in the *reverse order* (ie. a *3* x **2** matrix represents a linear transformation from a **2**-dimensional space into a *3*-dimensional space).

SUMMARY

linear transformation T $(T \mid V \rightarrow W)$	standard matrix A (m x n matrix)
T(X) = Y	[T(X)] = A[X] = [Y]
X in V, Y in W	[X] in n-space, [Y] in m-space
Y - actual image of X	[Y] - coordinates of Y
to find Y, use T	to find [Y], pre-multiply [X] by A

EXERCISES 4.2

Part A. Given the transformation T(x,y) = (2x, x + y, x - y)
1. T is a transformation from_____ into _____. ans. $\Re^2$, $\Re^3$
2. The domain of T is _____.
3. The range of T is contained in the vector space _____. ans. $\Re^3$
4. Prove that T is linear.
5. Find the 3 by 2 matrix A which represents the transformation T so that T(X) = A[X].

ans. $A = \begin{bmatrix} 2 & 0 \\ 1 & 1 \\ 1 & -1 \end{bmatrix}$

6. Find the pre-image(s) of Y = (8,9,-1).

Part B. Given the transformation T from 3-space into 2-space such that T(1,0,0) = (1,0), T(0,1,0) = (0,1) and T(0,0,1) = (0,0);

7. Find the standard matrix of T. ans. $A = \begin{bmatrix} 1 & 0 & 0 \\ 0 & 1 & 0 \end{bmatrix}$

8. Find T(1,2,3) using the standard matrix.
9. Can you find two different pre-images X_1 and X_2 such that $T(X_1) = T(X_2) = (3,4)$?
ans. yes; $X_1 = (3,4,5)$, $X_2 = (3,4,6)$
10. For any vector Y = (a,b) in 2-space, can you find a pre-image X such that T(X) = Y?

Part C. Given a linear transformation T such that $T(1,0,1) = (1,4)$ and $T(1,0,-1) = (-2,0)$ and $T(0,1,0) = (3,3)$;

11. Find the standard matrix of T. ans. $A = \begin{bmatrix} -\frac{1}{2} & 3 & \frac{3}{2} \\ 2 & 3 & 2 \end{bmatrix}$.

12. Use A to find the image of $X = (4,2,-2)$.

13. Check your answer by writing $X = (4,2,-2)$ as a linear combination of the vectors in $B = \{(1,0,1),(1,0,-1),(0,1,0)\}$ and using the linearity of T. ans. (1,10)

14. Find the pre-image(s) of $Y = (2,7)$

15. Find the pre-image of $Y = (a,b)$ using A. Hint: recall $[T(X)] = A[X] = [Y]$

ans. $X = (-2a/5+2b/5-45t, 4a/15+b/15-8t, t)$

Part D. Given $A = \begin{bmatrix} 1 & 2 & 2 & 0 \\ -4 & 4 & 0 & 8 \\ 0 & 1 & 0 & 1 \end{bmatrix}$;

16. A represents a transformation from _____ into _____.

17. $T(1,0,0,0) = $ _____, $T(0,1,0,0) = $ _____ ans. (1,-4,0), (2,4,1)

18. $T(2,3,0,0) = $

19. $T(0,0,1,0) = $ _____, $T(0,0,0,1) = $ _____ ans. (2,0,0),(0,8,1)

20. $T(1,2,3,4) = $

Part E. Given a linear transformation T such that $T(1,0,0,0) = (1/2,0,0,1/2)$, $T(0,1,0,0) = (0,1,0,0)$, $T(0,0,1,0) = (0,0,1,0)$, $T(0,0,0,1) = (1/2,0,0,1/2)$ find:

21. standard matrix A of T ans. $A = \begin{bmatrix} \frac{1}{2} & 0 & 0 & \frac{1}{2} \\ 0 & 1 & 0 & 0 \\ 0 & 0 & 1 & 0 \\ \frac{1}{2} & 0 & 0 & \frac{1}{2} \end{bmatrix}$

22. $T(x,y,z,w)$ using A

23. all pre-images of $Y = (5,4,6,5)$ ans. $X = (10-t,4,6,t)$

24. any pre-image of $Y = (2,3,4,5)$

25. if there exists $X = (x,y,z,w)$ such that $T(X) = (a,b,c,d)$ for any a,b,c & d.

ans. no; $a = d$

26. if T has a fixed point (ie. if there exists X in dmn(T) such that $T(X) = X$

ans. $X = (x,y,z,x)$

Part F. Given the linear transformation $T(x,y,z) = 1/3(x + y +z, x + y + z, x + y + z)$;

27. Find the standard matrix A of T. ans. $A = \dfrac{1}{3}\begin{bmatrix} 1 & 1 & 1 \\ 1 & 1 & 1 \\ 1 & 1 & 1 \end{bmatrix}$

28. Use the standard matrix to find $T(X)$ if $X = (4,6,10)$.
29. Use the standard matrix to find all pre-images of $Y = (2,2,2)$. ans. $X = (6\text{-s-t},s,t)$
30. If possible, find the pre-image of $Y = (2,3,4)$.
31. Using the fact that T is linear, show that if $X = t(1,1,1)$, then $T(X) = X$. Therefore X is a **fixed point** of the transformation T (ie. image = pre-image). NB. T represents a projection in 3-space onto the line $W = \{(x,y,z) = t(1,1,1)\}$ and A is idempotent (ie. $A^2 = A$)and symmetric.

Part G. Suppose T is a linear transformation from 4-space into 3-space such that
$T(1,0,1,0) = (0,1,0)$, $T(0,1,0,1) = (2,2,2)$, $T(1,0,-1,0) = (3,2,1)$ and $T(0,1,0,-1) = (0,4,0)$.
32. Find $T(1,2,3,4)$. Hint: Write $X = (1,2,3,4)$ as a combination of the pre-images given above and use the linearity of T.

33. Find the standard matrix A of T. ans.
$$A = \begin{bmatrix} \dfrac{3}{2} & 1 & -\dfrac{3}{2} & 1 \\ \dfrac{3}{2} & 3 & -\dfrac{1}{2} & -1 \\ \dfrac{1}{2} & 1 & -\dfrac{1}{2} & 1 \end{bmatrix}$$

34. Use the matrix A from # 33 to again find $T(1,2,3,4)$ this time by a simple matrix multiplication.
35. Find all pre-images of $Y = (5,9,3)$. ans. $X = (1+2t,1-t,2t,t)$
36. Find the pre-image(s) of $Y = (a,b,c)$.

Part H. Given the linear transformation $T(p) = x p$ where $T | P_1 \rightarrow P_2$;

37. Using the standard basis for P_1, find the image of each basis vector. ans. x, x^2
38. Find the standard matrix A for T.
39. Use the matrix A to find the image of $p(x) = 3 + 4x$ ans. $3x + 4x^2$
40. If $q(x) = a + b x + c x^2$, does there exist p(x) such that $T(p(x)) = q(x)$ for all q(x)?

Part I. Given the linear transformation $T(p(x)) = \int\limits_{0}^{x} p(t)dt \quad where\, T | P_1 \rightarrow P_2$;

41. Using the standard basis for P_1, find the image of each basis vector. ans. $x, \dfrac{x^2}{2}$

42. Find the standard matrix for T.
43. Use the matrix A to find the image of $p(x) = 3 + 4x$ ans. $3x + 2x^2$
44. If $q(x) = a + b x + c x^2$, does there exist p(x) such that $T(p(x)) = q(x)$ for all q(x)?

Part J. Given the linear transformation $T(p) = p - dp/dx$ where $T | P_2 \rightarrow P_2$;
45. Using the standard basis for P_2, find the image of each basis vector.

ans. $1, -1+x, -2x+x^2$

46. Find the standard matrix A for T.

47. Use the matrix A to find the image of $p(x) = 2 + 3x - 4x^2$. ans. $-1 + 11x - 4x^2$

48. If $q(x) = a + bx + cx^2$, does there exist p(x) such that T(p(x)) = q(x) for all q(x)?

Part K. Given the transformation $T \mid M_{2,2} \to M_{2,2}$ where $T(M) = M^t$;

49. Prove that T is indeed linear.

50. Find T(M) if $M = \begin{bmatrix} 1 & 2 \\ 2 & 4 \end{bmatrix}$.

51. Using the standard basis for $M_{2,2}$, find the standard matrix for T.

ans. $A = \begin{bmatrix} 1 & 0 & 0 & 0 \\ 0 & 0 & 1 & 0 \\ 0 & 1 & 0 & 0 \\ 0 & 0 & 0 & 1 \end{bmatrix}$

52. Find T(M) using A if $M = \begin{bmatrix} 0 & -3 \\ 3 & 0 \end{bmatrix}$.

53. Find all M satisfying T(M) = M. ans. $M = \begin{bmatrix} a & b \\ b & c \end{bmatrix}$ (ie. 2 by 2 symmetric matrices)

4.3 Kernel and Range

The Kernel of a Transformation

Suppose T is a linear transformation from the vector space V into the vector space W.

DEFINITION
The **kernel** of T is a subset of the domain of T (ie. V). It is defined by: $ker(T) = \{ X : X \in V \text{ and } T(X) = \theta_W \}$ where theta is the identity in W.

In other words, it is the set of all *pre-images* of the identity in W (or all vectors in V whose image is the identity in W). Now ker(T) is not empty, since if T is linear, $T(\theta) = \theta$ so that the identity vector must be in ker(T).

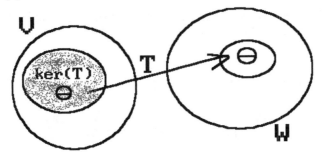

Let's prove that the ker(T) is actually a **subspace** of V:

$$X_1, X_2 \in ker(T) \Rightarrow T(X_1) = \theta_W \text{ and } T(X_2) = \theta_W \text{ so } T(X_1 + X_2) =$$
$$T(X_1) + T(X_2) = \theta_W + \theta_W = \theta_W$$

so ker(T) is closed under addition. ker(T) is also closed under scalar multiplication, since

$$X \in ker(T) \Rightarrow T(kX) = kT(X) = k\theta_W = \theta_W \Rightarrow kX \in ker(T)$$

The role of the kernel of a transformation is to determine if a transformation is one-to-one or not.

DEFINITION
A transformation $T|V \rightarrow W$ is **one-to-one**
iff $X_1 \neq X_2 \Rightarrow T(X_1) \neq T(X_2)$
(or equivalently $T(X_1) = T(X_2) \Rightarrow X_1 = X_2$).

This means that if T is one-to-one, different pre-images have different images (or two different pre-images cannot share an image). Clearly if ker(T) has more than one vector in it, then at least two vectors in V are mapped to the zero vector in W so then T **cannot** be one-to-one. We thus have:

THEOREM A linear transformation T is **one-to-one** iff $ker(T) = \{ \theta_V \}$.

Proof: $\Rightarrow$ Suppose ker(T) = $\{\theta\}$; let $T(X_1) = Y$ and $T(X_2) = Y$. Then

$$T(X_1) - T(X_2) = Y - Y = \theta_w$$

but

$$T(X_1) - T(X_2) = T(X_1 - X_2) = \theta_w$$

by the linearity of T so $X_1 - X_2 \in ker(T)$. However the only vector in ker(T) is θ so that $X_1 - X_2 = \theta$ which means $X_1 = X_2$. By definition this means that T is one-to-one. QED.

Thus if ker(T) is trivial (ie. if ker(T) = $\{\theta\}$, then T is one-to-one; if ker(T) is non-trivial, then T is not 1-1.

DEFINITION

The dimension of ker(T) is the **nullity** of T. That is dim(ker(T)) = nullity(T).

Using the above theorem then, we see that *a transformation T is one-to-one iff nullity(T) = 0.*

EXAMPLE 1 Determine if the transformation T(x,y,z) = (y,x,2z) is one-to-one.
T is one-to-one if T(x,y,z) = (0,0,0) implies (y,x,2z) = (0,0,0). There is an implied system:

$$\begin{matrix} y = 0 \\ x = 0 \\ 2z = 0 \end{matrix} \Rightarrow \begin{bmatrix} 0 & 1 & 0 & 0 \\ 1 & 0 & 0 & 0 \\ 0 & 0 & 2 & 0 \end{bmatrix} \Rightarrow \begin{bmatrix} 0 & 1 & 0 \\ 1 & 0 & 0 \\ 0 & 0 & 2 \end{bmatrix} \begin{bmatrix} x \\ y \\ z \end{bmatrix} = \begin{bmatrix} 0 \\ 0 \\ 0 \end{bmatrix}$$

The "matrix multiplication" version of this system shows how naturally the standard matrix of T arises is such a problem. The matrix equation above is just A[X] = θ which means we are just finding the null space of A (a familiar maneuver) where

$$A = \begin{bmatrix} 0 & 1 & 0 \\ 1 & 0 & 0 \\ 0 & 0 & 2 \end{bmatrix}$$

The rank of A is clearly 3 so that the nullity of A is ZERO, since rank(A) + nullity(A) = number of columns of A = 3. Thus the kernel of T is trivial (ie. ker(T) = $\{(0,0,0)\}$), hence nullity(T) = 0 and therefore T is one-to-one.

EXAMPLE 2 Find the kernel of T where $T \mid \Re^2 \to \Re^3$ such that T(1,0) = (1,3,1) and T(0,1) = (1,3,1). We need to find T(x,y). But since we know the images of the basis vectors, we can immediately write down the standard matrix of T:

$$A = \begin{bmatrix} 1 & 1 \\ 3 & 3 \\ 1 & 1 \end{bmatrix}$$

Now

$$[T(x,y)] = A[X] = \begin{bmatrix} 1 & 1 \\ 3 & 3 \\ 1 & 1 \end{bmatrix} \begin{bmatrix} x \\ y \end{bmatrix} = \begin{bmatrix} x+y \\ 3x+3y \\ x+y \end{bmatrix}$$

so to find ker(T) we wish to solve

$$\begin{aligned} x+y &= 0 \\ 3x+3y &= 0 \Rightarrow \begin{bmatrix} 1 & 1 \\ 3 & 3 \\ 1 & 1 \end{bmatrix} \begin{bmatrix} x \\ y \end{bmatrix} = \begin{bmatrix} 0 \\ 0 \\ 0 \end{bmatrix} \Rightarrow \begin{bmatrix} 1 & 1 & 0 \\ 3 & 3 & 0 \\ 1 & 1 & 0 \end{bmatrix} \sim \begin{bmatrix} 1 & 1 & 0 \\ 0 & 0 & 0 \\ 0 & 0 & 0 \end{bmatrix} \\ x+y &= 0 \end{aligned}$$

Row1 says x + y = 0 where y = t so x = -y= -t and ker(T) = {t(-1,1)}; geometrically this a line through the origin in the direction of (-1,1). Thus nullity(T) = 1 (since ker(T) = span{(-1,1)}). Hence T is NOT one-to-one since ker(T) is not trivial.

EXAMPLE 3 Determine if the transformation T(x,y,z) = (x,y,0) is one-to-one.
First let's find the standard matrix of T: T(1,0,0) = (1,0,0); T(0,1,0) = (0,1,0) and
T(0,0,1) = (0,0,0). Now use these images (in column form) to get the standard matrix of T:

$$A[X] = \begin{bmatrix} 1 & 0 & 0 \\ 0 & 1 & 0 \\ 0 & 0 & 0 \end{bmatrix} \begin{bmatrix} x \\ y \\ z \end{bmatrix} = \begin{bmatrix} 0 \\ 0 \\ 0 \end{bmatrix} \Rightarrow \begin{bmatrix} 1 & 0 & 0 & 0 \\ 0 & 1 & 0 & 0 \\ 0 & 0 & 0 & 0 \end{bmatrix}$$

The rank of A is clearly 2 so that nullity(A) is 1. The system is already in echelon form, we can see clearly that z = t so that ker(T) = {(0,0,t)} (ie. all vectors lying on the z-axis) so T is NOT one-to-one. Geometrically this transformation is a *projection* since a vector like X = (1,2,3) gets sent to its projection Y = (1,2,0) in the x-y plane. If you think about it, a projection transformation cannot be one-to-one!

The Range of a Transformation

DEFINITION

The **range** of transformation $T | V \rightarrow W$ is the set
$$rng(T) = \{ Y : Y \in W \text{ and } Y = T(X) = Y \text{ for some } X \in V \}$$
(ie. the set of all images of elements in the domain V).

THEOREM The range of a transformation T is a **subspace** of W.
Proof: Suppose $Y_1, Y_2 \in rng(T)$; then there must exist two pre-images $X_1, X_2 \in dmn(T)$ such

that $T(X_1) = Y_1$ and $T(X_2) = Y_2$. But since T is linear,

$$T(X_1) + T(x_2) = T(X_1 + X_2) = Y_1 + Y_2$$

so $Y_1 + Y_2 \in rng(T)$ whenever Y_1 and Y_2 are. Hence rng(T) is closed under addition. If $Y_1 \in rng(T)$ then $kY_1 = kT(X_1) = T(kX_1)$ so that $kY_1 \in rng(T)$ whenever Y_1 is.
So rng(T) is closed under scalar multiplication. Therefore rng(T) is a subspace of W.
We next investigate when rng(T) is **all** of W (and not just a proper subspace).

EXAMPLE 4: Given T(x,y,z) = (x - y, 2x - 3z); find rng(T).
Let $Y = (a, b)$ be in rng(T); then a pre-image X = (x,y,z) must exist such that T(X) = Y. This means that $(x - y, 2x - 3z) = (a, b)$ so we have a system in x,y and z:

$$x - y \qquad = a$$

$$2x \quad - 3z = b$$

$$\begin{bmatrix} 1 & -1 & 0 & a \\ 2 & 0 & -3 & b \end{bmatrix} \Rightarrow \begin{bmatrix} 1 & -1 & 0 \\ 2 & 0 & -3 \end{bmatrix} \begin{bmatrix} x \\ y \\ z \end{bmatrix} = \begin{bmatrix} a \\ b \end{bmatrix}$$

The "matrix multiplication" version of this system again is of the form A[X] = [B], where A is the standard matrix for T (ie. T(1,0,0) = (1,2), T(0,1,0) = (-1,0) and T(0,0,1) = (0,-3).
Writing down the augmented matrix and putting it in echelon form:

$$\begin{bmatrix} 1 & -1 & 0 & a \\ 2 & 0 & -3 & b \end{bmatrix} \sim \begin{bmatrix} 1 & -1 & 0 & a \\ 0 & 1 & -\dfrac{3}{2} & \dfrac{b-2a}{2} \end{bmatrix}$$

The echelon form of the augmented matrix shows that we can solve for x and y for any
Y = (a,b). This transformation is thus onto. Furthermore, we can see that the solution cannot be unique (column 3 is a non-pivot column) so we also learn that T *cannot be one-to-one*.

DEFINITION	The **rank** of T is the dimension of rng(T). Simply put, rank(T) = dim(rng(T)).

If rng(T) = W (every vector Y in W has a pre-image in V) then the transformation T is **onto**.

This means that if rank(T) = dim(W) then T is onto; if not, rng(T) is a proper subspace of W and at least one vector in W has no pre-image in V.
Looking back at **EX.4**, we see easily that rank(A) = 2 and since $T \mid \Re^3 \to \Re^2$ and
dim(W) = dim($\Re^2$) = 2, this transformation *must have been onto*.

190

Dimension Theorem

The kernel of T and the range of T appear to be unrelated subspaces of V and W, respectively; however they are constrained by the size of V. The theorem which gives this relationship is called the dimension theorem.

THEOREM (Dimension TH): Given the linear transformation $T \mid V \to W$ where dim(V) = n, dim(ker(T)) = k and dim(rng(T)) = r; then r + k = n (or rank plus nullity = dim(V))

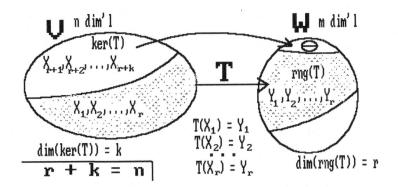

Note that this means for a linear transformation, dim(rng(T)) $\leq$ dim(dmn(T)).

Outline of proof: Step 1- show that if $K = \{ X_{r+1}, X_{r+2}, \cdots, X_{r+k} \}$ then there exists a set $P = \{ X_1, X_2, \cdots, X_r \}$ such that K and P are disjoint and together span V (since dim(V) = n).

Step 2- show that the set $R = \{ T(X_1), T(X_2), \cdots, T(X_r) \}$ spans rng(T).

Step 3- show that R is actually independent and therefore forms a basis for rng(T); thus dim(rng(T)) = r.

Now if we have the standard matrix A of T, then the vectors in rng(T) are represented uniquely by the column vectors in the **column space** of A. For suppose $Y \in rng(T)$; then there exists an X in dmn(T) (which is the vector space V) such that T(X) = Y. But if we have the standard matrix A of the transformation T, then [T(X)] = A[X] so we must be able to solve A[X] = [Y]. But A[X] = [Y] has at least one solution if [Y] is in the column space of A.

Thus **rng(T)** is determined in this fashion by the column space of the standard matrix A.

Now if the rank of A matches the dimension of the vector space W, then the range of T will be **all** of W (and T is **onto**); if not, the range of T will be a proper *subspace* of W (and then T is **NOT** onto). Thus if we can find the rank of A we can determine if T is onto.

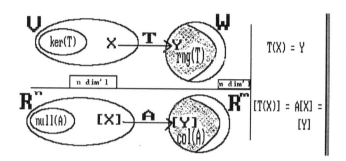

THEOREM Given A is the standard matrix for the linear transformation $T \mid V \to W$; then T is **onto** iff rng(T) = W iff rank(A) = dim (W).

EXAMPLE 5 Determine if T(x,y) = (x+y,x-y,2y) is onto where $V = \Re^2$, $W = \Re^3$. Writing the image of (x,y) in column matrix form gives us

$$
\begin{bmatrix} x+y \\ x-y \\ 2y \end{bmatrix} = \begin{bmatrix} x \\ x \\ 0 \end{bmatrix} + \begin{bmatrix} y \\ -y \\ 2y \end{bmatrix} = x \begin{bmatrix} 1 \\ 1 \\ 0 \end{bmatrix} + y \begin{bmatrix} 1 \\ -1 \\ 2 \end{bmatrix} = \begin{bmatrix} 1 & 1 \\ 1 & -1 \\ 0 & 2 \end{bmatrix} \begin{bmatrix} x \\ y \end{bmatrix} = A \ [\ X \ ]
$$

And the echelon form of A is

$$
A = \begin{bmatrix} 1 & 1 \\ 1 & -1 \\ 0 & 2 \end{bmatrix} \sim \begin{bmatrix} 1 & 1 \\ 0 & 1 \\ 0 & 0 \end{bmatrix}
$$

so rank(A) = 2; but dim(W) = 3 and dim(rng(T)) < dim(W) so T is NOT onto. Intuitively we may have been able to think our way out of this question- could a transformation from a smaller space into a larger space ever be onto???

EXAMPLE 6 Find rng(T) and determine if T is onto if T(x,y,z) = (x + y, 2y - z). Let Y = (a,b) be any element in rng(T); then there must exist X = (x,y,z) such that T(X) = (a,b) which means that (x + y,2y - z) = (a,b) .

192

We have a linear system whose augmented matrix is:

$$\begin{bmatrix} 1 & 1 & 0 & a \\ 0 & 2 & -1 & b \end{bmatrix} \sim \begin{bmatrix} 1 & 1 & 0 & a \\ 0 & 1 & -\dfrac{1}{2} & \dfrac{b}{2} \end{bmatrix}$$

Therefore $\text{rng}(T) = \Re^2$ and T is onto. Alternatively we could observe that the standard matrix of T is $A = \begin{bmatrix} 1 & 1 & 0 \\ 0 & 2 & -1 \end{bmatrix}$ and rank(A) = 2 means that $\text{col}(A) = \Re^2$ so T is onto.

EXAMPLE 7 Find the kernel and range of T if $T \mid P_2 \rightarrow P_1$ and T(p) = dp/dx.

Recall the standard bases for P_2 and P_1 are $B = \{ 1, x, x^2 \}$ and $B' = \{ 1, x \}$ so we need to find the coordinate matrices of the images of the basis vectors in B:

$$T(1) = 0 \Rightarrow [T(1)]_{B'} = \begin{bmatrix} 0 \\ 0 \end{bmatrix} \quad T(x) = 1 \Rightarrow [T(x)]_{B'} = \begin{bmatrix} 0 \\ 1 \end{bmatrix} \quad T(x^2) = 2x \Rightarrow [T(x^2)]_{B'} = \begin{bmatrix} 0 \\ 2 \end{bmatrix}$$

so the standard matrix A for T is

$$A = \begin{bmatrix} 0 & 1 & 0 \\ 0 & 0 & 2 \end{bmatrix}$$

Now rank(A) = 2 so nullity(A) = 1 (since rank(A) + nullity(A) = # of columns of A = 3. This means dim(ker(T)) = 1 so ker(T) is not trivial and thus T is not one-to-one. If you passed calculus you should know that d/dx(a) = 0 so any constant polynomial is in ker(T).

$$p = a = a \cdot 1 \Rightarrow ker(T) = \{ p : p = a \} = span \{ 1 \}$$

We know that rank(T) = rank(A) = 2 and dim(P_1) = 2; then since rng(T) = P_1, T must be onto. This means every degree one polynomial in P_1 has a pre-image in dmn(T).

In fact

$$T(c + ax + \frac{b}{2}x^2) = a + bx$$

EXAMPLE 8 Find the kernel and range of the transformation T if

$$T \mid P \rightarrow P_2 \quad \text{where } T(p) = \int_0^x p(t)\,dt$$

Clearly this transformation is linear (check this out for yourself). The standard basis for P_1 is $B = \{ 1, x \}$ and we need to find the T(1) and T(x) and write the components of the image in column format:

$$T(1)=\int_0^x 1\,dt=[\,t\,]_0^x=x-0=x=0\cdot 1+1\cdot x+0\cdot x^2$$

$$T(x)=\int_0^x t\,dt=[\,\frac{t^2}{2}\,]_0^x=\frac{1}{2}x^2-0=0\cdot 1+0\cdot x+\frac{1}{2}x^2$$

so the standard matrix is a 3 x 2 matrix of the form

$$A=\begin{bmatrix} 0 & 0 \\ 1 & 0 \\ 0 & \frac{1}{2} \end{bmatrix}$$

Clearly this matrix has rank 2; since dim(P2) = 3 this means that T is NOT onto. But since dim(dmn(T)) = dim(P1) = 2 we know that rank(T)+nullity(T) = 2 so nullity(T) = 0.
This means that T is *one-to-one* (ie. two different degree one polynomials have two different antiderivatives).

Row operations and the standard matrix

Consider the standard matrix A below:

$$A=\begin{bmatrix} 1 & 3 \\ -1 & 4 \\ 2 & 5 \end{bmatrix}$$

You should be able to easily show that the transformation hiding behind this matrix is:
T(x,y) = (x + 3y, -x + 4y, 2x + 5y) and we can answer all the important questions about this transformation by using the echelon form of A:

$$A=\begin{bmatrix} 1 & 3 \\ -1 & 4 \\ 2 & 5 \end{bmatrix} \sim \begin{bmatrix} 1 & 3 \\ 0 & 1 \\ 0 & 0 \end{bmatrix}$$

For example we know that rank(A) = rank(T) = 2 and since

$$T\mid\Re^2 \to \Re^3$$

T cannot be onto. Also nullity(T) = nullity(A) = 0 so T **is** 1-1. We can say that ker(T) is invariant under row operations done to A. You DON'T change ker(T) by doing row operations on A.

But the rng(T) could change by doing even one row operation on A! Looking at the columns of A, they are independent and thus form a basis for rng(T). If you look at the *echelon form* of A, the column vectors have no z-component. The span of those two column vectors certainly does NOT form rng(T). In fact rng(T) = (x,y,z): 13x −y − 7z =0} and the column vectors of the echelon form of A don't satisfy this equation.

Moral- if you do even one row operation on the standard matrix A, you in general will change col(A) and thus rng(T). The range of a transformation is NOT invariant under row operations. If you do row operations on the transpose of A, you will NOT change the range of T. Be careful - remember WHY you are doing the row operations!

Concluding comment: If $T | V \rightarrow W$ and V is smaller than W, clearly T cannot be *onto*; on the other hand if V is larger than W, T cannot be *one-to-one*. Thus if we want a transformation to be both one-to-one AND onto, we need dim(V) = dim(W). This is precisely our setting in the next section.

EXERCISES 4.3

Part A. Given the linear transformation T(x,y,z) = (x + y, 2z);
1. T is a transformation from_____ into _____. ans. $\mathfrak{R}^3, \mathfrak{R}^2$
2. Find the image of X = (1,2,3).
3. Find the pre-image(s) of Y = (2,2). ans. (2-t,t,1)
4. Find the standard matrix A of T.
5. Find the rank and nullity of A. ans. 2,1
6. Is T onto? Why or why not?
7. Find the range of T. ans. $\mathfrak{R}^2$
8. Find the kernel of T.

Part B. Given the linear transformation T(x,y) = (x + y, 2y, x - 3y)
9. T is a transformation from_____ into _____. ans. $\mathfrak{R}^2 \rightarrow \mathfrak{R}^3$
10. Find T(3,4).

11. Find the standard matrix of T. ans. $A = \begin{bmatrix} 1 & 1 \\ 0 & 2 \\ 1 & -3 \end{bmatrix}$

12. Find the pre-image(s) of Y = (3,4,-5).
13. Describe the kernel of T and the range of T. ans. ker(T) = {(0,0)},
rng(T) = span({(1,0,1),(0,1,-2)})
14. Is T one-to-one? Why or why not?
15. Is T onto? Why or why not? ans. NO; $rng(T) \neq \mathfrak{R}^3$

Part C. Given the linear transformation T(x,y) = (x,2x + 3y, 3x + 5y):
16. Find the standard matrix A for T.
17. Find the image of X = (4,5) and the pre-image(s) of Y = (2,1,1). ans.(4,23,37), (2,-1)

18. Determine rank(A) and nullity(A).

19. Determined if T is 1-1 and/or onto. ans. 1-1, NOT onto

20. Find ker(T) and rng(T).

Part D. Given the transformation T(p) = dp/dx where $T \mid P_2 \rightarrow P_1$;

21. Show that T is linear.

22. Find $T(x^2)$ and $T(x^2 + 5)$; your answer show that T is NOT ____.

23. Find the pre-image(s) of $q = a x + b$; your answer shows that T is _____ .

Hint: you are finding p such that T(p) = dp/dx = q. ans. $p = \dfrac{a}{2} x^2 + b x + c$, onto

24. Find the standard matrix A of T.

25. Find the rank and nullity of A. ans. 2,1

26. Using # 25, determine if T is one-to-one and/or onto.

27. Find ker(T) and rng(T). ans. ker(T) = span(1), rng(T) = P1

Part E. Given the linear transformation T(x,y,z) = 1/3(x + y +z, x + y + z,x + y + z);

28. Find the standard matrix for T.

29. Find ker(T) and rng(T). ans. ker(T) = span{(-1,1,0),(-1,0,1)}, rng(T) = span{(1,1,1)}

30. Find rank(A) and nullity(A). ans. 1, 2

31. Determine whether T is 1-1 and/or onto. ans. neither

Part F. Given the transformation T(p) = xp where $p \in P_1$;

32. Show that T is a linear transformation from P_1 into P_2.

33. Find the standard matrix A of T. ans. $A = \begin{bmatrix} 0 & 0 \\ 1 & 0 \\ 0 & 1 \end{bmatrix}$

34. Find the rank and nullity of A.

35. What is ker(T)? Is T one-to-one? ans. ker(T) = {0},YES

36. What is rng(T)? Is T onto?

Part G. Given the transformation $T(f) = \int\limits_0^x f(t)\,d t$ on the vector space P_1;

37. Show that T is a linear transformation from P_1 into P_2.

38. Find the standard matrix A for T (your matrix A will do integration).

39. Find the rank of A and nullity of A. ans. 2, 0

40. Determine if T is one-to-one and/or onto.

41. Find ker(T) and rng(T). ans. ker(T) = {(0,0)}, $r n g(T) = span(\{ x, x^2 \})$

Part H. Given the linear transformation T(M) = BM
where $B = \begin{bmatrix} 1 & 2 \\ 0 & 3 \end{bmatrix}$ and $M = \begin{bmatrix} a & b \\ c & d \end{bmatrix}$ so that T is a linear operator on $M_{2,2}$.

42. Find A, the standard matrix of T.

43. Find its rank and nullity. ans. 4, 0

44. Is T one-to-one? It is onto? Explain why/why not.

Part I. Given the transformation $T(M) = M^t$ where $M \in M_{2,2}$;

45. Show that T is linear so T is a linear operator on $M_{2,2}$.

46. Find the standard matrix A of T; ie. find A so that T(M) = AM.

47. Find the rank and nullity of A. ans. rank(A) = 4, nullity(A) = 0

48. Determine if T is one-to-one and/or onto.

Part J. Given the transformation $T \mid M_{2,2} \to \Re$ where $T(M) = tr(M)$

and $M \in M_{2,2}$ (NB.tr(M) = trace(M) = sum of diagonal entries of M);

49. Show that T is a linear transformation.

50. Find ker(T). NB. ker(T) consists of matrices called *traceless*.

51. Find the standard matrix of T. ans. $A = \begin{bmatrix} 1 & 0 & 0 & 1 \end{bmatrix}$

52. Find rank(A) and nullity(A).

53. Determine if T is 1-1 and/or onto. ans. onto, NOT 1-1

Part K. Prove (assume T is linear) :

54. If $T \mid V \to W$ is one-to-one, then ker(T) = { θ } where θ is the identity in V.

55. If $T \mid V \to W$ and dim(V) > dim(W), then T cannot be one-to-one.

56. If $T \mid V \to W$ and dim(V) < dim(W), then T cannot be onto.

57. Step 1 in dimension theorem.

58. Step 2 in dimension theorem.

59. Step 3 in dimension theorem.

4.4 Linear Operators

Recall that a linear operator is a linear transformation from one vector space V back into itself- we are "staying inside" one vector space. The key idea of this chapter is that any linear transformation can be represented by a matrix(its standard matrix); if the vector space V is n-dimensional, then the matrix which represents the transformation is a **square** n by n matrix. What is significant about square matrices is that they can be multiplied and also may be invertible; m x n matrices (if $m \neq n$) can be added but *cannot be multiplied*. In $\Re^2$ or $\Re^3$ a linear operator may have a simple geometric interpretation.

EXAMPLE 1 In $\Re^2$ consider the linear operator T(x,y) = (3x,3y). Clearly this transformation simply *lengthens* each vector by a factor of 3 and preserves its direction. Now the standard matrix can by found by find the images of the standard basis vectors for $\Re^2$:

$$T(1,0)=(3,0) \quad T(0,1)=(0,3)$$

Using these images in column matrix format, we have the standard matrix for T:

$$A = \begin{bmatrix} 3 & 0 \\ 0 & 3 \end{bmatrix}$$

so that the operator T can by represented by a matrix multiplication by A:

$$[T(X)]= A [X]= \begin{bmatrix} 3 & 0 \\ 0 & 3 \end{bmatrix} \begin{bmatrix} x \\ y \end{bmatrix} = \begin{bmatrix} 3x \\ 3y \end{bmatrix}$$

It is clear that rank(A) = 2 and since $dim(\Re^2)= 2$, the dimension of the range of T = 2 and rng(T) = col(A) is all of $\Re^2$ so T must be onto. By the rank plus nullity theorem, rank(A) + null(A) = 2 and rank(A) = 2 so null(A) = 0 and T is one-to-one (remember that null(A) = 0 implies T is one-to-one). All of this information can be deduced from the fact that the rank of A is maximal (ie. it is equal to 2, its maximum possible value here).

EXAMPLE 2 In $\Re^2$ consider the linear operator T(x,y) = (x,0). Clearly this transformation is a projection- every vector is projected onto the x-axis. Now the standard matrix can by found by getting the images of the standard basis vectors for $\Re^2$:

$$T(1,0)=(1,0) \quad T(0,1)=(0,0)$$

Using these images in *column matrix* format, we have the standard matrix for T:

$$A = \begin{bmatrix} 1 & 0 \\ 0 & 0 \end{bmatrix}$$

so that the operator T can by represented by a matrix multiplication by A:

$$[T(X)] = A[X] = \begin{bmatrix} 1 & 0 \\ 0 & 0 \end{bmatrix} \begin{bmatrix} x \\ y \end{bmatrix} = \begin{bmatrix} x \\ 0 \end{bmatrix}$$

It is clear that rank(A) = 1 (it is already in row-echelon form) and since $dim(\Re^2) = 2$, the dimension of the range of T = 1 and rng(T) = col(A) is a *proper subspace* of $\Re^2$ so T cannot be onto. By the rank plus nullity theorem, rank(A) + null(A) = 2 but rank(A) = 1 so null(A) = 1 and T is not one-to-one (remember T is one-to-one iff null(A) = 0). All of this information can be deduced from the fact that the rank of A is not maximal (ie. it is not equal to 2, its maximum value here).

EXAMPLE 3 In $\Re^2$ consider the linear operator T(x,y) = (-y,x). One can show that this transformation represents a rotation of 90 degrees. Check out the fact that T(1,0) = (0,1) and T(0,1) = (-1,0). Each of these vectors is rotated 90 degrees with no change of length. Now the standard matrix can by found by find the images of the standard basis vectors for $\Re^2$:

$$T(1,0) = (0,1) \qquad T(0,1) = (-1,0)$$

Using these images in column matrix format, we have the standard matrix for T:

$$A = \begin{bmatrix} 0 & -1 \\ 1 & 0 \end{bmatrix}$$

so that the operator T can by represented by a matrix multiplication by A:

$$[T(X)] = A[X] = \begin{bmatrix} 0 & -1 \\ 1 & 0 \end{bmatrix} \begin{bmatrix} x \\ y \end{bmatrix} = \begin{bmatrix} -y \\ x \end{bmatrix}$$

It is clear that rank(A) = 2 and since $dim(\Re^2) = 2$, the dimension of the range of T = 2 and rng(T) = col(A) = $\Re^2$ so T is onto. By the rank plus nullity theorem, rank(A) + null(A) = 2 but rank(A) = 2 so null(A) = 0 and T is one-to-one (remember that null(A) = 0 implies T is one-to-one).
All of this information can be deduced from the fact that the rank of A is maximal (ie. it is equal to 2, its maximum value here).

Composition of Operators

Suppose that $S(x,y) = (2x, 2y)$ and $T(x,y) = (y,-x)$. Each of these linear operators is represented by a unique matrix and from the previous section we know how to find the standard matrices which represent S and T:

$$A_1 = \begin{bmatrix} 2 & 0 \\ 0 & 2 \end{bmatrix} \qquad A_2 = \begin{bmatrix} 0 & 1 \\ -1 & 0 \end{bmatrix}$$

If we want to combine the effect of both of these operators, we are finding the *composition* of the operator S with the operator T.

$$S \circ T(x,y) = S(T(x,y)) = S(y,-x) = (2y,-2x)$$

199

Using matrix multiplication to accomplish this means:

$$[S \circ T(X)] = (A_1 A_2)[X] = \begin{bmatrix} 2 & 0 \\ 0 & 2 \end{bmatrix} \begin{bmatrix} 0 & 1 \\ -1 & 0 \end{bmatrix} \begin{bmatrix} x \\ y \end{bmatrix} = \begin{bmatrix} 2 & 0 \\ 0 & 2 \end{bmatrix} \begin{bmatrix} y \\ -x \end{bmatrix} = \begin{bmatrix} 2y \\ -2x \end{bmatrix}$$

It is important here that we observe that the standard matrix of S(T(X)) is calculated by multiplying X on the left by A_2 FIRST and then multiplying that on the left by A_1. Geometrically we are **rotating first** (since we applied T **first**) and then stretching by a factor of 2. The reader can verify that, in this case, composing in the reverse order gives $T \circ S(x, y) = (2y, -2x)$ so that the composition here was commutative (this is unusual!). But we know that in general, composition is NOT commutative so something special must have been going on here- in fact the matrix of S was a *scalar* matrix (which commutes with all other square matrices). If an operator has a matrix which is not scalar, that operator will not commute (in general) with other operators.

EXAMPLE 4 Find the matrix C which represents $S \circ T$ if $S(x, y) = (x + 2y, 3y)$ and $T(x, y) = (-2y, 2x)$.

We need to find the standard matrices for S and T; let's deal with S first:

$$[S(x, y)] = \begin{bmatrix} x + 2y \\ 3y \end{bmatrix} = \begin{bmatrix} x \\ 0 \end{bmatrix} + \begin{bmatrix} 2y \\ y \end{bmatrix} = x \begin{bmatrix} 1 \\ 0 \end{bmatrix} + y \begin{bmatrix} 2 \\ 1 \end{bmatrix} = \begin{bmatrix} 1 & 2 \\ 0 & 1 \end{bmatrix} \begin{bmatrix} x \\ y \end{bmatrix} = A_1$$

Now it's T's turn:

$$[T(x, y)] = \begin{bmatrix} -2y \\ 2x \end{bmatrix} = \begin{bmatrix} 0 \\ 2x \end{bmatrix} + \begin{bmatrix} -2y \\ 0 \end{bmatrix} = x \begin{bmatrix} 0 \\ 2 \end{bmatrix} + y \begin{bmatrix} -2 \\ 0 \end{bmatrix} = \begin{bmatrix} 0 & -2 \\ 2 & 0 \end{bmatrix} \begin{bmatrix} x \\ y \end{bmatrix} = A_2$$

Therefore the standard matrix C for the composition of S with T is given by:

$$C = A_1 A_2 = \begin{bmatrix} 1 & 2 \\ 0 & 3 \end{bmatrix} \begin{bmatrix} 0 & -2 \\ 2 & 0 \end{bmatrix} = \begin{bmatrix} 4 & -2 \\ 6 & 0 \end{bmatrix}$$

The reader can check that here order is important (ie. composition is NOT commutative in general) by changing the order of the factors and finding $A_2 A_1$. We could have predicted this by simply observing that neither matrix is a scalar matrix. The actual image can be then found by a simple matrix multiplication: $C \begin{bmatrix} x \\ y \end{bmatrix} = \begin{bmatrix} 4 & -2 \\ 6 & 0 \end{bmatrix} \begin{bmatrix} x \\ y \end{bmatrix} = \begin{bmatrix} 4x - 2y \\ 6x \end{bmatrix}$ so S(T(x,y)) = (4x - 2y,6x), which can be checked by simply composing S with T without using matrices.

EXAMPLE 5 Find the inverse operator of the linear operator T(x,y) = (x + 2y, 3x + 4y)

The standard matrix of T is given by: $A = \begin{bmatrix} 1 & 2 \\ 3 & 4 \end{bmatrix}$. Thus to get the inverse of T, all we need to do is find the inverse of A:

$$A^{-1} = \frac{1}{\det(A)} \begin{bmatrix} 4 & -2 \\ -3 & 1 \end{bmatrix} = \frac{1}{-2} \begin{bmatrix} 4 & -2 \\ -3 & 1 \end{bmatrix} = \begin{bmatrix} -2 & 1 \\ \frac{3}{2} & -\frac{1}{2} \end{bmatrix}$$

Then we can get the inverse of T:

$$A^{-1} \begin{bmatrix} x \\ y \end{bmatrix} = \begin{bmatrix} -2 & 1 \\ \frac{3}{2} & -\frac{1}{2} \end{bmatrix} \begin{bmatrix} x \\ y \end{bmatrix} = \begin{bmatrix} -2x + y \\ \frac{3}{2}x - \frac{1}{2}y \end{bmatrix}$$

$$T^{-1}(x,y) = (-2x + y, \frac{3}{2}x - \frac{1}{2}y)$$

The reader should check this all out by finding the image of a particular vector (say T(1,2) = (5,11)) and show that under the *inverse operator* (5,11) gets sent to (1,2). Or better yet, compose T and its inverse and show that the result is the identity operator I(x,y) = (x,y).

$$T(T^{-1}(x,y)) = T(-2x + y, \frac{3}{2}x - \frac{1}{2}y) =$$

$$(-2x + y + 2(\frac{3}{2}x - \frac{1}{2}y), 3(-2x + y) + 4(\frac{3}{2}x - \frac{1}{2}y)) =$$

$$(-2x + y + 3x - y), -6x + 3y + 6x - 2y)) = (x,y)$$

We can now add a couple results to our fundamental theorem if the linear operator T has standard matrix A.

FUNDAMENTAL THEOREM OF LINEAR ALGEBRA

Given a square matrix of size n by n; the following are equivalent:

1) A is *invertible*
2) $\det(A) \neq 0$
3) rank(A) = n and nullity(A) = 0
4) rows(columns) of A are independent
5) *null(A) = { θ }*

6) $row(A) = col(A) = \Re^n$

7) $AX = B$ has a unique solution

8) T is both one-to-one and onto

9) $ker(T)=\{ \theta \}$ and $rng(T)=\Re^n$

10) T^{-1} exists

Keep in mind that the flip side of this theorem would start out with A is *singular* iff $det(A) = 0$ etc; in other words one can just negate every condition.

EXERCISES 4.4

Part A. Each matrix A represent a linear operator on 2-space. Find T(x,y) if [T(X)] = A[X] where X = (x,y). Determine if T is one-to-one and/or onto.

1. $A = \begin{bmatrix} 1 & -1 \\ 2 & -2 \end{bmatrix}$ ans. T(x,y) = (x-y,2x-2y), NOT 1-1, NOT onto

2. $A = \begin{bmatrix} 1 & -1 \\ 0 & 3 \end{bmatrix}$

3. $A = \begin{bmatrix} 0 & 1 \\ -1 & 0 \end{bmatrix}$ ans. T(x,y) = (y,-x), 1-1, onto

4. $A = \begin{bmatrix} 1 & -3 \\ 0 & 1 \end{bmatrix}$

5. $A = \begin{bmatrix} 3 & 0 \\ 0 & 3 \end{bmatrix}$ ans. T(x,y) = (3x,3y), 1-1, onto

6. $A = \begin{bmatrix} 0 & 1 \\ 1 & 0 \end{bmatrix}$

7. $A = \begin{bmatrix} 1 & 2 \\ 0 & 3 \end{bmatrix}$ ans. T(x,y) = (x+2y,3y), 1-1, onto

8. $A = \begin{bmatrix} 1 & 0 \\ 1 & 4 \end{bmatrix}$

202

9. $A = \begin{bmatrix} 0 & 2 \\ -2 & 0 \end{bmatrix}$ ans. T(x,y) = (2y,-2x), 1-1, onto

10. $A = \begin{bmatrix} 3 & 0 \\ 0 & 0 \end{bmatrix}$

Part B. Given T(x,y) = 1/2(x + y, x + y);
11. Draw the images of X = (3,3),(0,4), (0,-4), (-4,4),(6,0) and (0,2); observe that T is the projection onto the line y = x.
12. Find the standard matrix A of T.
13. What is rank(A) and nullity(A)? ans. 1, 1
14. Is T 1-1 and/or onto?
15. Find ker(T) and rng(T). ans. ker(T) = {s(-1,1)}, rng(T) = {t(1,1)}

Part C. Given S(x,y) = (-y,x), T(x,y) = (2x, 3x + y) find:
16. standard matrix A_1 for S and standard matrix A_2 for T
17. draw a diagram of X = (2,1) (in Quadrant I) and what S(X) and T(X) look like
18. find $S \circ T (x, y)$ using matrix multiplication
19. find $T \circ S (x, y)$ using matrix multiplication ans. (-2y,x-3y)
20. find $T^{-1} (x, y)$ using the inverse matrix
21. Do the operators S and T commute under composition? How could you tell this using only their standard matrices? ans. NO, neither is a scalar matrix
22. find $(T \circ S)^{-1} (x, y)$. Hint: look back at # 19.

Part D. Given the linear operator T(x,y) = 1/5(x+2y,2x+4y);

23. Find the standard matrix A for T. ans. $A = \begin{bmatrix} \dfrac{1}{5} & \dfrac{2}{5} \\ \dfrac{2}{5} & \dfrac{4}{5} \end{bmatrix}$

24. Find rank(T) and nullity(T); using these, determine if T is 1-1 and/or onto.
25. Find ker(T) and rng(T), thus verifying conclusions made in #24. ans. {s(-2,1)}, {t(1,2)}
26. Draw the vectors X_1 = (3,1), X_2 = (-1,3), and X_3 = (2,4) together with their images; can you determine the effect of this transformation from your art work?

Part E. Given the linear operator on 3-space whose standard matrix is $A = \begin{bmatrix} \frac{1}{2} & 0 & \frac{1}{2} \\ 0 & 1 & 0 \\ \frac{1}{2} & 0 & \frac{1}{2} \end{bmatrix}$;

27. Find T(x,y,z). ans. T(x,y,z) = 1/2(x + z, 2y, x + z)
28. Find rank(T) and nullity(T); thereby determine if T is onto/1-1.
29. Show that this transformation represents a *projection* onto the subspace W = span{(1,0,1),(0,1,0)}. Hint: rewrite the image of (x,y,z) in terms of a basis.
30. Find all pre-images of the vector Y = (2,3,2).

31. Show that $[T(T(X))] = A(A[X]) = A^2[X] = A[X]$; such an operator is called *idempotent*, since $A^2 = A$ and also symmetric. T is a projection operator.

Part F. Given the linear operator T(x,y,z,w) = 1/2(2x,y + w,2z, y + w);
32. Find the standard matrix A for T.
33. Get rank(A) and nullity(A). ans. 3, 1
34. Determine if T is 1-1 and/or onto.
35. Show that each vector in W is a fixed point of the operator (ie. T(X) = X), where W = span(S), S = {(1,0,1,0),(1,0,-1,0),(0,1,0,1)}. NB. T is the projection in 4-space onto the hyperplane spanned by the 3 vectors in S.
36. Could you find T^{-1} ? Why or why not?
37. Verify that T is indeed a projection by showing that A is *idempotent*; also show A is symmetric.

Part G. Given the linear operator on 3-space whose standard matrix is
$A = \begin{bmatrix} 1 & 0 & 0 \\ 0 & \cos(\theta) & -\sin(\theta) \\ 0 & \sin(\theta) & \cos(\theta) \end{bmatrix}$; this operator represents a rotation through the angle θ
about the x-axis.
38. Find T(1,2,3) if $\theta = \frac{\pi}{4}$.

39. Find the rank and nullity of A and thereby determine if T is onto/1-1. You should know the answer to these questions just by knowing what T does! ans. 3, 0, both 1-1 and onto
40. Show that T has an inverse by inverting A.

Part H. A **fixed point** of a linear operator T is a vector X in dmn(T) such that T(X) = X.
41. Show that the set of fixed points of a linear operator T is a *subspace* of dmn(T).
42. Find the fixed points of the operator T(X) = T(x,y) = (1/2x+1/4y,1/2x+3/4y).
43. Find the fixed points of the operator T(X) = T(x,y) = (y,x). ans. FP = {(x,x)}
44. Show that the operator whose standard matrix A is given by

$$A = \begin{bmatrix} \dfrac{1}{\sqrt{2}} & \dfrac{-1}{\sqrt{2}} \\ \dfrac{1}{\sqrt{2}} & \dfrac{1}{\sqrt{2}} \end{bmatrix}$$ has no fixed points (except the origin). NB. It represents a *rotation*.

Part I. Given the linear operator T whose standard matrix is $A = \begin{bmatrix} 1 & 2 \\ 2 & 5 \end{bmatrix}$ (an invertible matrix);

45. Show that if S = {(1,1),(3,4)} is independent, the set of its images is also independent.

46. Show that if $A = \begin{bmatrix} 1 & 2 \\ 2 & 4 \end{bmatrix}$ (a singular matrix) with S as given in # 45, the set of its images is dependent. This means an invertible transformation *preserves independence*, a singular one does not.

Part J. Given B = { e^x, sin (x), cos (x)} and T(f) = df/dx where V = span(B);

47. Find the matrix A for T with respect to the basis B. ans. $A = \begin{bmatrix} 1 & 0 & 0 \\ 0 & 0 & -1 \\ 0 & 1 & 0 \end{bmatrix}$

48. Find rank(A) and nullity(A).
49. Is T 1-1 and/or onto? Give a reason. ans. 1-1 since nullity(T) = 0, onto since rank(T) = 3 and co-domain of T is $\Re^3$
50. Is T invertible? If so, find the matrix A which represents T^{-1}.
51. Find $T (2 e^x + 3 \sin (x) + 4 \cos (x))$ using A.ans. $2 e^x - 4 \sin (x) + 3 \cos (x)$

4.5 Changing Bases

Let $B = \{ (1,0), (0,1) \} \subseteq \Re^2$ which is a basis for 2-space. Pick an arbitrary vector in 2-space, suppose we pick X = (3,4). Clearly

$$(3,4) = 3(1,0) + 4(0,1)$$

To emphasize the fact that we are using the basis B, we can use the notation

$$(X)_B = (3,4)$$

so that everyone knows that we are using the given basis B.

To get matrices into this discussion, we say the **coordinate matrix** of the vector X = (3,4) is

$$[X]_B = \begin{bmatrix} 3 \\ 4 \end{bmatrix}$$

Now what if we desire to change the basis, say we pick

$$B' = \{ (2,3), (-1,4) \} \subseteq \Re^2$$

Our question would be, what are the coordinates of the same vector X wrt. this new basis? We'd like $(X)_{B'}$. We need to find constants k_1 and k_2 such that

$$(3,4) = k_1(2,3) + k_2(-1,4) \Rightarrow (3,4) = (2k_1 - k_2, 3k_1 + 4k_2)$$

$$2k_1 - k_2 = 3$$

$$3k_1 + 4k_2 = 4$$

If we write down the corresponding augmented matrix we have:

$$\begin{bmatrix} 2 & -1 & 3 \\ 3 & 4 & 4 \end{bmatrix}$$

In matrix multiplication form, things are a little more exciting:

$$\begin{bmatrix} 2 & -1 \\ 3 & 4 \end{bmatrix} [X]_{B'} = [X]_B$$

The matrix on the left clearly is formed by the components of the new basis (ie. B') wrt. the old basis B; it is called a **transition matrix** and we will represent it by P:

206

$$P \left[\ X\ \right]_{B'} = \left[\ X\ \right]_B$$

Thus P is the transition matrix from B' to B (from **new** to old); however, to solve the problem at hand, we need to multiply both sides by the inverse of P. Since the columns of P must be independent (why??), we can always do this and

$$\left[\ X\ \right]_{B'} = P^{-1} \left[\ X\ \right]_B$$

If we reverse the order, we have:

$$P^{-1} \left[\ X\ \right]_B = \left[\ X\ \right]_{B'}$$

Hence *P inverse* is also a *transition matrix* but it takes you in the **reverse direction**:

transition matrix P	trans. matrix $Q = P^{-1}$
$P \left[\ X\ \right]_{B'} = \left[\ X\ \right]_B$	$P^{-1} \left[\ X\ \right]_B = \left[\ X\ \right]_{B'}$
new to old (NO)	old to **new** (ON)

In our case we have

$$P^{-1} = \frac{1}{11} \begin{bmatrix} 4 & 1 \\ -3 & 2 \end{bmatrix} = \begin{bmatrix} \dfrac{4}{11} & \dfrac{1}{11} \\ -\dfrac{3}{11} & \dfrac{2}{11} \end{bmatrix}$$

so that the coordinate matrix of X wrt. B' is given by:

$$\left[\ X\ \right]_{B'} = \begin{bmatrix} \dfrac{4}{11} & \dfrac{1}{11} \\ -\dfrac{3}{11} & \dfrac{2}{11} \end{bmatrix} \begin{bmatrix} 3 \\ 4 \end{bmatrix} = \begin{bmatrix} \dfrac{16}{11} \\ -\dfrac{1}{11} \end{bmatrix}$$

To verify our results:

$$\frac{16}{11}(2,3) - \frac{1}{11}(-1,4) = \left(\frac{32+1}{11}, \frac{48-4}{11}\right) = \left(\frac{33}{11}, \frac{44}{11}\right) = (3,4)$$

so that it appears we have found the correct coordinates of X = (3,4) wrt. the new basis B'.

Linear Transformations and Similar Matrices

Consider the transformation represented by the matrix $A = \begin{bmatrix} 1 & 2 \\ 2 & 4 \end{bmatrix}$; recall we view this in the form

$[T (X)] = A [X]$ so A is the standard matrix of T. We know certainly that T is linear and is an operator on $\Re^2$. The basis that we are actually using with this transformation is the standard basis B = {(1,0),(0,1)} and the first column of A is the image of (1,0) and the second column of A is the image of (0,1). Fine- now what if we choose to change to a different (non-standard) basis? Let's say we use B' = {(1,0),(1,1)}. The situation in general is that if we want to find an image (output) we need to be careful as to what basis we are using. If we use the standard matrix A, it is assumed we are using the standard basis B. Let us be very careful with the notation and write

$$[Y]_B = [T (X)]_B = A [X]_B \; where \, Y \in rng (T) \, \& \, X \in dmn (T) \tag{1}$$

If we want to change to the new basis, remember the fundamental equation which accomplishes this:

$$P [X]_{B'} = [X]_B \tag{2}$$

In equ.(1), let's replace [X]B from equ.(2)

$$[Y]_B = [T (X)]_B = A (P [X]_{B'})$$

But we need to change the coordinates of the image Y also:

$$[Y]_B = P [Y]_{B'}$$

so that

$$P [Y]_{B'} = A (P [X]_{B'}) \tag{3}$$

Now just solve equ.(3) for the coordinates of Y wrt. B' (the new basis) by multiplying both sides equ. (3) by P^{-1}:

$$[Y]_{B'} = P^{-1} A P [X]_{B'}$$

This means that the matrix which represents the same transformation wrt. the new basis B' is the matrix $P^{-1} A P$ where P is the transition matrix from B' to B. Now this matrix is just another 2 x 2 so let's call it C; then we have

$$[Y]_{B'} = C [X]_{B'}$$

DEFINITION	Two square matrices A and C are **similar** iff there exists an invertible matrix P such that $C = P^{-1} A P$.

From our example, the matrices A and C were related to each other- they represent the *same linear transformation* T with wrt. to **different bases** B and B'. This is always true for two matrices A and C which are similar.

THEOREM If two matrices A and C are similar, they represent the same linear transformation with respect to two (different) bases.

EXAMPLE 1 If X = (5,3); find T(X) using A and then the similar matrix C if the new basis is B' = {(1,0),(1,1)}.
To use A, we write X as a column matrix and multiply by A to get its image:

$$[Y]_B = \begin{bmatrix} 1 & 2 \\ 0 & 3 \end{bmatrix} \begin{bmatrix} 5 \\ 3 \end{bmatrix} = \begin{bmatrix} 11 \\ 9 \end{bmatrix}$$

so T(5,3) = (11,9). Now let's find the matrix C which is similar to A- remember that C is related to A via the equation

$$C = P^{-1} A P$$

where the columns of P come from the new basis B'. So we need P and its inverse:

$$P = \begin{bmatrix} 1 & 1 \\ 0 & 1 \end{bmatrix} \Rightarrow P^{-1} = \begin{bmatrix} 1 & -1 \\ 0 & 1 \end{bmatrix}$$

so that

$$C = P^{-1} A P = \begin{bmatrix} 1 & -1 \\ 0 & 1 \end{bmatrix}\begin{bmatrix} 1 & 2 \\ 0 & 3 \end{bmatrix}\begin{bmatrix} 1 & -1 \\ 0 & 1 \end{bmatrix} = \begin{bmatrix} 1 & 0 \\ 0 & 3 \end{bmatrix}$$

If we are going to use C to get the image of X though, we must find the coordinates of X wrt. the basis B' ; just use P again:

$$P[X]_{B'} = [X]_B \Rightarrow [X]_{B'} = P^{-1}[X]_B = \begin{bmatrix} 1 & -1 \\ 0 & 1 \end{bmatrix}\begin{bmatrix} 5 \\ 3 \end{bmatrix} = \begin{bmatrix} 2 \\ 3 \end{bmatrix}$$

Now multiply the new coordinate matrix by C on the left:

$$[Y]_{B'} = C[X]_{B'} = \begin{bmatrix} 1 & 0 \\ 0 & 3 \end{bmatrix}\begin{bmatrix} 2 \\ 3 \end{bmatrix} = \begin{bmatrix} 2 \\ 9 \end{bmatrix}$$

This means that $[Y]_B = [T(X)]_B = 2\begin{bmatrix} 1 \\ 0 \end{bmatrix} + 9\begin{bmatrix} 1 \\ 1 \end{bmatrix} = \begin{bmatrix} 11 \\ 9 \end{bmatrix}$ and of course this is the image

that we found directly using the old basis and the standard matrix of the transformation A. Why would we want to go through this work?? Well, check out the matrix C. It is a much simpler matrix than A- it is a *diagonal* matrix. The new basis B' was chosen so that its matrix C would be diagonal. Then it is really easy to deal with the transformation T if the basis B' is used, since its matrix representative C is a diagonal matrix.

EXERCISES 4.5

Part A. Given the basis B', which is a non-standard basis; find the coordinates of the vector X wrt. the basis B'.

1. B' = {(1,2),(2,1)}, X = (3,3) ans. $(X)_{B'} = (1,1)$

2. B' = {(1,1),(1,-1)}, X = (5,6)

3. B' = {(1,0,1),(0,1,0),(1,1,2)}, X = (1,2,3) ans. $(X)_{B'} = (-1,0,2)$

4. B' = {(0,1,1),(1,1,0),(1,1,1)}, X = (3,4,5)

5. B' = {(0,2,2,0),(2,0,0,2),(0,1,-1,0),((1,0,0,-1)}, X = (3,4,5,6) ans. $(X)_{B'} = (\frac{9}{4}, 3, \frac{-1}{2}, -3)$

Part B. Given the non-standard basis B' = {(1,0,1),(1,0,-1),(0,2,0)} for 3-space, find:

6. Transition matrix P from B' to B = {(1,0,0),(0,1,0),(0,0,1)}.

]7. Transition matrix Q from B to B'. ans. $P = \begin{bmatrix} 1 & 1 & 0 \\ 0 & 0 & 2 \\ 1 & -1 & 0 \end{bmatrix}$

8. Coordinate matrix of X = (5,5,5) wrt. B' using a transition matrix.

9. Check your answer to # 8 with a simple vector addition.

Part C. Given the basis B' = {(2,1),(1,1)} for 2-space and the linear transformation T whose standard matrix is $A = \begin{bmatrix} 1 & 2 \\ -1 & 4 \end{bmatrix}$;

10. find the image of X = (11,7) using the matrix A

11. find the coordinate matrix $[X]_{B'}$ using a transition matrix ans. $[X]_{B'} = \begin{bmatrix} 4 \\ 3 \end{bmatrix}$

12. find the matrix C which represents the transformation T wrt. the basis B' (ie. find C such that $[T (X)]_{B'} = C [X]_{B'}$).

13. find $[T (X)]_{B'}$ using C if X = (11,7) ans. $[T (X)]_{B'} = \begin{bmatrix} 8 \\ 9 \end{bmatrix}$

14. find $[T (X)]_B$ using your answer to # 13 and the basic interpretation of coordinates as multipliers in a linear combination of basis vectors

15. find $[T (X)]_B$ using your answer to # 13 and a transition matrix (NB. your answer to #15 of

course is the same as your answer to # 14!) $[T (X)]_B = \begin{bmatrix} 25 \\ 17 \end{bmatrix}$

Part D. Given the basis B' = {(-1,1,0),(1,0,1),(-1,-1,1)} for 3-space and the linear transformation T

whose standard matrix is $A = \begin{bmatrix} 0 & -1 & 1 \\ -1 & 0 & 1 \\ 1 & 1 & 0 \end{bmatrix}$;

16. find the image of X = (-2,-2,5) using the matrix A

17. find the coordinate matrix $[X]_{B'}$ using a transition matrix ans. $[X]_{B'} = \begin{bmatrix} 1 \\ 2 \\ 3 \end{bmatrix}$

18. find the matrix C which represents the transformation T wrt. the basis B (ie. find C such that $[T (X)]_{B'} = C [X]_{B'}$. Why is the matrix C much simpler than the matrix A?

19. find $[T (X)]_{B'}$ using C ans. $[T (X)]_{B'} = \begin{bmatrix} 1 \\ 2 \\ -6 \end{bmatrix}$

20. find $[T (X)]_B$ using your answer to # 19 and the basic interpretation of coordinates as multipliers in a linear combination of basis vectors
21. find $[T (X)]_B$ using your answer to # 19 and a transition matrix (NB. your answer to #21 of

course is the same as your answer to # 20!) ans. $[T (X)]_B = \begin{bmatrix} 7 \\ 7 \\ 4 \end{bmatrix}$

Part E. Given the transformation T(p) = dp/dx on the vector space P_2 with standard basis $B = \{ 1, x, x^2 \}$; suppose the basis is changed to $B' = \{ 1, 2x, 4x^2 - 1 \}$
22. find T(p) if $p = 1 + 2x + 4x^2$

23. find the standard matrix A for the linear operator T ans. $A = \begin{bmatrix} 0 & 1 & 0 \\ 0 & 0 & 2 \\ 0 & 0 & 0 \end{bmatrix}$

24. find $[T (p)]_B$ using A

25. find $[p]_{B'}$ ans. $[p]_{B'} = \begin{bmatrix} 2 \\ 1 \\ 1 \end{bmatrix}$

26. find the matrix C which represents T wrt. the basis B'

27. use C to find $[T(p)]_{B'}$ ans. $[T(X)]_{B'}=\begin{bmatrix} 2 \\ 4 \\ 0 \end{bmatrix}$

28. what is $[T(p)]_B$ using your answer to # 27 and the basic interpretation of coordinates as multipliers in a linear combination of basis vectors

29. repeat # 28 using a transition matrix ans. $[T(p)]_B=\begin{bmatrix} 2 \\ 8 \\ 0 \end{bmatrix}$

Part F. Given the basis B' = {(1,2,0),(1,1,0),(0,0,1)} for 3-space and the linear transformation T whose standard matrix is $A=\begin{bmatrix} 3 & -2 & 0 \\ 4 & -3 & 0 \\ 0 & 0 & 1 \end{bmatrix}$;

30. find the image of X = (2,3,1) using the matrix A

31. find the coordinate matrix $[X]_{B'}$ using a transition matrix ans. $[X]_{B'}=\begin{bmatrix} 1 \\ 1 \\ 1 \end{bmatrix}$

32. find the matrix C which represents the transformation T wrt. the new basis B' (ie. find C such that $[T(X)]_{B'}=C[X]_{B'}$. Why is the matrix C much simpler than the matrix A?

33. find $[T(X)]_{B'}$ using C ans. $\begin{bmatrix} -1 \\ 1 \\ 1 \end{bmatrix}$

34. find $[T(X)]_B$ using your answer to # 33 and the basic interpretation of coordinates as multipliers in a linear combination of basis vectors

35. find $[T(X)]_B$ using your answer to # 33 and a transition matrix (NB. your answer to #35 of course is the same as your answer to # 34!) ans. $[T(X)]_B=\begin{bmatrix} 0 \\ -1 \\ 1 \end{bmatrix}$

36. What is T(x,y,z) ?

4.6 Chapter Four Outline and Review

4.1. Linear Transformations

Given T is a transformation (function,mapping) from a vector space V into a vector space W; we write $T \mid V \rightarrow W$ where V is the domain of T and W is the co-domain of T.

If T(X) = Y then Y is the *image* and X is the pre-image of Y

dmn(T) = V and $rng \, (\, T \,) \subseteq W$ where

$$rng \, (\, T \,) = \{ \, Y : T (\, X \,) = Y \text{ for some } X \in V \, \}$$

(set of all images of vectors from dmn(T) = V)

T is **linear** iff

1) $T (\, X_1 + X_2 \,) = T (\, X_1 \,) + T (\, X_2 \,)$

2) T(kX) = k T(X)

for all $X \in V \text{ and } k \in \Re$

4.2 Matrices and Transformations

Every m x n matrix A represents a linear transformation from $\Re^n$ into $\Re^m$

Conversely, every *linear* transformation (from an n-dimensional space into an m-dimensional space) can be represented by an m x n matrix A where $A [\, X \,]_B = [\, T (\, X \,) \,]_{B'}$ and the bases of V and W are B and B', respectively

To find A- map every basis vector (from B) and use the *coordinates of the image* (with respect to B') to form the corresponding columns in A

$$A = \left[\, [\, T (\, X_1 \,) \,]_{B'} \vdots [\, T (\, X_2 \,) \,]_{B'} \vdots ... \vdots [\, T (\, X_n \,) \,]_{B'} \, \right]$$

where B = $\{ \, X_1, X_2, ..., X_n \, \}$

If B and B' are the standard bases for V and W (resp.) then A is the **standard matrix** of T.

4.3 Kernel and Range

Given T is a linear transformation from V into W:

1) $ker (\, T \,) = \{ \, X : T (\, X \,) = \theta_W \, \} \subseteq V$ (set of all pre-images of the identity from W)

2) $rng (\, T \,) = \{ \, Y : T (\, X \,) = Y \text{ for some } X \in V \, \} \subseteq W$

(set of all images of vectors in V)

ker(T) is a subspace of V; rng(T) is a *subspace* of W

One-to-one and onto transformations:

1) T is one-to-one iff $X_1 \neq X_2 \Rightarrow T(X_1) \neq T(X_2)$ (or different pre-images have different images)

2) T is onto iff rng(T) = W (every element Y in W has a pre-image X in V so T(X) = Y)

THEOREM T is 1-1 iff ker(T) = $\{ \, \theta \, \}$ (ie. ker(T) is lonely)

THEOREM T is onto iff rng(T) = W (ie. T fills up W)

DIMENSION THEOREM Given $T \mid V \rightarrow W$ such that T is linear, then

$$rank(T) + nullity(T) = dim(V) \text{ where } V = dmn(T)$$

Thus T is <u>onto</u> if rank(T) = dim(W) and T is **1-1** if nullity(T) = 0

4.4 Linear Operators

If $T|V \rightarrow V$ such that T is linear, then T is a linear *operator* on V.

Standard matrix A for T is square: $[T(X)] = A[X] = [Y]$,

[X] and [Y] are coordinate matrices of X and Y

T is invertible iff A is nonsingular; then the standard matrix for T^{-1} is A^{-1}

Composition of operators: if S and T are two linear operators on V (with standard matrices A and B resp.) then $S \circ T (X) = S (T (X))$ where standard matrix of $S \circ T$ is AB

T(S(X)) has standard matrix BA; in general, composition is *NOT commutative*

Know <u>fundamental theorem</u>: if A is invertible then...

4.5 Changing Bases

Given vector space V with two bases: B(standard basis) and B'(non-standard).

If we wish to translate between B and B',
we use the matrix equation $P \ [X \ ]_{B'} = [X \ ]_B$.

transition matrix P	transition matrix Q = P^{-1}
$P \ [X \ ]_{B'} = [X \ ]_B$	$P^{-1} \ [X \ ]_B = [X \ ]_{B'}$
new to old	old to new

Suppose T(X) = Y; then $[T (X)]_B = [Y \ ]_B = A [X \ ]_B$. If the linear operator T is represented by two matrices A(wrt. standard basis B) and C(wrt. non-standard basis B') then A and C are **similar**, which means $C = P^{-1} A P$ and $[Y \ ]_{B'} = C \ [X \ ]_{B'}$.

Review Problems

Part A: Determine whether T is a linear transformation or not.

1. $T (x) = 2 x - 5 \quad T | \Re \rightarrow \Re$ ans. NOT

2. $T (x) = 2 x \quad T | \Re \rightarrow \Re$ ans. linear

3. $T (A) = \det (A) \quad T | M_{2,2} \rightarrow \Re$ ans. NOT

4. $T (A) = \begin{bmatrix} 1 & 0 \\ 0 & 2 \end{bmatrix} A \quad T | M_{2,2} \rightarrow M_{2,2}$ ans. linear

5. $T (x, y) = (x - y, x + y, 3 x) \quad T | \Re^2 \rightarrow \Re^3$ ans. linear

6. [T(X)] = A[X] where $A = \begin{bmatrix} 1 & 2 \\ 2 & -1 \end{bmatrix}$ and $T | \Re^2 \rightarrow \Re^2$ ans. linear

Part B: Given the linear transformation T(x,y) = (x - 2y, 3y, y - 5x) find:

7. image of X = (1,2) ans. (-3,6,-3)

8. pre-image(s) of Y = (-5,12,-11) ans. X = (3,4)

9. standard matrix A for T so that $[T(X)] = A[X]$ ans. $A = \begin{bmatrix} 1 & -2 \\ 0 & 3 \\ -5 & 1 \end{bmatrix}$

10. rank(A) and nullity(A) ans. 2, 0
11. if T is 1-1 and/or onto ans. 1-1, not onto
12. ker(T) and rng(T) ans. ker(T) = {(0,0)}, rng(T) = span({1,0,-5),(0,1,-3)})

Part C: Given the linear transformation T(p) = dp/dx where $T \mid P_3 \rightarrow P_2$ with the standard bases for P3 and P2;

13. find the image of $p = -3 + 4x - 5x^2 + 6x^3$ ans. $4 - 10x + 18x^2$

14. find the pre-image(s) of $q = -4 + 12x - 30x^2$ ans. $-4x + 6x^2 - 10x^3 + C$

15. find the standard matrix A for T so that $[T(p)]_{B'} = A[p]_B$

ans. $A = \begin{bmatrix} 0 & 1 & 0 & 0 \\ 0 & 0 & 2 & 0 \\ 0 & 0 & 0 & 3 \end{bmatrix}$

16. find ker(T) and rng(T) ans. ker(T) = {a}, $rng(T) = P_2$

17. find rank(A) and nullity(A) ans. 3, 1
18. determine whether T is 1-1 and/or onto ans. NOT 1-1, onto

Part D: Given the linear transformation T(x,y,z) = (x - 2y, 3y - 4z)
19. Find the image of X = (1,2,3). ans. (-3,-6)
20. Find the pre-image(s) of Y = (5,6) ans. X = (9+8/3t,2+4/3t,t)

21. Find the standard matrix A for T. ans. $A = \begin{bmatrix} 1 & -2 & 0 \\ 0 & 3 & -4 \end{bmatrix}$

22. Find rank(A) and nullity(A). ans. 2, 1
23. Find ker(T) and rng(T). ans. {t(8/3,4/3,1)}, $\mathfrak{R}^2$
24. Determine if T is 1-1 and/or onto. ans. NOT 1-1, onto

Part E: Given $T \mid M_{2,2} \rightarrow \mathfrak{R}$ where T(M) = tr(M) where tr(M) = sum of entries on main diagonal of M;
25. Show T is linear.
26. Find ker(T) and nullity(T). ans. $ker(T) = \{ \begin{bmatrix} a & b \\ c & -a \end{bmatrix} \}, 3$

27. Find rng(T) and rank(T); show dim(ker(T)) + dim(rng(T)) = dim(dmn(T)).
ans. $rng(T) = \mathfrak{R}, 3 + 1 = 4$
28. Find the standard matrix A for T. ans. $A = \begin{bmatrix} 1 & 0 & 0 & 1 \end{bmatrix}$
29. Find rank(A) and nullity(A); show rank(A) + nullity(A) = number of columns of A.
ans. rank(A) = 1, nullity(A) = 3, 1 + 3 = 4
30. Show that T is NOT 1-1 but IS onto. ans. ker(T) is NOT trivial, $rng(T) = \mathfrak{R}$

Part F: Given the linear operator T(x,y,z) = 1/3(x+y+z,x+y+z,x+y+z);

31. find the standard matrix A for T ans. $A = \dfrac{1}{3}\begin{bmatrix} 1 & 1 & 1 \\ 1 & 1 & 1 \\ 1 & 1 & 1 \end{bmatrix}$

32. get rank(A) and nullity(A). ans. 1, 2
33. find ker(T) and rng(T). span{(-1,1,0),(-1,0,1)}, span{(1,1,1)}
34. determine if T is 1-1 and/or onto. ans. neither
35. Show that A is idempotent (ie. $A^2 = A$). NB. T represents a projection in 3-space of a vector onto the line L = {t(1,1,1)}. Show that if X lies on this line, then T(X) = X.

Part G: Given the linear operator T(x,y,z) = (2x-y,3x-2y,z);

36. Find the standard matrix for T. ans. $A = \begin{bmatrix} 2 & -1 & 0 \\ 3 & -2 & 0 \\ 0 & 0 & 1 \end{bmatrix}$

37. Find the rank and nullity of T. ans. 3,0
38. Determine if T is 1-1 and/or onto. ans. both
39. Show that T is its own inverse.

Part H: Given the linear operator $T(M) = CM$ where $C = \begin{bmatrix} 1 & 2 \\ 0 & 3 \end{bmatrix}$ and $T \vert M_{2,2} \to M_{2,2}$; using the standard basis for $M_{2,2}$;

40. find T(M) if $M = \begin{bmatrix} 0 & 1 \\ 2 & 3 \end{bmatrix}$ ans. $\begin{bmatrix} 4 & 7 \\ 6 & 9 \end{bmatrix}$

41. find the pre-image(s) of $N = \begin{bmatrix} 3 & 3 \\ 3 & 3 \end{bmatrix}$ ans. $M = \begin{bmatrix} 1 & 1 \\ 1 & 1 \end{bmatrix}$

42. find the standard matrix A for T so that $[T(M)]_B = A[M]_B$ ans. $A = \begin{bmatrix} 1 & 0 & 2 & 0 \\ 0 & 1 & 0 & 2 \\ 0 & 0 & 3 & 0 \\ 0 & 0 & 0 & 3 \end{bmatrix}$

43. find rank(A) and nullity(A) ans. 4, 0
44. find ker(T) and rng(T) ans. $\{\begin{bmatrix} 0 & 0 \\ 0 & 0 \end{bmatrix}\}$, $rng(T) = M_{2,2}$

45. determine whether T is 1-1 and/or onto ans. both

Part I: Given the linear operators S(x,y) = (-y,x) and T(x,y) = (0,y);
46. describe geometrically what S and T do to a typical vector in the plane ans. rotation of 90 degrees anti-clockwise, projection onto y-axis

47. find the standard matrices A and B for S and T, respectively ans. $A = \begin{bmatrix} 0 & -1 \\ 1 & 0 \end{bmatrix}$ $B = \begin{bmatrix} 0 & 0 \\ 0 & 1 \end{bmatrix}$

48. determine if S is 1-1 and/or onto ans. both
49. determine if T is 1-1 and/or onto ans. neither
50. find S(T(X)) and T(S(X)) ans. S(T(X)) = (-y,0), T(S(X)) = (0,x)
51. find $[S \circ T (X)]$ and $[T \circ S (X)]$ using A and B ans. (-y,0), (0,x)

52. find ker(S) and rng(S) ans. $\{(0,0)\}, \Re^2$
53. find ker(T) and rng(T) ans. $\{(t,0)\}, \{(0,t)\}$
54. find $ker(S \circ T)$ and $rng(S \circ T)$ ans. $\{(t,0)\}, \{(t,0)\}$
55. find $ker(T \circ S)$ and $rng(T \circ S)$ ans. $\{(0,t)\}, \{(0,t)\}$

56. find S(S(X)) and T(T(X)) and their corresponding standard matrices
57. describe geometrically what the operators in #56 do to a typical vector in the plane
ans. rotation of 180 degrees anticlockwise, projection onto y-axis
58. Find S^{-1} and T^{-1} if possible; check your answer(s) via composition. ans. $S^{-1}(x, y)=(y, - x)$,
T has no inverse

Part J. Given the linear operator T(x,y) = (x+y,4x+y);

59. find the standard matrix A for T ans. $A = \begin{bmatrix} 1 & 1 \\ 4 & 1 \end{bmatrix}$

60. if B' = $\{(1,-2),(1,2)\}$ find the transition matrix P from B' to B ans. $P = \begin{bmatrix} 1 & 1 \\ -2 & 2 \end{bmatrix}$

61. find the transition matrix Q from B to B'. ans. $Q = \frac{1}{4} \begin{bmatrix} 2 & -1 \\ 2 & 1 \end{bmatrix}$

62. find the matrix A' that represents T wrt. the new basis B'. ans. $A' = \begin{bmatrix} -1 & 0 \\ 0 & 3 \end{bmatrix}$

Part K (understanding the dimension theorem): Let $A = \begin{bmatrix} 1 & 2 & 3 \\ 2 & 4 & 6 \\ 3 & 6 & 9 \end{bmatrix}$ where A is the standard matrix

of T where $T | \Re^3 \to \Re^3$;
63. Find ker(T) and show it is 2-dimensional. ans. ker(T) = span($\{(-2,1,0),(-3,0,1)\}$)
64. Using the dimension theorem, what is dim(rng(T))? ans. 3 - 2 = 1
65. Put A in row-echelon form; show row1 only remains and col1 is the only pivot column. Then
show that T(row1) is a basis for rng(T) which spans the same subspace as
span{col 1 of A}. ans. T(1,2,3) = (14,28,42) = 14 col 1
66. Show that if K = basis of ker(T) and P = {row1}, then K and P are disjoint and together form a
basis for V.
67. Let X_p be a vector in span(P); show T(X_p) is in rng(T). Let X_k be a vector in ker(T); show
T(X_k) = θ. Where does T($X_k + X_p$) lie? ans. in rng(T)

Part L: Given the transformation $T | \Re^3 \to P_2$ $T(a,b,c,)= a + b x + c x^2$;
68. Find the standard matrix A of T.
69. Show rank(A) = 3 and nullity(A) = 0.

70. Prove that T is both one-to-one and onto. Here T is called an *isomorphism*; it shows that $\Re^3$ and P_2 are structurally the same.

71. Let $X_1 = (2,3,4)$ $X_2 = (-5,6,7)$ and let $T(X_1) = p_1$ $T(X_2) = p_2$; show $T(X_1 + X_2) = p_1 + p_2$ and $T(6X_1) = 6p_1$.

Part M: Let L = {T: $T|V \to W$, T linear} where V and W are fixed vector spaces; define $(T_1 + T_2)(X) \equiv T_1(X) + T_2(X)$ and $(kT)(X) \equiv kT(X)$, show:

72. Vector space axioms 1,2, & 3 hold for L.

73. Vector space axioms 4 & 5 hold for L. Thus *L itself* is a *vector space*, the set of all linear transformations from V into W. This is not a surprise, since L is isomorphic to the set of all m x n matrices (ie. the standard matrices of all transformations in L).

CHAPTER 5

ORTHOGONALITY

CHAPTER FIVE OVERVIEW

In this chapter we take the concept of vector space further- we add a geometric element to it which enables us to answer questions like what is the **length** of a vector and what is the **angle** between two vectors, in particular when are two vectors **orthogonal** (ie. perpendicular). Again our model is the familiar *dot product* (or scalar product) from 2-space or 3-space; this scalar product has 4 basic properties which we take as our axioms for the general concept called an **inner product**. In section 2 we will show that we can define an inner product on many of our familiar vector spaces (like polynomials or continuous functions). This will enable us to do calculations on polynomials or continuous functions like we do in 2-space; find the length of a vector, find the angle between two vectors or even project one vector onto another.

In section 3 we discover why a basis consisting of *mutually perpendicular unit vectors* is so convenient; in section 4 we show how to take an arbitrary basis and force it to be orthogonal.

In Section 5 we learn how the concept of orthogonality allows us to find the best possible solution to a system which is *inconsistent*- and this has a lot of practical uses. Finally we study how the concept of an *infinite orthogonal basis* can be used in the space of continuous functions to find the best trigonometric representative of a particular continuous function; this leads into the realm of Fourier series which also has a host of practical applications (like approximating solutions to partial differential equations).

5.1 Scalar Product

Euclidean n-space

Consider $V = \Re^2$; the reader is probably familiar with the Pythagorean theorem. If $X = (\ x_1, x_2\ )$ then we know geometrically that the *length* of the vector X (ie. the hypotenuse of the corresponding right triangle) is given by

$$\|X\| = \sqrt{x_1^2 + x_2^2}$$

where the notation $\|X\|$ means the **norm** (geometrically the length) of the vector X.

This function (ie.the **norm** of X) has three properties:

1) $\|X\| \geq 0\ and\ \|X\| = 0\ iff\ X = \theta$ (positivity)

2) $\|kX\| = |k|\|X\|$ (homogeneity)

3) $\|X + Y\| \leq \|X\| + \|Y\|$ (triangle inequality)

This can be generalized to n-dimensional space in a natural fashion, where for $X = (\ x_1, x_2, ..., x_n\ ) \in \Re^n$, the norm is given by

$$\|X\| = \sqrt{x_1^2 + x_2^2 + ... + x_n^2}$$

and has the same three properties given above. The vector space $\Re^n$ together with the norm described above is called **n-dimensional Euclidean space**[1]. It is one example of a *normed vector space*. We would like to do more than just be able to find the length of a vector though- we would like to find the angle between two vectors, we would especially like to be able to determine if two vectors are orthogonal. This will be accomplished by the scalar (dot) product.

The Scalar Product

Given two vectors in n-space, we can multiply the two vectors and get a scalar by simply *multiplying corresponding components and adding*:

$$X = (\ x_1, x_2, ..., x_n\ ), Y = (\ y_1, y_2, ..., y_n\ ) \Rightarrow X \cdot Y = x_1 y_1 + x_2 y_2 + ... + x_n y_n = \sum_{i=1}^{n} x_i\ y_i$$

This is called the **scalar product** (or dot product) of the two vectors X and Y. It is a **real number** and can be positive, negative or zero. You might be familiar with the dot product from a physics course or some other math course. It is easy to prove the following properties of the scalar product, simply by writing out the components and carrying out the arithmetic:

---

[1] Here I follow the definition given in G. Simmon's *Topology and Modern Analysis*

PROPERTY	NAME
1) $X \cdot Y = Y \cdot X$	symmetry
2) $(X + Y) \cdot Z = X \cdot Z + Y \cdot Z$	additivity
3) $(kX \cdot Y) = k(X \cdot Y)$	homogeneity
4) $X \cdot X \geq 0$ and $X \cdot X = 0$ iff $X = \theta$	positivity

The scalar product allows us to do more *geometry* than we can in just a vector space. For example, we can now define **length** in terms of the scalar product and then get the **distance** between two vectors, and also the **angle** between two vectors. Suppose we dot a vector X in n-space with itself:

$$X = (x_1, x_2, ..., x_n) \Rightarrow X \cdot X = x_1^2 + x_2^2 + ... + x_n^2$$

The number on the right is the length of X squared (in 2-space this is just the *Pythagorean theorem*) so we have:

$$X \cdot X = x_1^2 + x_2^2 + ... + x_n^2 \Rightarrow \|X\| = \sqrt{x_1^2 + x_2^2 + ... + x_n^2}$$

where $\| X \|$ is the **norm** (magnitude or length) of the vector X.

EXAMPLE 1 Given X = (1,2,3,4) find $\|X\|$ Dotting X with itself:

$$X \cdot X = 1^2 + 2^2 + 3^2 + 4^2 \Rightarrow \|X\| = \sqrt{X \cdot X} = \sqrt{30}$$

DEFINITION

The **distance** between two vectors $X = (x_1, x_2, ..., x_n)$ and $Y = (y_1, y_2, ..., y_n)$ is given by

$$d(X,Y) = \|X - Y\| = \|(x_1 - y_1, x_2 - y_2, ..., x_n - y_n)\| =$$

$$\sqrt{(X - Y) \cdot (X - Y)} = \sqrt{(x_1 - y_1)^2 + (x_2 - y_2)^2 + ... + (x_n - y_n)^2}$$

EXAMPLE 2 Given X = (1,2,3,4) and Y = (5,6,-2,-3) find d(X,Y).

$$d(X,Y) = \|X - Y\| = \|(1 - 5, 2 - 6, 3 - (-2), 4 - (-3))\| = \sqrt{(-4)^2 + (-4)^2 + 5^2 + 7^2} = \sqrt{106}$$

Suppose neither X nor Y is the zero vector; if we want to find the angle[2] between two vectors, we can also use the scalar product:

$$\cos(\phi) = \frac{X \cdot Y}{\|X\|\|Y\|} \qquad 0 \le \phi \le \pi$$

The most important special case of angle, we say that X and Y are **orthogonal** if $\phi = \frac{\pi}{2}$. But this means that $\cos(\phi) = \cos(\frac{\pi}{2}) = 0$ so

$$\cos(\phi) = \frac{X \cdot Y}{\|X\|\|Y\|} = 0 \Rightarrow X \cdot Y = 0$$

ie. X dot Y must be zero if X and Y are orthogonal. Turning this around means that if $X \cdot Y = 0$ then $X \perp Y$ (ie. X and Y are orthogonal). By the way, the zero vector is orthogonal to every vector (and is the only vector orthogonal to itself).

EXAMPLE 3 Show that X = (1,2,0,-3) and Y = (1,1,5,1) are orthogonal.

Just dot X and Y: $X \cdot Y = 1 + 2 + 0 - 3 = 0$ thus $\cos(\phi) = 0 \Rightarrow \phi = \frac{\pi}{2}$.

EXAMPLE 4 Find all vectors X in 4-space that are orthogonal to Y = (1,2,3,4). Let $W = \{ X : X = (x_1, x_2, x_3, x_4) \}$ Then we must have:

$$(x_1, x_2, x_3, x_4) \cdot (1, 2, 3, 4) = 0$$

which means that $x_1 + 2x_2 + 3x_3 + 4x_4 = 0$. But this is a hyperplane through the origin and is therefore a *subspace* of $\Re^4$. We can find a basis for this subspace W:

$$x_1 = -2x_2 - 3x_3 - 4x_4 \Rightarrow X = (x_1, x_2, x_3, x_4) = (-2x_2 - 3x_3 - 4x_4, x_2, x_3, x_4)$$

$$X = (-2x_2 - 3x_3 - 4x_4, x_2, x_3, x_4) = (-2x_2, x_2, 0, 0) + (-3x_3, 0, x_3, 0) + (-4x_4, 0, 0, x_4)$$

$$X = x_2(-2, 1, 0, 0) + x_3(-3, 0, 1, 0) + x_4(-4, 0, 0, 1)$$

Thus we've found a basis for the set of all vectors orthogonal to the given vector Y; let S = {(-2, 1,0, 0),(-3, 0, 1,0),(-4, 0,0,1)} Certainly S spans W and is also independent (this is easy to verify); thus it forms a *basis* for W. Since W has a basis of three vectors, it must be 3-dimensional (so it is a hyperplane in 4-space).

---

[2] The Cauchy-Schwarz inequality guarantees that this is well-defined.

One can prove that the set of vectors orthogonal to a given vector (or set of vectors) *always* will form a subspace. This concept is important enough to make an official definition.

Orthogonal Complements

Given a subspace W of a vector space V. The set of all vectors

DEFINITION

$$W^{\perp} = \{ \ X : X \in V, X \cdot Y = 0 \ for \ all \ Y \in W \ \}$$

is called the **orthogonal complement** of W (or W perp).

One can prove easily that W perp is actually a *subspace* of V. We need to show closure under addition and scalar multiplication.

Proof: 1) Let X_1 & X_2 be in $W^{\perp}$; then $X_1 \cdot Y = 0$ & $X_2 \cdot Y = 0$ for all Y in W. But

$$X_1 \cdot Y + X_2 \cdot Y = 0 = (\ X_1 + X_2 \ ) \cdot Y$$

so $X_1 + X_2 \in W^{\perp}$.

2) Now for scalar multiplication:
If $X_1 \cdot Y = 0$ then

$$k (\ X_1 \cdot Y \ ) = 0 \Rightarrow (\ k \ X_1 \ ) \cdot Y = 0 \Rightarrow k \ X_1 \in W^{\perp}$$

for any k which means any scalar multiple of a vector in $W^{\perp}$ is in $W^{\perp}$. Therefore $W^{\perp}$ is closed under both addition and scalar multiplication so it is a subspace of V. QED

EXAMPLE 5 Suppose we are given two vectors $X_1 = (1,0,1)$ and $X_2 = (0,2,-4)$; let W = span($\{ X_1, X_2 \}$). We know that this is a plane thorough the origin. Find the orthogonal complement $W^{\perp}$.

Let's let X = (x,y,z) be in $W^{\perp}$. Then $X \cdot X_1 = 0$ & $X \cdot X_2 = 0$; but this implies two equations in x,y and z:

$$(\ 1,0,1 \ ) \cdot (\ x, y, z \ ) = 0 \qquad (\ 0, 2, -4 \ ) \cdot (\ x, y, z \ ) = 0$$

So we have a linear system to solve:

$$x \ + \ z = 0$$

$$2 \ y - 4 \ z = 0$$

The augmented matrix is:

$$\begin{bmatrix} 1 & 0 & 1 & 0 \\ 0 & 2 & -4 & 0 \end{bmatrix} \sim \begin{bmatrix} 1 & 0 & 1 & 0 \\ 0 & 1 & -2 & 0 \end{bmatrix}$$

so that y = 2z and x = -z. If z = t, then X = (-t,2t,t); this is a line through the origin.

But what we have done is to find the null space of $A = \begin{bmatrix} 1 & 0 & 1 \\ 0 & 2 & -4 \end{bmatrix}$; null(A) must be *orthogonal* to the row space of A. We actually get more than we bargained for though, since we can show that we can write any vector in $\Re^3$ (in a unique fashion) as a sum of two vectors, one in W and a second in its orthogonal complement. The reason is that $\{ X_1, X_2 \}$ forms a basis for W and $X_3 = (-1,2,1)$ forms a basis for its complement; together $\{ X_1, X_2, X_3 \}$ is an independent set and must therefore span $\Re^3$.

THEOREM: For any matrix A, $(row(A))^\perp = null(A)$.

This theorem allows us to get the orthogonal complement of any vector subspace W of $\Re^n$ by just finding the null space of the matrix whose ROWS span the subspace W.

EXAMPLE 6 Find the orthogonal complement of W if W = span(S) where $S = \{(1,0,1,0),(0,1,1,1)\}$. Let $A = \begin{bmatrix} 1 & 0 & 1 & 0 \\ 0 & 1 & 1 & 1 \end{bmatrix}$ so W = row(A). If we solve $A X = \theta$ we will find null(A).

$$\begin{bmatrix} 1 & 0 & 1 & 0 & 0 \\ 0 & 1 & 1 & 1 & 0 \end{bmatrix} \Rightarrow x_2 + x_3 + x_4 = 0 \Rightarrow x_2 = -x_3 - x_4$$

so $x_2 = -s - t$. Using the first row of the augmented matrix, we have $x_1 + x_3 = 0 \Rightarrow x_1 = -x_3 = -s$ so null(A) = $\{(-s,-s-t,s,t)\}$. If we separate s and t and let X = $(-s,-s-t,s,t)$ then we have

$$X = s(-1,-1,1,0) + t(0,-1,0,1).$$

The reader can verify that $(-1,-1,1,0)$ and $(0,-1,0,1)$ are orthogonal to the rows of A. Thus $W^\perp$ = span($\{(-1,-1,1,0),(0,-1,0,1)\}$) which is a *plane* through the origin. In 4-space, the orthogonal complement of a plane is a plane!

Projection

Consider the problem of projecting a vector X onto a vector Y in 2-space. Let $X = (x_1, x_2)$, $Y = (y_1, y_2)$ be the two vectors involved, where we want to **project Y onto X**. Let's indicate this projection by $proj_X Y$; we must agree that this projection vector does two things:

1) it is a multiple of X so that $proj_X Y = k X = k (x_1, x_2)$

2) it is related to an orthogonal vector, call it Z, dropped from the head of Y so that $Z \cdot X = 0$ (see Fig. 1).

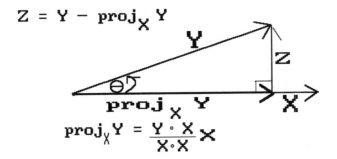

$$proj_X Y = \frac{Y \cdot X}{X \cdot X} X$$

By simple vector addition, $proj_X Y + Z = Y$ so that $Z = Y - proj_X Y$:

$$Z = Y - proj_X Y = (\ y_1, y_2\ ) - k(\ x_1, x_2\ ) = (\ y_1 - k\ x_1, y_2 - k\ x_2\ )$$

Now using 2) if Z is orthogonal to X we have:

$$(\ y_1 - k\ x_1, y_2 - k\ x_2\ ) \cdot (\ x_1, x_2\ ) = 0$$

This means that

$$(\ y_1 - k\ x_1\ )\ x_1 + (\ y_2 - k\ x_2\ )\ x_2 = 0 \Rightarrow y_1 x_1 - k\ x_1^2 + y_2 x_2 - k\ x_2^2 = 0$$

$$y_1 x_1 + y_2 x_2 = k(\ x_1^2 + x_2^2\ )$$

so we found k:

$$k = \frac{y_1 x_1 + y_2 x_2}{x_1^2 + x_2^2} = \frac{Y \cdot X}{X \cdot X}$$

and thus the formula for $proj_X Y$ (the projection of Y onto X):

$$proj_X Y = (\ \frac{Y \cdot X}{X \cdot X}\ )\ X = (\ \frac{Y \cdot X}{\|X\|^2}\ )\ X$$

projection of Y onto X

This formula is not difficult to remember if we think that $proj_X Y$ is a scalar multiple of X. It is the vector in span({X}) which is closest to Y. Clearly the length of the projection must be *less than* the length of the vector Y. Of course, this formula is valid in $\Re^n$; we did the work in 2-space for convenience.

We will generalize this procedure in the future, and actually be able to project one vector onto a subspace of V. This one simple idea has **many** important practical applications. See *Part G* in the exercises for a **calculus approach** to this formula.

EXAMPLE 7 Find the projection of the vector Y = (1,2,3) onto the vector X = (1,-1,1).
Using the above formula for the projection, we get:

$$proj_X Y = (\frac{Y \cdot X}{X \cdot X})X = \frac{1 \cdot 1 + 2 \cdot (-1) + 3 \cdot 1}{1^2 + (-1)^2 + 1^2}(1,-1,1) = \frac{2}{3}(1,-1,1) = (\frac{2}{3}, \frac{-2}{3}, \frac{2}{3})$$

Here the reader can verify that $Z = Y - proj_X Y$ = (1,2,3) - (2/3,-2/3,2/3) = (1/3,8/3,7/3) is really orthogonal to X = (1,-1,1). Always verify that Z is *orthogonal* after having found the projection. Clearly the *length* of the projection is always *less than the* length of the vector Y – see Fig. 1 presuming Y in not in span({X}).

EXERCISES 5.1

Part A. Given X = (1,2), Y = (3,-4) and k = -6, using the norm in $\Re^2$ given by $\|X\| = \sqrt{x_1^2 + x_2^2}$ show:
1. $\|kX\| = |k|\|X\|$.
2. $\|X\| \geq 0 \ and \ \|X\| = 0 \ iff \ X = \theta$.
3. $\|X + Y\| \leq \|X\| + \|Y\|$.
4. $\|X + Y\|^2 + \|X - Y\|^2 = 2\|X\|^2 + 2\|Y\|^2$.

Part B. Given X = (1,2,3), Y = (-4,0,5) & k = -10 show:
5. $\|kX\| = |k|\|X\|$
6. $\|X\| \geq 0 \ and \ \|X\| = 0 \ iff \ X = \theta$
7. $\|X + Y\| \leq \|X\| + \|Y\|$
8. $\|X + Y\|^2 + \|X - Y\|^2 = 2\|X\|^2 + 2\|Y\|^2$

Part C. Given X = (1,2,3,4) and Y = (1,-1,1,1) find:
9. $X \cdot Y$ ans. 6
10. $\|X\|, \|Y\|$
11. $\|X + Y\|$ ans. $\sqrt{46}$
12. $\|X + Y\|$

226

13. dist(X,Y) ans. $\sqrt{22}$

14. $\cos(\phi)$ where phi is the angle between X and Y

15. $proj_X Y$ ans. 1/5(1,2,3,4)

16. a vector Z perpendicular to X such that $proj_X Y + Z = Y$ (show your answer is correct by dotting Z with X)

17. $proj_Y X$ ans. 3/2(1,-1,1,1)

18. a vector Z perpendicular to Y such that $proj_Y X + Z = X$ (show your answer is correct by dotting Z with Y)

Part D. Lines, planes and hyperplanes: find (and describe geometrically)

19. set of all vectors X = (x,y,z) which are perpendicular to Y = (0,1,4). Find a basis for this subspace also. ans. y + 4z = 0, plane,{(1,0,0),(0,-4,1)}

20. set of all vectors X = (x,y,z) which are perpendicular to Y = (1,0,1) and Z = (2,-2,0)

21. set of all vectors X = (x,y,z,w) which are perpendicular to S = {(1,2,2,1)}. Find a basis for this subspace also. ans. x + 2y + 2z + w = 0 (hyperplane),
B = {(1,0,0,-1),(0,1,0,-2),(0,0,1,-2)}

22. set of all vectors X = (x,y,z,w) which are perpendicular to S = {(1,2,2,1),(1,1,-1,-1)}. Find a basis for this subspace also.

23. set of all vectors X = (x,y,z,w) which are perpendicular to
S = {(1,2,2,1),(1,1,-1,-1,),(2,3,1,2)} ans. X = t(4,-3,1,0), line

Part E. Given the subspace W, find W perp by using a matrix A whose row space = W;
verify that the subspaces are indeed perpendicular by dotting each vector in a basis for W with each vector in a basis for W perp.

24. W = span({(1,2)})

25. W = span({(1,2,3),(-1,-1,1)}) ans.(5,-4,1)

26. W = span({(1,0,1),(0,1,1)})

27. W = span({(1,2,3,4),(-4,-2,0,2)}) ans.{(1,-2,1,0),(2,-3,0,1)}

28. W = span({(1,0,2,3)}) ans. {(-2,0,1,0),(0,1,0,0),(-3,0,0,1)}

29. W = span({(1,0,1,0,1),(0,1,0,1,0), (2,2,2,2,2),(3,3,3,3,4)}) ans. {(0,-1,0,1,0),(-1,0,1,0,0)}

30. W = span({(1,0,1,0),(0,1,0,-1),(1,1,1,1)})

31. W = span({1,0,1,0),(0,1,1,1),(1,1,2,1)}) ans. {(-1,-1,1,0),(0,-1,0,1)}

32. W = span({(1,0,0,1,1),(0,1,2,-2,0), (1,1,2,-1,1)})

Part F. Given the matrix $A = \begin{bmatrix} 2 & 0 & 0 & 6 \\ 0 & 1 & 3 & 1 \\ 2 & 1 & 3 & 7 \end{bmatrix}$;

33. Using W = row(A), find $W^{\perp}$. Describe geometrically W and its orthogonal complement. ans. $W^{\perp}$ = span({(-3,-1,0,1),(0,-3,1,0)}), 2 orthogonal planes

34. Find null(A); show it is the same as $W^{\perp}$.

35. Find a basis for W; show each vector in this basis is orthogonal to the basis of $W^{\perp}$.
ans. S = {(1,0,0,3), (0,1,3,1)}

36. Using V = col(A), find $V^{\perp}$. Describe geometrically V and its orthogonal complement.

Part G: Projection and calculus; let X and Y be two unrelated given vectors in $\Re^2$;
if we project Y onto X, we produce a vector P of the form P = kX.

37. Let Z = Y - P; find the norm of Z and then square it.

ans. $\|Z\|^2 = \|Y\|^2 - 2k\,X \cdot Y + k^2 \|X\|^2$

38. Now differentiate both sides wrt. k (remember X and Y are constant vectors!).

39. Set this derivative equal to zero; solve for k. Claim that the k-value you found minimizes the

norm of Z squared and thus the norm of Z. ans. $k = \dfrac{X \cdot Y}{\|X\|^2}$

40. Verify this all works for X = (3,4) and Y = (5,7).

Part H. Let Y = (1,2,3);

41. Let W = span(1,0,0) and find the projection of Y onto W. ans. (1,0,0)

42. Let $Z = Y - proj_W Y$; show that Z is orthogonal to the projection of Y onto W.

43. Now let W = span({(1,0,0),(0,1,0)}); find the projection of Y onto W.
Hint: since the vectors given are orthogonal, simply project onto each basis vector and add.
ans. (1,2,0)

44. Again let $Z = Y - proj_W Y$; show that Z is orthogonal to the projection of Y onto W.

45. Find the magnitude of Z from #42 & #44; what is happening to these magnitudes as W gets larger?? Show that the magnitude of the projections found in #41 & #43 is increasing as W gets larger. Moral- as the dimension of W increases, the length of the projection **increases** while the length of Z *decreases*. ans. $\sqrt{13}, 3$

Part I. Given the line L = span({(1,-2,1)}); find a linear operator T which projects onto this line by:

46. Using the basic projection formula with Y = (x,y,z) and X = (1,-2,1); let T(x,y,z) = $proj_X$ Y.

47. By writing the image T(x,y,z) in column matrix format and separating x,y and z, find the

standard matrix A of T. ans. $A = \dfrac{1}{6}\begin{bmatrix} 1 & -2 & 1 \\ -2 & 4 & -2 \\ 1 & -2 & 1 \end{bmatrix}$

48. Show rank(T) < 3 and nullity(T) is NOT zero, so T is neither 1-1 nor onto.

49. Show that the matrix A here is idempotent (ie. $A^2 = A$).

50. Find all *fixed points* (ie. T(X) = X) of this transformation; show they form a subspace.
Hint: If [T(X)] = A[X] = [X] then A[X] - [X] = zero matrix so (A - I)[X] = zero matrix.

Part J. Given $X = (x_1, x_2)$, $Y = (y_1, y_2)$ in $\Re^2$ using the scalar product;

51. Show that $\|X + Y\|^2 + \|X - Y\|^2 = 2\|X\|^2 + 2\|Y\|^2$ for any X and Y in $\Re^2$; this is called the *parallelogram law*.

52. Show $\|X - Y\|^2 = \|X\|^2 + \|Y\|^2 - 2\|X\|\|Y\|\cos(\phi)$. Hint: expand $(X - Y) \cdot (X - Y)$.

5.2 Inner Products

Introduction- Normed Vector Spaces

Consider $V = \Re^2$ and given X = (x,y), we know from Section 5.1 that we can find the length of the vector X using $\|X\| = \sqrt{x^2 + y^2}$. This is just the Pythagorean theorem in the plane. We can think of "length" as a function on V of the form $\|\cdot\| | V \to \Re$. This function has three important properties:

N1) $\|X\| \geq 0$ & $\|X\| = 0$ iff $X = \theta$

N2) $\|kX\| = |k| \|X\|$

N3) $\|X + Y\| \leq \|X\| + \|Y\|$ (triangle inequality)

These are easy to prove in $\Re^2$; for example, to prove 2):

$$\|kX\| = \|(kX, kY)\| = \sqrt{(k\,x\,)^2 + (k\,y\,)^2} = \sqrt{k^2\,x^2 + k^2\,y^2} = \sqrt{k^2}\sqrt{x^2 + y^2} = |k|\|(X,Y)\| = |k|\|X\|$$

Given an arbitrary vector space V, we would like to be able to define something analogous to this "length function", which is called a norm. A norm is a generalization of the length function on $\Re^2$. Just like we created vector spaces by making axioms out of important properties of vectors in 2-space, we now can do the same thing with the properties of length in 2-space.

DEFINITION	Given a vector space V and a function $\|\cdot\| | V \to \Re$; Then $\|\cdot\|$ is a **norm** on V if for all X,Y in V and real k:

1) $\|X\| \geq 0$ & $\|X\| = 0$ iff $X = \theta$

2) $\|kX\| = |k| \|X\|$

3) $\|X + Y\| \leq \|X\| + \|Y\|$ (triangle inequality)

For example, if we let V = C[a,b] then we can define a norm[3] on V by letting

$$\| f \| = \max_{x \in [a,b]} |f(\,x\,)|$$

Proving all three axioms is not easy but the first one is not difficult to see- by taking the absolute value of a function f on [a,b], its graph cannot dip below the x-axis (so $|f| \geq 0$) and by a theorem from calculus, every continuous function has a maximum value on a closed interval.

Consider for example the funtion f(x) = $x^2 - 2x$ in V = C[0,3]; below is the graph of its absolute value:

---

[3] This type of norm is important in a branch of mathematics called *functional analysis*.

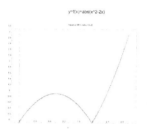
y=f(x)=abs(x^2-2x)

Clearly the absolute max of abs(f(x)) = 3 (which occurs when x = 3) so we write $\| f \| = 3$.
Alternatively, one may define a norm on C[0,3] by:

$$\| f \| = \int_0^3 | f | \, d x$$

It is not difficult to show that this definition satisfies all three axioms for an inner product. Note that in this IPS, the function given above has a different value for the norm:

$$\| f \| = \int_0^3 | x^2 - 2 x | \, d x = \frac{8}{3}$$

If a vector space V has a norm defined on it, it is called a **normed vector space** (NVS). In a NVS we now can find the **length** of a vector or the distance between two vectors. The distance is defined simply by

$$d (X, Y) = \| X - Y \|$$

However in a NVS we still cannot determine the *angle* between two vectors. The most important angle is $\frac{\pi}{2}$ - in other words, we'd like to be able to determine when two vectors X and Y are orthogonal (ie. perpendicular). So we need something else- something like the scalar product in n-space. We cannot call it a dot product, since this refers specifically to $\mathfrak{R}^n$ so we call this function an inner product. It must have all of the properties of the scalar product defined on n-space. The definition is given below.

DEFINITION

An **inner product** is a function $f | V \, x V \to \mathfrak{R}$ defined on a vector space V where we write <X,Y> to mean the **value** of the function f at the two vectors X and Y, ie. f(X,Y) = <X,Y>. The function f must satisfy four axioms:

IP 1) <X,Y> = <Y,X> (symmetry)
IP 2) <X + Y,Z> = <X,Z> + <Y,Z> (additivity)
IP 3) <kX,Y> = k<X,Y> (homogeneity)
IP 4) $< X, X > \geq 0$ and $< X, X > = 0$ *iff* $X = \theta$ (positivity)

Note that these are precisely the properties of the scalar product in n-space. It is important to note that *every inner product gives rise to a norm* by letting $\|X\| = \sqrt{< X, X >}$.

Once we have a vector space V with an inner product, we have what is called an inner product space.

DEFINITION

An **inner product space**(IPS) is a vector space on which is defined an inner product.

By the above comment, every IPS is also a NVS.

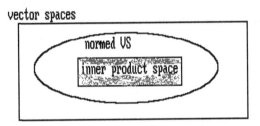

EXAMPLE 1 Let V = C[a,b] and define $< f, g > = \int_a^b f \, g \, d x$. Certainly it makes sense that the integral will produce a real number since continuity implies the integral exists. Show that this definition satisfies all the axioms for an inner product.

$$< f, g > = \int_a^b f \, g \, d x = \int_a^b g \, f \, d x = < g, f > \qquad (1)$$

$$< f + g, h > = \int_a^b (f + g) h \, d x = \int_a^b f \, h \, d x + \int_a^b g \, h \, d x = < f, h > + < g, h > \qquad (2)$$

$$<k\,f,g\,> = \int_a^b k\,f\,g\ dx = k \int_a^b f\,g\ dx = k < f,g\,> \qquad (3)$$

$$<f,f\,> = \int_a^b f \cdot f\ dx = \int_a^b f^2\ dx \qquad (4)$$

Clearly $\int_a^b f^2\ dx \geq 0$ since $f^2(x) \geq 0$; also

$$\int_a^b f^2\ dx = 0$$

only if f = 0 since f is continuous. Thus the vector space C[a,b] together with the integral inner product $<f,g\,> = \int_a^b f\,g\ dx$ forms an *inner product space*.

EXAMPLE 2 Consider the function

$$<A,B\,> = tr(A\,B^t) = a_1 a_2 + b_1 b_2 + c_1 c_2 + d_1 d_2$$

if $A = \begin{bmatrix} a_1 & b_1 \\ c_1 & d_1 \end{bmatrix}, B = \begin{bmatrix} a_2 & b_2 \\ c_2 & d_2 \end{bmatrix}$ defined on the vector space of all 2 by 2 matrices $V = M_{2,2}$.
axiom 1) Now

$$A\,B^t = \begin{bmatrix} a_1 & b_1 \\ c_1 & d_1 \end{bmatrix}\begin{bmatrix} a_2 & c_2 \\ b_2 & d_2 \end{bmatrix} = \begin{bmatrix} a_1 a_2 + b_1 b_2 & a_1 c_2 + b_1 d_2 \\ c_1 a_2 + d_1 b_2 & c_1 c_2 + d_1 d_2 \end{bmatrix}$$

so $tr(A\,B^t) = a_1 a_2 + b_1 b_2 + c_1 c_2 + d_1 d_2$ and the definition makes sense as stated. Not trying to be overly slick, one may remember that $(C\,D)^t = D^t\,C^t$. Using this idea now, we have $(A\,B^t)^t = (B^t)^t A^t = B\,A^t$ and also by definition $<B,A> = tr(B\,A^t)$. Since $tr(G) = tr(G^t)$ we must have $tr(B\,A^t) = tr(A\,B^t) = <A,B>$ so the function is symmetric. I'll leave axioms 2 and 3 to you and prove axiom 4. Calculating <A,A>:

$$A\,A^t = \begin{bmatrix} a_1^2 + b_1^2 & a_1 c_1 + b_1 d_1 \\ c_1 a_1 + d_1 b_1 & c_1^2 + d_1^2 \end{bmatrix} \Rightarrow tr(A\,A^t) = a_1^2 + b_1^2 + c_1^2 + d_1^2$$

But $tr(A\,A^t)$ cannot be negative so $<A,A> = tr(A\,A^t) \geq 0$ and
clearly <A,A> = 0 iff $A = \theta$.
We can now find the length of a matrix, the distance between two matrices, or the angle between two matrices. This may sound peculiar, but sometimes you have to divorce yourself from geometric

232

intuition!

EXAMPLE 3 Consider the determinant function on the vector space of all 2 by 2 matrices $V = M_{2,2}$. Let's try $<A, B> = \det(AB)$; certainly $\det | V \times V \to \Re$ so again we are at least in the ball park. Now for the four axioms:

1)

$$<A, B> = \det(AB) = \det(A)\det(B) = \det(B)\det(A) = <B, A> \tag{1}$$

so symmetry holds.

$$<A + B, C> = \det((A + B)C) = \tag{2}$$

$$\det(AC + BC) \neq \det(AC) + \det(BC) = <A, C> + <B, C>$$

so the determinant function does NOT have the additivity property so it was a good try but it cannot be an inner product.

Now that we know what an inner product space is, we want to know what good does it do us to have an IPS? Since every IPS is a NVS, we can find the length of a vector or the distance between two vectors. Compare and contrast - in an IPS we will also be able to find the *angle* between two vectors; in particular we will be able to figure out if two vectors are orthogonal to each other. Let's make the definitions for these concepts now.

DEFINITION

1) Length (magnitude or norm): $\|X\| = \sqrt{<X, X>}$

2) Distance $d(X, Y) = \|X - Y\| = \sqrt{<X - Y, X - Y>}$

3) Angle:

$$\cos(\phi) = \frac{<X, Y>}{\|X\|\|Y\|} \quad 0 \leq \phi \leq \pi$$

This angle is well-defined due to the *Cauchy-Schwarz* inequality, which says that in any inner product space,

$$|<X, Y>| \leq \|X\|\|Y\|$$

Thus the fraction on the right in 3) will always be between -1 and 1 (ie. a good cosine number).

4) Orthogonality:

$$X \perp Y \text{ iff } <X, Y> = 0$$

since that means that $\phi = \frac{\pi}{2}$. Therefore if the inner product of X with Y is zero, X is **orthogonal** to Y. Believe it or not, this idea is the essence of a lot of applications of inner product spaces (eg. least squares, Fourier series).

EXAMPLE 4 Find the norm of the matrix $A = \begin{bmatrix} 1 & 2 \\ 3 & 4 \end{bmatrix}$ using $<A, B> = tr(A B^t)$.

We need to multiply A times its transpose:

$$A A^t = \begin{bmatrix} 1 & 2 \\ 3 & 4 \end{bmatrix} \begin{bmatrix} 1 & 3 \\ 2 & 4 \end{bmatrix} = \begin{bmatrix} 5 & 11 \\ 11 & 25 \end{bmatrix}$$

Thus

$$tr\left(\begin{bmatrix} 5 & 11 \\ 11 & 25 \end{bmatrix}\right) = 5 + 25 = 30 \Rightarrow \|A\| = \sqrt{<A, A>} = \sqrt{30}$$

EXAMPLE 5 Find the norm of the function f(x) = 2x + 1 in the inner product space C[0,1] where $<f, g> = \int_0^1 f \, g \, d x$.

We use

$$\|f\| = \sqrt{<f, f>} = \sqrt{\int_0^1 f^2 \, d x}$$

Here then we can find the length of the function f with respect to the given inner product:

$$\|f\| = \sqrt{<f, f>} = \sqrt{\int_0^1 (2x+1)^2 \, d x} = \sqrt{\int_1^3 \frac{1}{2} u^2 \, d u} = \sqrt{\frac{1}{2}[\frac{u^3}{3}]_1^3} = \sqrt{39}/3$$

EXAMPLE 6 Find the distance between the two functions $f = x, \ g = x^2$ using the integral inner product $<f, g> = \int_0^1 f \, g \, d x$. The distance between the two is the norm of the difference of the two functions:

$$d(f, g) = \| <f - g, f - g> \| = \sqrt{\int_0^1 (f - g)^2 d x} = \sqrt{\int_0^1 (x - x^2)^2 d x} \Rightarrow$$

$$d(f, g) = \sqrt{\int_0^1 (x^2 - 2 x^3 + x^4) d x} = \sqrt{[\frac{x^3}{3} - 2\frac{x^3}{4} + \frac{x^5}{5}]_0^1} = \sqrt{\frac{1}{3} - \frac{1}{2} + \frac{1}{5}} = \frac{1}{\sqrt{30}}$$

EXAMPLE 7 Find the angle between the two functions $f = x, \ g = x^2$ using the IP

$$<f, g> = \int_0^1 f \, g \, d x$$

Recall that

$$\cos(\phi) = \frac{<f,g>}{\| f \| \| g \|}$$

So

$$\cos(\phi) = \frac{<f,g>}{\|f\|\|g\|} = \frac{<x,x^2>}{<x,x><x^2,x^2>} = \frac{\int_0^1 x \cdot x^2 \, dx}{\int_0^1 x^2 \, dx \sqrt{\int_0^1 x^4}} = \frac{\frac{1}{4}}{\sqrt{\frac{1}{3} \cdot \frac{1}{5}}} = \frac{\sqrt{15}}{4}$$

Therefore

$$\phi = \text{Cos}^{-1}(\frac{\sqrt{15}}{4}) \approx 0.2527$$

Properties of an inner product

Let V be an IPS, then:

1) $\|X + Y\| \leq \|X\| + \|Y\|$ (triangle inequality)

Proof: By Cauchy-Schwarz we know that

$$<X,Y> = \|X\| \|Y\| \cos(\phi) \Rightarrow <X,Y> \leq \|X\| \|Y\|$$

since the cosine function is always less than or equal to one.
Thus clearly

$$2 <X,Y> \leq 2\| X \| \| Y \|$$

Certainly

$$\left\| X \right\|^2 + \left\| Y \right\|^2 = \| X \|^2 + \| Y \|^2$$

Adding implies:

$$\left\| X \right\|^2 + 2 <X,Y> + \left\| Y \right\|^2 \leq \| X \|^2 + 2\|X\| \|Y\| + \left\| Y \right\|^2$$

But the left hand side is just <X+Y,X+Y> and the right hand side is:

$$(\left\| X \right\| + \left\| Y \right\|)^2$$

Thus we have:

$$\left\|X + Y\right\|^2 \leq (\left\|X\right\| + \left\|Y\right\|)^2$$

Since norms cannot be negative finally we have:

$$\left\|X + Y\right\| \leq \left\|X\right\| + \left\|Y\right\|$$

which is the triangle inequality.   QED

2) $\left\|X + Y\right\|^2 = \left\|X\right\|^2 + \left\|Y\right\|^2$ iff $<X,Y> = 0$ (Pythagorean theorem)

This is also a nice verification of the orthogonality of X and Y.

EXAMPLE 8 Use the inner product $<f, g> = \int_0^2 f\, g\, d\,x$ to show that prop(1) is true for $f = 2 - x$, $g = x^2$ in the IPS C[0,2].

$$f + g = (2 - x) + x^2 \Rightarrow \left\|f + g\right\| = \sqrt{\int_0^2 (2 - x + x^2)^2 d\,x} = \sqrt{\frac{176}{15}}$$

Now find the norms of each separately:

$$\left\|2 - x\right\| = \sqrt{\int_0^2 (2 - x)^2 d\,x} = \sqrt{\frac{8}{3}} \quad \text{and} \quad \left\|x^2\right\| = \sqrt{\int_0^2 (x^2)^2 d\,x} = \sqrt{\frac{32}{5}}$$

One can verify that

$$\sqrt{\frac{176}{15}} \leq \sqrt{\frac{8}{3}} + \sqrt{\frac{32}{5}} \qquad 3.4254 \leq 4.1628$$

NB. f and g were NOT orthogonal in this example

EXAMPLE 9 Show that prop(2) works for f = 1 and g = sin(x) in the IPS $C\,[-\pi, \pi\,]$ with inner product given by $<f,g> = \int_{-\pi}^{\pi} f\, g\, d\,x$.

$$\left\|\,1\,\right\| = \sqrt{<1\ 1>} = \sqrt{\int_{-\pi}^{\pi} [\ 1^2\ ]\, d\,x} = \sqrt{2\,\pi}$$

$$\left\|\sin(x)\right\| = \sqrt{< \sin(x), \sin(x)>} = \sqrt{\int_{-\pi}^{\pi} [\ \sin^2(x)\,]\, d\,x} = \sqrt{\pi}$$

$$\| \ 1 + \sin(x) \ \| = \sqrt{< 1 + \sin(x), 1 + \sin(x) >} = \sqrt{\int_{-\pi}^{\pi} [\ 1 + 2\sin(x) + \sin^2(x)\ ]\ dx} = \sqrt{3\pi}$$

You can verify then that

$$(\sqrt{2\pi} \ )^2 + (\sqrt{\pi} \ )^2 = 2\pi + \pi = (\sqrt{3\pi} \ )^2$$

Here f and g are like the legs of a right triangle (and 1 + sin(x) is the hypotenuse).
This all works since

$$< f, g > = \int_{-\pi}^{\pi} f\ g\ dx = \int_{-\pi}^{\pi} 1 \cdot \sin(x)\ dx = 0$$

ie. f and g were orthogonal wrt. the given inner product.

EXERCISES 5.2

Part A. Given the Euclidean norm on $\Re^2$, $\| \ (x, y) \ \| = \sqrt{x^2 + y^2}$ where X = (1,2),
Y = (3,4) & k = 5:

1. Find $\| \ X \ \| \ \& \|Y \|$. ans. $\| \ X \ \| = \sqrt{5},\ \| \ Y \ \| = 5,$

2. Show $\|kX\| = |k| \|X\|$.

3. Show $\|X + Y\| \le \|X\| + \|Y\|$. ans. $\| \ X + Y \ \| = 2\sqrt{13}$

4. Show $\| \ X + Y \ \|^2 + \| \ X - Y \ \|^2 = 2\| \ X \ \|^2 + 2\| \ Y \ \|^2$.

Part B. Given V = C[0,2]; define a norm on V by letting $\| \ f \ \| = \max_{x \in [0,2]} |f(x)|$. If
f = x(x - 2), g = 2x - 1 and k = 5 (NB. Graphs would be helpful here!):

5. Find $\| \ f \ \| \ \& \|g \|$. ans. $\| \ f \ \| = 1,\ \| \ g \ \| = 3$

6. Show $\|kX\| = |k| \|X\|$.

7. Show $\|f + g\| \le \|f\| + \|g\|$. ans. $\| \ f + g \ \| = 3$

8. Show that the parallelogram law $\| \ f + g \ \|^2 + \| \ f - g \ \|^2 = 2\| \ f \ \|^2 + 2\| \ g \ \|^2$ does NOT
hold here.

Part C. Given $V = P_1 = \{ a + bx : a, b \in \Re \}$ with a function on V where
$< p, q > = a_0 a_1 + b_0 b_1$ if $p = a_0 + b_0 x,\ q = a_1 + b_1 x$; show that this definition gives rise to an
inner product on V by proving:

9. axiom 1

10. axiom 2

11. axiom 3

12. axiom 4

Part D. Using the inner product defined in part C with p = 1 + 3x, q = -4 + 5x find:

237

13. <p,q> ans. 11

14. $\| p \|$, $\| q \|$

15. dist(p,q) ans. $\sqrt{29}$

16. cos(phi) between p and q

17 a function r in P_1 such that r is orthogonal to p ans. r = b(-3 + x)

18. two functions formed from p and q which are unit vectors

Part E. Given two matrices $A = \begin{bmatrix} a & b \\ c & d \end{bmatrix}$, $B = \begin{bmatrix} e & f \\ g & h \end{bmatrix}$ from the vector space V $M_{2,2}$; show that

$< A, B > = a e + b f + c g + d h$ gives an inner product for V by proving:

19. axiom 1

20. axiom 2

21. axiom 3

22. axiom 4

Part F. Use the inner product given in part E to find the requested quantity if

$A = \begin{bmatrix} 1 & 2 \\ 3 & 4 \end{bmatrix}$, $B = \begin{bmatrix} -2 & 1 \\ -4 & 4 \end{bmatrix}$, $C = \begin{bmatrix} -1 & 1 \\ 1 & -1 \end{bmatrix}$;

23. <A,B> ans. 4

24. <A, B + C>

25. <A,B> + <A,C> ans. 4

26. $\| A \|$

27. a vector formed from A whose norm is one. ans. $\dfrac{1}{\sqrt{30}} A$

28. d(A,B)

29. angle ϕ between A and B ans. $\phi = \text{Cos}^{-1} (\dfrac{4}{\sqrt{30} \sqrt{37}})$

30. determine which pair of matrices from {A,B,C} is orthogonal

31. show that $\|A + B\| \le \|A\| + \|B\|$ (this is the *triangle inequality*, which holds in any inner product space)

32. show that $|< A, B >| \le \| A \| \| B \|$ (this is the *Cauchy-Schwarz* inequality)

33. Find a 2 by 2 matrix D which is orthogonal to A,B and C simultaneously.

ans. $D = d \begin{bmatrix} -5 & -6 \\ 3 & 2 \end{bmatrix}$

Part G. Given the set of functions $S = \{ f_1 = 1, f_2 = x, f_3 = x^2 \} \subseteq C[-1,1]$ with inner product

$< f, g > = \displaystyle\int_{-1}^{1} f \, g \, d x$ find the requested quantity;

34. <f₁,f₂>

35. <f₂,f₃> ans. 0

36. <f₁,f₃>

37. Is every function in S orthogonal to the others? ans. NO; f1 & f3 are not orthogonal

38. Find the magnitude of each function in S; which is the largest?

39. Find the cosine of the angle ϕ between f1 and f3. ans. $\dfrac{\sqrt{5}}{3}$

40. show that $\|f_1 + f_2\| \le \|f_1\| + \|f_2\|$ (this is the *triangle* inequality)

41. show that $|< f_1, f_3 >| \le \| f_1 \| \| f_3 \|$ (this is the *Cauchy-Schwarz* inequality)

42. find d(f1,f2)

43. find d(f2,f3) ans. $\dfrac{4\sqrt{15}}{15}$

44. Show d(f1,f3) $\le$ d(f1,f2) + d(f2,f3).

45. Find a polynomial p of degree three which is simultaneously orthogonal to f1, f2 and f3. Hint: What should the inner product of p with respect to f1,f2 and f3 be?

ans. $p = c(5x^3 - 3x)$

Part H. Given the set $S = \{ f_1 = 1, f_2 = \cos(x), f_3 = \sin(x) \} \subseteq C[-\pi, \pi]$ with inner product

$$< f, g > = \int_{-\pi}^{\pi} f\, g\, dx;$$

46. Show that the functions in S are pairwise orthogonal (you have 3 integrals to do!).

47. Find the magnitude of each function in S. ans. $\| f_1 \| = \sqrt{2\pi},\ \| f_2 \| = \| f_3 \| = \sqrt{\pi}$

48. Find d(1,cos(x))

49. Find d(cos(x),sin(x)). ans. $\sqrt{2\pi}$

50. Show that every function in T = {cos(2x), sin(2x)} is orthogonal to each function in S.

51. Find a set S' formed from S in which each function is a unit vector.

ans. $S' = T = \{ \dfrac{1}{\sqrt{2\pi}}, \dfrac{\cos(x)}{\sqrt{\pi}}, \dfrac{\sin(x)}{\sqrt{\pi}} \}$

Part I. Prove:

52. In any inner product space: $< kX + jY, Z > = k < X, Z > + j < Y, Z >$ NB. One says that an inner product is *bilinear*.

53. In any IPS: $X \perp Y \Rightarrow \| X + Y \|^2 = \| X - Y \|^2 = \| X \|^2 + \| Y \|^2$.

54. In any IPS: $\| X + Y \|^2 + \| X - Y \|^2 = 2\| X \|^2 + 2\| Y \|^2$. This is called the *parallelogram law*.

55. For functions f and g in C[a,b]: $\|kf_1\| \le |k| \|f_1\|$ (this is homogeneity)

Use the integral inner product $< f, g > = \int_{a}^{b} f\, g\, dx$.

J. Using the norm $\| f \| = \max_{x \in [0,2]} |f(x)|$ with f = 1 + x & $g = -x^2$:

56. Find $\| f \|$.

57. Find $\left\| \; g \; \right\|$. ans. 4

58. Show $\left\| f_1 + f_2 \right\| \leq \left\| f_1 \right\| + \left\| f_2 \right\|$.

NB. Graphs are helpful here!

5.3 Orthonormal Bases

Recall from Chapter 3 that every finite-dimensional vector space has a basis. The vectors in the basis must span the vector space and be independent. If there is an inner product defined on the vector space, we can check out another property that a set of vectors can have: orthogonality. Keep in mind that orthogonality is a "pairwise" concept- in other words we always check out **pairs** of vectors to see if they are orthogonal to each other. It does not make sense for example to say that the vector v = (1,2,3) is orthogonal. A definition is in order.

DEFINITION	A set of vectors $B = \{ X_1, X_2, \cdots , X_n \} \subseteq V$ is **orthogonal** iff $i \neq j \Rightarrow \langle X_i, X_j \rangle = 0$ *for all possible i & j*.

In other words, if we form all possible inner products of vectors in B (except the inner product $\langle \alpha_i, \alpha_j \rangle$ (why do we exempt such an inner product?) we must get *zero*. The vectors in B must be pairwise orthogonal in order for the whole set B to be called orthogonal. A familiar set will be a good example to keep in mind. Recall that the standard basis for $\Re^3$ is the set
B = {(1,0,0), (0,1,0), (0,0,1)}. What do we know about this set? It must be independent and it must span $\Re^3$. But it also has another property; if you form all possible dot products (there should be 3 different possibilities), the result should be zero.

$$(1,0,0)\cdot(0,1,0)=0, \quad (1,0,0)\cdot(0,0,1)=0, \quad (0,1,0)\cdot(0,0,1)=0$$

That is, B is composed of 3 mutually orthogonal vectors. Knowing that a set is orthogonal also implies that the set is independent. A proof for a set of 3 vectors is given below.
Let $B = \{ X_1, X_2, X_3 \} \in \Re^3$ be an orthogonal set. Write down the independence equation:

$$k_1 X_1 + k_2 X_2 + k_3 X_3 = \theta$$

If we dot this equation with X_1 we have:

$$X_1 \cdot (k_1 X_1 + k_2 X_2 + k_3 X_3) = X_1 \cdot \theta$$
$$X_1 \cdot (k_1 X_1) + X_1 (k_2 X_2) + X_1 \cdot (k_3 X_3) = 0$$
$$k_1 (X_1 \cdot X_1) + k_2 (X_1 \cdot X_2) + k_3 (X_1 \cdot X_3) = 0$$

But since B is orthogonal, we have $X_1 \cdot X_2 = 0$ *and* $X_1 \cdot X_3 = 0$ so therefore

$$k_1 | X_1 |^2 + k_2 (0) + k_3 (0) = 0 \Rightarrow k_1 = 0$$

In a similar fashion we could show $k_2 = 0$ and $k_3 = 0$ so the set B is independent. **Orthogonality implies independence!**

THEOREM: Given $B = \{ X_1, X_2, \cdots , X_n \} \subseteq V$, where V is an IPS and no vector in B is the zero vector. If set B is orthogonal, then the set B is an independent set.

EXAMPLE 1 Show that the set B is independent if

$$B = \{ \ (\ 1,0,1,0 \ ),(0,1,0,1 \ ),(\ 1,0,-1,0 \ ) \ \} \subseteq \mathfrak{R}^4$$

Performing all possible dot products, we see easily that they are all zero. Therefore the set is orthogonal and hence independent.

EXAMPLE 2 Find a third vector $X = (x,y,z)$ such that the set $B = \{(1,2,4),(2,1,-1),(x,y,z)\}$ is orthogonal.
We need to have both the first and second vectors in B orthogonal to X:

$$(\ 1,2,4 \ ) \cdot (\ x,y,z \ ) = 0 \quad and \quad (\ 2,1,-1 \ ) \cdot (\ x,y,z \ ) = 0$$

This implies a system of linear equations:

$$x + 2\,y + 4\,z = 0$$

$$2\,x + y - z = 0$$

$$\begin{bmatrix} 1 & 2 & 4 & 0 \\ 2 & 1 & -1 & 0 \end{bmatrix} \sim \begin{bmatrix} 1 & 2 & 4 & 0 \\ 0 & 1 & 3 & 0 \end{bmatrix}$$

So we have $y + 3\,z = 0 \Rightarrow y = -3\,z$ and then from the first row:

$$x + 2\,y + 4\,z = 0 \Rightarrow x = -2\,y - 3\,z = -2\,(-3\,z) - 4\,z = 2\,z$$

so we have the form of the orthogonal vector: $X = (2z,-3z,z) = z(2,-3,1)$. Any multiple of $(2,3,-1)$ is orthogonal to the given two vectors. Thus the set of 3 vectors

$$B = \{ \ (\ 1,2,4 \ ),(\ 2,1,-1),(\ 2,-3,1) \ \}$$

is orthogonal. Since it is orthogonal, it is also independent. It could serve as a basis for 3-space.

Normal Vectors

Getting back to the set $B = \{(1,0,0), (0,1,0), (0,0,1)\}$; another nice thing that you might not think of right off the top of your head is that each vector in B has magnitude **one** (each vector is a *unit vector*). If every vector in a set B has magnitude "one", it is called a *normal* set of vectors. If a set is both orthogonal and normal, it is called orthonormal.

Orthonormal means orthogonal and normal

Now, why is it nice to have a orthogonal set of unit vectors? The answer is - coordinates!
Let's suppose that we have a set of three vectors that form the basis of some IPS V:

$$B = \{ \ X_1, \ X_2, X_3 \ \}$$

Any vector Y in V must be expressible as a linear combination of these three vectors:

$$Y = k_1 X_1 + k_2 X_2 + k_3 X_3$$

In this context, the scalars $k_1, k_2,$ & k_3 are unique for a given vector and are the **coordinates** of Y *with respect to the basis B*. A point to keep in mind is that if I change the basis B, then the coordinates of Y will also change , even though Y stays the same. More on this later. Let's suppose that B is an orthogonal set of vectors and let's try to find k_1 without solving the corresponding linear system using Gaussian elimination. Let's take the inner product of Y with X_1:

$$<Y, X_1> = <k_1 X_1 + k_2 X_2 + k_3 X_3, X_1> = <k_1 X_1, X_1> + <k_2 X_2, X_1> + <k_3 X_3, X_1> \Rightarrow$$

$$<Y, X_1> = k_1 < X_1, X_1> + k_2 < X_2, X_1> + k_3 < X_3, X_1>$$

$$<Y, X_1> = k_1 \| X_1 \|^2 + k_2 \cdot 0 + k_3 \cdot 0 = k_1 \| X_1 \|^2$$

Therefore to find k_1:

$$k_1 = \frac{<Y, X_1>}{\| X_1 \|^2}$$

Similarly:

$$k_2 = \frac{<Y, X_2>}{\| X_2 \|^2} \qquad k_3 = \frac{<Y, X_3>}{\| X_3 \|^2}$$

A further simplification occurs if B is a **normal**- then all the denominators become "1" (ie. the magnitude squared of any normal vector is one) and we've got

$$k_i = <Y, X_i>$$

In other words, to find the component of Y with respect to X_i, just take the inner product of Y with X_i (you must agree that is a fairly simple thing to do!).

For example, consider the basis B = {(1,0,1), (1,0,-1)} and the vector space W = span(B) using the ordinary dot product as the inner product. B is orthogonal but not normal; but we could make it normal with no problem. Simply divide each vector by its magnitude:

$$B' = \{ (\frac{1}{\sqrt{2}}, 0, \frac{1}{\sqrt{2}}), (\frac{1}{\sqrt{2}}, 0, \frac{-1}{\sqrt{2}}) \}$$

THEOREM Given $B = \{ X_1, X_2, \cdots, X_n \} \subseteq V$, where V is an IPS and no vector in V is the zero vector; if B is orthogonal and X is in V, then $X = k_1 X_1 + k_2 X_2 + \ldots + k_n X_n \Rightarrow k_i = \frac{<X, X_i>}{<X_i, X_i>}$.

<u>Corollary</u>: If B is orthonormal, then $k_i = <X, X_i>$.

EXAMPLE 3 Find the coordinates of the vector X = (5,0,1) with respect to the basis B, where $B = \{ (1,0,1), (1,0,-1), (0,1,0) \}$. First note that B is orthogonal but not normal. Thus we must find three scalars $k_1, k_2,$ & k_3 which are the coordinates of X with respect to the basis B.

$$k_1 = \frac{Y \cdot X_1}{X_1 \cdot X_1} = \frac{(5,0,1) \cdot (1,0,1)}{(1,0,1) \cdot (1,0,1)} = \frac{5+0+1}{1+1} = \frac{6}{2} = 3$$

$$k_2 = \frac{X \cdot X_2}{X_2 \cdot X_2} = \frac{(5,0,1) \cdot (1,0,-1)}{(1,0,-1) \cdot (1,0,-1)} = \frac{5+0-1}{1+1} = \frac{4}{2} = 2$$

$$k_3 = \frac{X \cdot X_3}{X_3 \cdot X_3} = \frac{(5,0,1) \cdot (0,1,0)}{(0,1,0) \cdot (0,1,0)} = \frac{0}{1} = 0$$

Therefore $(X)_B = (3,2,0)$; to verify we are happy to know that

$$X = (5,0,1) = 3(1,0,1) + 2(1,0,-1) + 0(0,1,0)$$

EXAMPLE 4 Find the coordinate matrix of the vector X = (3,4) with respect to the basis

$$B = \{ (\frac{1}{\sqrt{2}}, \frac{1}{\sqrt{2}}), (\frac{1}{\sqrt{2}}, -\frac{1}{\sqrt{2}}) \}$$

It is easy to check that this basis is both orthogonal and normal- it is *orthonormal*. Therefore, to find the coordinates of X with respect to B, we just need to dot X with each vector in B:

$$k_1 = X \cdot X_1 = (3,4) \cdot (\frac{1}{\sqrt{2}}, \frac{1}{\sqrt{2}}) = \frac{3}{\sqrt{2}} + \frac{4}{\sqrt{2}} = \frac{7}{\sqrt{2}}$$

$$k_2 = X \cdot X_2 = (3,4) \cdot (\frac{1}{\sqrt{2}}, \frac{-1}{\sqrt{2}}) = \frac{3}{\sqrt{2}} - \frac{4}{\sqrt{2}} = \frac{-1}{\sqrt{2}}$$

Thus we have

$$[X]_B = \begin{bmatrix} \dfrac{7}{\sqrt{2}} \\ \dfrac{-1}{\sqrt{2}} \end{bmatrix}$$

EXAMPLE 5 Let B = {1, cosh(x) - sinh(1), sinh(x)} where V = C[-1,1]; show B is *orthogonal* but not normal using the integral inner product $<f,g>=\int_{-1}^{1}f\,g\,d\,x$.

Since we are integrating on a *symmetric* interval [-1,1], we can use properties of even/odd functions nicely here. Notice that $f_1 = 1$ and $f_2 = \cosh(x) - \sinh(1)$ are both even and $f_3 = \sinh(x)$ is *odd* so the only integral we need to actually do is

$$<f_1, f_2>=\int_{-1}^{1}1\cdot(\cosh(x)-\sinh(1))\,d\,x=\int_{-1}^{1}[\sinh(x)-\sinh(1)\,x\,]_{-1}^{1}$$

$$<f_1, f_2>=[\sinh(1)-\sinh(1)]-[\sinh(-1)-\sinh(1)(-1)\,]=0$$

Thus B is orthogonal; hence it is also independent. But is it normal? Let's start out with f_1:

$$<f_1, f_1>=\int_{-1}^{1}f_1^2\,d\,x=\int_{-1}^{1}1\,d\,x=2\Rightarrow\|\,f_1\,\|=\sqrt{2}$$

so f_1 is NOT a unit vector so B cannot be normal.

Orthogonal matrices

If the columns of a matrix form an orthogonal set and the norm of each column (as a vector in n-space) is *one*, the matrix is called **orthogonal**. For example, the two by two matrix

$A=\begin{bmatrix}\dfrac{1}{\sqrt{2}} & \dfrac{-1}{\sqrt{2}}\\[2mm]\dfrac{1}{\sqrt{2}} & \dfrac{1}{\sqrt{2}}\end{bmatrix}$ is an orthogonal matrix, since if we dot the two columns we get zero and the norm

of each column (as a 2-vector) is one; for example, the magnitude of column one is

$$\|\,C_1\,\|=\sqrt{(\frac{1}{\sqrt{2}})^2+(\frac{1}{\sqrt{2}})^2}=\sqrt{\frac{1}{1}+\frac{1}{2}}=1$$

Orthogonal matrices have the special property that $A^{-1}=A^t$ or $A\,A^t=I$; this means to get the inverse of A, all you have to do is take its tranpose.

EXAMPLE 6 Show that the operator T preserves the length of a vector, where $[T(X)] = A[X]$ and

$A=\begin{bmatrix}\dfrac{1}{\sqrt{2}} & \dfrac{-1}{\sqrt{2}}\\[2mm]\dfrac{1}{\sqrt{2}} & \dfrac{1}{\sqrt{2}}\end{bmatrix}$.

Now

$$[T(x,y)] = A[X] = \begin{bmatrix} \dfrac{1}{\sqrt{2}} & \dfrac{-1}{\sqrt{2}} \\ \dfrac{1}{\sqrt{2}} & \dfrac{1}{\sqrt{2}} \end{bmatrix} \begin{bmatrix} x \\ y \end{bmatrix}$$

so the length of the image is given by

$$\left\| \left(\frac{1}{\sqrt{2}}x - \frac{1}{\sqrt{2}}y, \frac{1}{\sqrt{2}}x + \frac{1}{\sqrt{2}}y \right) \right\| =$$

$$\sqrt{\left(\frac{1}{2}x^2 - \frac{2}{\sqrt{2}}xy + \frac{1}{2}y^2 \right) + \left(\frac{1}{2}x^2 + \frac{2}{\sqrt{2}}xy + \frac{1}{2}y^2 \right)} = \sqrt{x^2 + y^2}$$

which is the length of the pre-image X =(x,y). Such a transformation (which preserves the length of a vector) is called an *isometry*.

.

EXERCISES 5.3

Part A. Below is a given a set of vectors from 3-space. Determine if the set is a) orthogonal, b) normal c) orthonormal.

1. B = {(1,0,1),(0,1,0),(1,0,-1)} ans. orthogonal
2. B = {(1,0,1),(1,0,-1),(2,2,2)}
3. B = {(0,1,1),(2,0,0),(0,1,-1)} ans. orthogonal
4. B = {(1,2,2),(0,1,-1),(-4,1,1)}
5. $B = \{ (\frac{1}{\sqrt{2}}, 0, \frac{1}{\sqrt{2}}), (0,1,0), (\frac{1}{\sqrt{2}}, 0, \frac{-1}{\sqrt{2}}) \}$ ans. orthonormal
6. $B = \{ (\frac{1}{\sqrt{2}}, \frac{1}{\sqrt{2}}, 0), (0,1,0), (\frac{1}{\sqrt{2}}, 0, \frac{-1}{\sqrt{2}}) \}$

Part B. Find the coordinates of the vector X wrt. the given orthogonal basis B. The basis B may or may not be normal.

7. X = (5,6), $B = \{ (\frac{1}{\sqrt{2}}, \frac{1}{\sqrt{2}}), (\frac{1}{\sqrt{2}}, -\frac{1}{\sqrt{2}}) \}$ ans. $(X)_B = (\frac{11}{\sqrt{2}}, \frac{-1}{\sqrt{2}})$

8. X = (2,3,4), $B = \{ (\frac{1}{\sqrt{2}}, \frac{1}{\sqrt{2}}, 0), (\frac{1}{\sqrt{2}}, -\frac{1}{\sqrt{2}}, 0)(0,0,1) \}$

9. X = (1,2,3), $B = \{ (1,1,0), (1,-1,0)(0,0,1) \}$ ans. (X)_B = (3,-1,3)

10. X = (1,2,3,4), $B = \{ (1,1,0,0), (1,-1,0,0), (0,0,1,0), (0,0,0,1) \}$

11. X = (1,2,3,4), $B = \{ (1,0,1,0), (0,1,0,1), (1,0,-1,0), (0,1,0,-1) \}$

ans. (X)B = (4,6,-2,-2)

12. re-do problem # 11 having normalized the basis B first

$$< p, q >= a_1 a_2 + b_1 b_2 + c_1 c_2$$

Part C. Given the inner product   ;

$$where \ p = a + b \ x + c \ x^2 \in P_2$$

13. Show the set $B = \{ 1 + x, 1 - x, x^2 \}$ is orthogonal with respect to the given inner product. Is B independent? Why or why not? Can it serve as a basis for P_2? ans. YES, it's orthogonal, YES

14. Find the coordinates of the vector $r = 1 + 2 x + 3 x^2$ with respect to the basis given in # 13.

15. Normalize the basis B given in # 13. ans. $B' = \{ \frac{1}{\sqrt{2}} (1 + x), \frac{1}{\sqrt{2}} (1 - x), x^2 \}$

16. Re-do problem # 14 with this new basis from # 15.

17. Show the set $B = \{ 1 - x^2, 2 x, 1 + x^2 \}$ is orthogonal wrt. the given inner product. Is B independent? Why or why not? Can it serve as a basis for P_2? ans. YES, it's orthogonal, YES

18. Find the coordinates of the vector $r = 1 + 2 x + 3 x^2$ wrt. the basis given in # 17.

19. Normalize the basis B given in # 17. ans. $B' = \{ \frac{1}{\sqrt{2}} (1 - x^2), x, \frac{1}{\sqrt{2}} (1 + x^2) \}$

20. Re-do problem # 18 with this new basis from # 19.

Part D. Given the set $B = \{ 1 - x^2, 2 x + x^3, 1 + x^2, x - 2 x^3 \} \subseteq P_3$ with the inner product

$$< p, q >= a_1 a_2 + b_1 b_2 + c_1 c_2 + d_1 d_2 \ \ where \ p = a + b x + c \ x^2 + d \ x^3 \in P_3$$

21. Is B an orthogonal set wrt. the given inner product? Is B independent? Why or why not? Can it serve as a basis for P_3? ans. YES, it's orthogonal, YES

22. Normalize B to get B'.

23. Find the coordinates of the vector $p = 1 + 2 x + 3 x^2 + 4 x^3$ wrt. B.

ans. (p)B = (-1,8/5,2,-6/5)

24. Find the coordinates of the same vector (in #23) wrt. the normalized basis B'.

Part E. Given the set of vectors S = {1,2,2,1),(1,1,-1,-1)};

25. Show that S is orthogonal.

26. Find all vectors in 4-space (x,y,z,w) which are orthogonal to W = span{S}; find $W^{\perp}$.

27. Find a basis for the subspace W perp (you should need two vectors).

ans. {(4,-3,1,0),(3,-2,0,1)}

28. Make S normal.

29. Show that any vector in 4-space can be written as a linear combination of the vectors in S and the two basis vectors from W perp. Choose the vector X = (1,2,3,4) to write as a combination of these vectors.

ans. X = 3/2(1,2,2,1)-1(1,1,-1,-1)-1(4,-3,1,0)+3/2(3,-2,0,1)

Part F. Using the inner product $< f, g >= \int_{-1}^{1} f \, g \, d \, x$ in C[-1,1], show that

30. $B = \{ 1, x, 3 x^2 - 1 \}$ is an orthogonal set but not normal.
31. Find a basis B' formed from B which is orthonormal.

ans. $B' = \{ \dfrac{1}{\sqrt{2}}, \dfrac{\sqrt{6}}{2} x, \dfrac{\sqrt{10}}{4} (3 x^2 - 1) \}$

32. Is B independent? Why or why not?

Part G. Using the inner product $< f, g >= \int_{-\pi}^{\pi} f \, g \, d \, x$ in $C [\pi, \pi]$, show that

33. $B = \{ 1, \cos(x), \sin(x) \}$ is an orthogonal set but not normal.
34. Find a basis B' formed from B which is orthonormal.
35. Is B independent? Why or why not? ans. YES, it is orthogonal
36. Show that the set $\{ \dfrac{1}{\sqrt{2\pi}}, \dfrac{\cos(x)}{\sqrt{\pi}}, \dfrac{\sin(x)}{\sqrt{\pi}}, \dfrac{\cos(2 x)}{\sqrt{\pi}}, \dfrac{\sin(2 x)}{\sqrt{\pi}} \}$ is orthonormal.

Part H. Given the matrix $A = \begin{bmatrix} \dfrac{1}{\sqrt{2}} & -\dfrac{1}{\sqrt{2}} \\ \dfrac{1}{\sqrt{2}} & \dfrac{1}{\sqrt{2}} \end{bmatrix}$, which is orthogonal;

37. Show that the linear operator [T(X)] = A[X] represents a rotation in the plane through an angle $\dfrac{\pi}{4}$. Hint: pick any easy vector to draw like X = (1,1)

38. Show that under this operator, the image Y = T(X) has the *same magnitude* as the pre-image (ie. this operator preserves the *length* of a vector).

39. Prove that this operator is 1-1 and onto. Intuitively this is clear.
ans. rank(A) = 2 so nullity(A) = 0

I. Given the matrix $A = \begin{bmatrix} \dfrac{1}{\sqrt{3}} & \dfrac{-1}{\sqrt{2}} & \dfrac{1}{\sqrt{6}} \\ \dfrac{1}{\sqrt{3}} & 0 & \dfrac{-2}{\sqrt{6}} \\ \dfrac{1}{\sqrt{3}} & \dfrac{1}{\sqrt{2}} & \dfrac{1}{\sqrt{6}} \end{bmatrix}$;

40. Show that A is *orthogonal*. Hint: recall $A^{-1} = A^t$ if A is orthogonal.
41. If [T(X)] = A[X] show that T is a linear operator on $\Re^3$ that is both 1-1 and onto.
42. Show that T preserves the length of a vector- ie. the length of the image = the length of the pre-image (ie. T is an isometry). Just pick some 3-vector like X = (3,4,5) and get its image using A.

5.4 Projection and Gram-Schmidt

Projection onto a subspace in n-space

The basis of this section is the projection formula using the dot product:

$$proj_X Y = \frac{Y \cdot X}{X \cdot X} X$$

where the notation means the **projection of Y onto X**. This actually gives us two vectors:
1) one vector parallel to X (ie. the projection) and
2) a second vector *orthogonal to X*.

Suppose we are given Y and X in n-space; then we can calculate the projection of Y onto X using $proj_X Y = (\frac{Y \cdot X}{X \cdot X}) X$. Firstly, if we let

$$Z = Y - proj_X Y$$

then this vector Z must be orthogonal to X. Subtracting away the projection of Y onto X produces an orthogonal vector.

Let's see why:

$$X \cdot Z = X \cdot [Y - \frac{Y \cdot X}{X \cdot X} X] = X \cdot Y - X \cdot (\frac{Y \cdot X}{X \cdot X}) X = X \cdot Y - (\frac{Y \cdot X}{X \cdot X}) X \cdot X = X \cdot Y - Y \cdot X = 0$$

Not only is Z orthogonal to the projection but clearly

$$proj_X Y + Z = Y$$

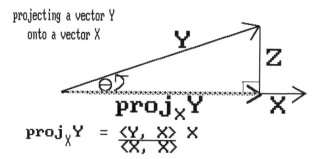

This allows us to decompose the vector Y into two vectors, one *parallel* to X (the projection vector) and one perpendicular to X (vector Z).

If we set W = span(X), then the projection of Y onto X is in W and Z is in $W^{\perp}$. Now, what if we wish to project onto a *subspace* instead of just one vector?

It turns out that we can use this same formula to project onto a subspace. Suppose W = span{S} where S = { X_1, X_2 } and S is an **orthogonal** set. Since the set { X_1, X_2 } is orthogonal, it is automatically independent and therefore forms a basis for W. We wish to decompose a given vector Y in n-space into the sum of two vectors, one in W and one orthogonal to W (ie. orthogonal to every vector in S). We want this decomposition to satisfy

$$Y = proj_W Y + Z \qquad proj_W Y \in W, \quad Z \in W^\perp$$

where

$$Y = (c_1 X_1 + c_2 X_2) + Z \Rightarrow Z = Y - (c_1 X_1 + c_2 X_2)$$

But how do we project onto a subspace??

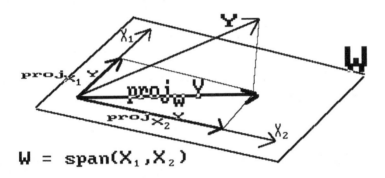

W = span(X_1, X_2)

Claim: Z is perpendicular to W only if

$$proj_W Y = proj_{X_1} Y + proj_{X_2} Y = \frac{Y \cdot X_1}{X_1 \cdot X_1} X_1 + \frac{Y \cdot X_2}{X_2 \cdot X_2} X_2$$

In other words, if we have an *orthogonal* basis, we can just separately project Y onto each basis vector for W and then add the results.

Proof: Let $proj_W Y = c_1 X_1 + c_2 X_2$ so that $Z = Y - (c_1 X_1 + c_2 X_2)$ is orthogonal to W = span({ X_1, X_2 }). Now if Z is orthogonal to X_1, then we must have

$$X_1 \cdot Z = X_1 \cdot [Y - (c_1 X_1 + c_2 X_2)] = 0$$

$$\Rightarrow X_1 \cdot Y - X_1 \cdot (c_1 X_1 + c_2 X_2) = 0 \Rightarrow X_1 Y - c_1 X_1 \cdot X_1 - c_2 X_1 \cdot X_2 = 0$$

But X_1 is orthogonal to X_2 so $X_1 \cdot X_2 = 0$; thus

$$X_1 \cdot Y - c_1 X_1 \cdot X_1 = 0 \Rightarrow X_1 \cdot Y = c_1 X_1 \cdot X_1 \Rightarrow c_1 = \frac{Y \cdot X_1}{X_1 \cdot X_1}$$

So c_1 is calculated via the ordinary projection formula for one vector onto another. The story for c_2 is basically a rerun; if Z is orthogonal to X_2 we must have $c_2 = \dfrac{Y \cdot X_2}{X_2 \cdot X_2}$. Thus we see that to project onto a subspace spanned by two **orthogonal** vectors, all we need to do is to *separately project onto each basis vector* and then add the results. This idea can be extended to a subspace spanned by 3 or more vectors, as long as the basis vectors form an orthogonal set.

All of this can be generalized and works in any inner product space- all we need to do in change the notation. To project the vector Y onto the vector X in any inner product space V, we use the formula

$$proj_X Y = \frac{<Y, X>}{<X, X>} X$$

THEOREM If S is an orthogonal set of m vectors, W = span(S) where W is a subspace of the inner product space V, then the projection of the vector Y onto W (for Y in V) is given by:

$$proj_W Y = proj_{X_1} Y + proj_{X_2} Y + \ldots + proj_{X_m} Y$$

Thus to project onto a subspace W use the above theorem- and we can simply project onto each basis vector in S (since S is orthogonal). Intuitively this projection $proj_W Y$ is the vector in W closest to Y; thus we have

THEOREM If $proj_W Y$ = P and U is any vector in W (not equal to $proj_W Y$), then $\|Y - P\| < \|Y - U\|$.

Proof: Assume $\|Y - U\| < \|Y - P\|$ since clearly $\|Y - P\| \neq \|Y - U\|$ if U is not P.

Now $Y - U = (Y - P) + (P - U)$ and also $P - U \in W$ (since both are in W) and $Y - P \in W^{\perp}$ so the Pythagorean theorem implies that

$$\|Y - U\|^2 = \|Y - P\|^2 + \|P - U\|^2$$

But this means that $\|Y - U\| > \|Y - P\|$ (contradiction). Therefore

$$\| Y - proj_W Y \| < \|Y - U\|$$

which means that $proj_W Y$ is **THE** vector *in W* closest to Y. QED

EXAMPLE 1 Given the basis $S = \{1, x\}$ in the inner product space $V = C[-1,1]$ with inner product $<f, g> = \int_{-1}^{1} f\, g\, d\,x$; find the projection of the function $y = \exp(x)$ on $W = \text{span}(S) = P_1$. First verify that S is indeed orthogonal. But the product of an even function (ie. f1) and an odd function (f2) is ODD; and the integral of an **odd** function on a symmetric interval equals zero. So S is indeed orthogonal. Now we can simply project y separately onto f1 = 1 and f2 = x and add.

$$proj_{f_1} y = \frac{<y, f_1>}{<f_1, f_1>} f_1 = \frac{\int_{-1}^{1} e^x \cdot 1\, d\,x}{\int_{-1}^{1} 1 \cdot 1\, d\,x} f_1 = \frac{[\,e^x\,]_{-1}^{1}}{[\,x\,]_{-1}^{1}} f_1 = \frac{e^1 - e^{-1}}{2} \cdot 1 = \sinh(\,1\,)$$

$$proj_{f_2} y = \frac{<y, f_2>}{<f_2, f_2>} f_2 = \frac{\int_{-1}^{1} e^x \cdot x\, d\,x}{\int_{-1}^{1} x \cdot x\, d\,x} f_2 = \frac{[\,x e^x - e^x\,]_{-1}^{1}}{\int_{-1}^{1} x^2\, d\,x} f_2 = \frac{2 e^{-1}}{\frac{2}{3}} f_2 = 3 e^{-1} x$$

Thus our projection of y onto $W = \text{span}(S)$ is

$$proj_W y = \sinh(\,1\,) \cdot 1 + 3\, e^{-1} \cdot x = \sinh(\,1\,) + 3\, e^{-1} x$$

We can view this projection as *the function in W which is* **closest to y**. As a vindication of our work, we know that $z = y - proj_W y = e^x - [\,\sinh(\,1\,) + 3\, e^{-1} x\,]$ is *orthogonal* to W and we could verify this by finding the inner products of z with both f1 and f2; these will both be *zero*. The norm of z is minimized by using

$$proj_W y = \sinh(\,1\,) + 3\, e^{-1} x$$

This means that

$$\|\,z\,\| = \|\,e^x - [\,\sinh(\,1\,) + 3\, e^{-1} x\,]\,\| = \sqrt{\int_{-1}^{1} (\,e^x - [\,\sinh(\,1\,) + 3\, e^{-1} x\,]\,)^2 d\,x}$$

which measures the distance between y and its projection onto W, is a minimum.

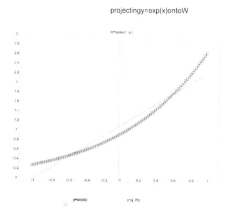

projectingy=exp(x)ontoW

EXAMPLE 2 Project $y = \sin(x)$ onto the subspace W of C[-1,1] , where W = span(S) and

$S = \{ \; 1, x, x^2 - \dfrac{1}{3} \; \}$, using the integral inner product $< f, g > = \displaystyle\int_{-1}^{1} f \; g \; d \; x$.

First verify for yourself that the three polynomials here form an orthogonal set; this is important!
Thus we can project $y = \sin(x)$ separately onto each basis vector and then add the results. Clearly
$y = \sin(x)$ is NOT in W! Note that y is odd and$\{ \; 1, x^2 - \dfrac{1}{3} \}$ are even so that the product of y and any
of these two functions is odd (odd times even is odd) and the integrals of the inner product would
disappear. Hence the projection of y onto these two functions is *zero* so all we need do is project y
onto x (since odd times odd = even).

$$proj_W \; y = proj_x \; y = \frac{<y, x>}{<x, x>} \; x = \frac{\displaystyle\int_{-1}^{1} x \sin(x) \, dx}{\displaystyle\int_{-1}^{1} x \cdot x \, dx} \; x = \frac{[\sin(x) - x\cos(x)]_{-1}^{1}}{[\frac{x^2}{3}]_{-1}^{1}} \; x$$

$$proj_W \; y = \frac{2 \sin(1) - 2 \cos(1)}{\dfrac{2}{3}} \; x = [\; 3 \sin(1) - 3 \cos(1) \;] \; x$$

Note that we can interpret this projection as a polynomial approximation to the function $y = \sin(x)$
on [-1,1]. This means that $proj_W \; y$ is the *closest* function to y in $W = P_2$. How good is this
approximation? Substituting x = 1/2 into $y = \sin(x)$ and its projection onto W gives:

$$\sin(\frac{1}{2}) = 0.4794, \quad proj_W \; y = 0.4518$$

Even though S only has three functions in it, this approximation is not bad. As you increase the
number of orthogonal functions in S, this approximation gets better and better.

As a check on our work, it must be that $proj_w \, y + z = y$ so that we get a function orthogonal to W, which here is $z = y - proj_w = \sin(x) - [3\sin(1) - 3\cos(1)]x$

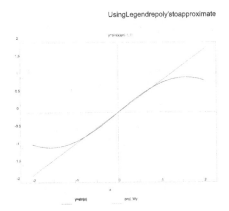

If indeed $proj_w \, y$ is the closest function in W to y, then $\| z \| = \sqrt{\int_{-1}^{1} z^2 \, dx}$ (which measures the discrepancy between y and its projection) is as small as possible. Here $\|z\| \approx 0.0337$; one can observe from the graph that the projection does a remarkable job on the "playground" [-1,1]. Outside this region, the projection no longer does a very good job of approximating y =sin(x).

Gram-Schmidt process

The Gram-Schmidt process takes any old basis B for the IPS V and shows how to produce an **orthonormal basis** E. The key is to recall that subtracting away a projection produces an orthogonal vector.

Given the basis $B = \{ \ X_1, X_2, X_3 \ \}$;

Step 1- rename X_1; let $Y_1 = X_1$.

Step 2- find Y_2 such that $\{ \ Y_1, Y_2 \ \}$ is orthogonal:

$$Y_2 = X_2 - proj_{Y_1} X_2 = X_2 - \frac{<X_2, Y_1>}{\| Y_1 \|^2} Y_1$$

Step 3- find Y_3 such that $\{ Y_1, Y_2, Y_3 \}$ is orthogonal:

$$Y_3 = X_3 - proj_{Y_1} X_3 - proj_{Y_2} X_3 = X_3 - \frac{<X_3, Y_1>}{\| Y_1 \|^2} Y_1 - \frac{<X_3, Y_2>}{\| Y_2{}^2 \|} Y_2$$

Now the set $\{ Y_1, Y_2, Y_3 \}$ is guaranteed to be orthogonal; it is easy to normalize it- just divide each vector by its norm.

$$E = \{ \frac{Y_1}{\| Y_1 \|}, \frac{Y_2}{\| Y_2 \|}, \frac{Y_3}{\| Y_3 \|} \} = \{ Z_1, Z_2, Z_3 \}$$ is thus **orthonormal**. This process can be extended in an inductive fashion to any (finite) number of vectors.

EXAMPLE 3 Produce an orthonormal basis from the set B = $\{(1,0,1), (1,0,-1), (1,2,3)\}$. Use the usual dot product here.
Let $Y_1 = X_1 = (1,0,1)$; then

$$Y_2 = (1,0,-1) - proj_{Y_1} X_2 = (1,0,-1) - \frac{(1,0,-1)\cdot(1,0,1)}{\| (1,0,1) \|^2} Y_1 = (1,0,-1) - \frac{0}{\sqrt{2}^2} Y_1 = (1,0,-1)$$

Thus $Y_2 = X_2$, since $\{ X_1, X_2 \}$ was already an orthogonal set. Now we get Y_3:

$$Y_3 = X_3 - proj_{Y_1} X_3 - proj_{Y_2} X_3 = (1,2,3) - \frac{X_3 \cdot Y_1}{\| (1,0,1) \|^2} Y_1 - \frac{X_3 \cdot Y_2}{\| (1,0,-1) \|^2} Y_2 \Rightarrow$$

$$Y_3 = (1,2,3) - \frac{4}{(\sqrt{2})^2}(1,0,1) - \frac{-2}{(\sqrt{2})^2}(1,0,-1)$$

$$Y_3 = (1,2,3) - (2,0,2) + (1,0,-1) = (0,2,0)$$

Check that the set $\{ Y_1, Y_2, Y_3 \}$ is an orthogonal set. Dividing each vector by its magnitude then produces an orthonormal set:

$$E = \{ \frac{1}{\sqrt{2}}, 0, \frac{1}{\sqrt{2}}), (\frac{1}{\sqrt{2}}, 0, -\frac{1}{\sqrt{2}}), (0,1,0) \}$$

EXAMPLE 4 Find an orthogonal basis in V = C[-1,1] with IP $<f, g> = \int_{-1}^{1} f\, g\, d\, x$ for

$W = \text{span}\{1,\ x,\ x^2\}$.

The Gram-Schmidt process can be used on $B = \{1,\ x,\ x^2\}$ to get an orthogonal basis for W.

$$p_1 = 1 \qquad p_2 = x - \text{proj}_1\ x = x - \frac{<x,1>}{<1,1>}\ 1 = x - \frac{\int_{-1}^{1} 1 \cdot x\, dx}{\int_{-1}^{1} 1^2\, dx} = x - 0 = x$$

ie. 1 and x were already orthogonal so no subtraction was needed.

$$p_3 = x^2 - \text{proj}_1\ x^2 - \text{proj}_x\ x^2 = x^2 - \frac{<x^2,1>}{<1,1>}\ 1 - \frac{<x^2,x>}{<x,x>}\ x = x^2 - \frac{\int_{-1}^{1} 1 \cdot x^2\, dx}{\int_{-1}^{1} 1^2\, dx} \cdot 1 - \frac{\int_{-1}^{1} x^2 \cdot x\, dx}{\int_{-1}^{1} x^2\, dx} \cdot x$$

The last integral is zero (an odd integrand!) so we finally have:

$$p_3 = x^2 - \frac{\left[\frac{x^2}{3}\right]_{-1}^{1}}{\left[x\right]_{-1}^{1}}\ 1 = x^2 - \frac{\frac{2}{3}}{2} = x^2 - \frac{1}{3}$$

Therefore the set of functions $\{\ 1,\ x,\ x^2 - \frac{1}{3}\ \}$ is an orthogonal set- we can continue in this fashion to create other higher degree basis functions; when these functions are multiplied by a certain constant, they are called *Legendre polynomials* and they are orthogonal on the interval $[-1,1]$ wrt. the inner product defined by the definite integral.

Orthogonal Complements in Inner Product Spaces

THEOREM Given the IPS V, where $\dim(V) = n$ and W, a proper subspace of V such that $\dim(W) = r$. Then there exists a set of r orthogonal vectors S, such that $S = \{Y_1, Y_2, ..., Y_r\}$ and a set of $n - r$ vectors $T = \{Y_{r+1}, Y_{r+2}, ..., Y_n\}$ such that $W = \text{span}(S)$ and $W^{\perp} = \text{span}(T)$. Furthermore, any vector Y in V can be written in a unique way as $Y = X + Z$, where

$$X \in W \ and \ Z \in W^{\perp}$$

Claim 1: S and T exist such that $S \cap T = \{\ \}$

Since W is r-dimensional, there must exist a set of r independent vectors which span W. Using Gram-Schmidt, this set can be made orthogonal, thus giving S. Find $W^{\perp}$ and get a basis for it. Now apply Gram-Schmidt to this set of of $n - r$ vectors; this set will consist of independent vectors (why?) which can be used to form T.

Adjoining these vectors to S will thus produce a set of n independent vectors which will span V. Claim 2: Since V is n-dimensional, any vector in V can be written as a combination of the vectors in $S \cup T$;

$$Y = (c_1 Y_1 + \ldots + c_r Y_r) + (c_{r+1} Y_{r+1} + \ldots + c_n Y_n) = X + Z$$

Now the first vector is in W and the second in $W^\perp$; thus we have found X and Z. In fact, once we have X, we can get Z by subtraction: $Z = Y - X$. And this is the essential lesson of Gram-Schmidt: subtracting away a projection produces an orthogonal vector. In fact, it must be that $X = proj_W Y$. Thus X and Z must be unique, since X is **the** vector in W *closest* to Y.

EXAMPLE 5 Let B = $\{1, x, x^2\}$ with inner product $<f, g> = \int_0^1 f\, g\, d x$; let

V = span(B) and W = span(S) where S = {1,x}. Find $W^\perp$ in V.
Apply Gram-Schmidt to the basis S for W; let $f_1 = g_1 = 1$, $f_2 = x$ then we calculate:

$$g_2 = f_2 - \frac{<f_2, g_1>}{<g_1, g_1>} g_1 = x - \frac{\int_0^1 f_2\, g_1\, d x}{\int_0^1 g_1^2\, d x} f_1 = x - \frac{\int_0^1 x \cdot 1\, d x}{\int_0^1 1^2\, d x} 1 = x - \frac{1}{2}$$

Now the set $\{1, x-1/2\}$ is orthogonal and we can continue with Gram-Schmidt to find a third function which will be orthogonal to W. Apply Gram-Schmidt again to $f_3 = x^2$:

$$g_3 = f_3 - \frac{<f_3, g_1>}{<g_1, g_1>} g_1 - \frac{<f_3, g_2>}{<g_2, g_2>} g_2 = x^2 - \frac{\int_0^1 x^2 \cdot 1\, d x}{\int_0^1 1^2\, d x} 1 - \frac{\int_0^1 x^2 \cdot (x-\frac{1}{2})\, d x}{\int_0^1 (x-\frac{1}{2})^2\, d x} (x-\frac{1}{2}) = x^2 - x + \frac{1}{6}$$

Thus we have W perp; it is the set

$$W^\perp = span(T) \quad T = \{\frac{1}{6} - x + x^2\}$$

It would be good for you to verify that this third function is orthogonal to S. Any quadratic polynomial in V could then be written as the sum of two functions, one in W and one in W perp. For example, if I chose $y = 1 + 2x + 3x^2$ the reader may verify that

$$proj_W y = \frac{1}{2} + 5x \quad z = y - proj_W y = \frac{1}{2} - 3x + 3x^2 \in W^\perp$$

We also have produced an orthogonal basis for V, namely $B' = \{1, -\frac{1}{2} + x, \frac{1}{6} - x + x^2\}$.

258

EXERCISES 5.4

Part A. For each basis in n-space B, find a basis B' which is orthogonal using the Gram-Schmidt process.

1. $B = \{ (1,2), (0,3) \}$ ans. $\{(1,2),(-6/5,3/5\}$
2. $B = \{ (1,0), (1,-1) \}$
3. $B = \{ (1,0,0), (1,1,0), (1,1,1) \}$ ans. $\{(1,0,0),(0,1,0),(0,0,1)\}$
4. $B = \{ (1,1,0), (1,-1,0), (1,1,1) \}$

Part B. Construct an orthonormal basis B' from each basis B.

5. $B = \{ (1,2), (0,3) \}$ ans. $\{ (\frac{1}{\sqrt{5}}, \frac{2}{\sqrt{5}}), (\frac{-2}{\sqrt{5}}, \frac{1}{\sqrt{5}}) \}$

6. $B = \{ (1,0), (1,-1) \}$
7. $B = \{ (1,0,0), (1,1,0), (1,1,1) \}$ ans. same as #3
8. $B = \{ (1,1,0,0), (1,-1,0,0), (1,0,1,0), (2,0,1,2) \}$

Part C. Determine if the basis B is orthogonal; if not make it orthogonal using Gram-Schmidt. Use the inner product $<f, g> = \int_{-1}^{1} f \, g \, dx$.

9. $B = \{1, \cosh(x)\}$ ans. NO, $B' = \{1, \cosh(x) - \sinh(1)\}$
10. $B = \{1, \sinh(x)\}$
11. $B = \{1, \cosh(x), \sinh(x)\}$ ans. NO, $B' = \{1, \cosh(x) - \sinh(1), \sinh(x)\}$
12. $B = \{1, \exp(x)\}$

Part D. Project the vector Y onto the subspace W; also find the vector Z orthogonal to W such that $proj_W Y + Z = Y$. Use the usual dot product.

13. $Y = (1,2)$, $W = span((4,1))$ ans. $(24/17,6/17),(-7/17,28/17)$
14. $Y = (1,2,3)$, $W = span((1,0,1),(0,1,-1))$
15. $Y = (1,2,3)$, $W = span\{(1,0,2),(2,0,-1)\}$ ans. $(1,0,3),(0,2,0)$
16. $Y = (1,2,3,4)$, $W = span\{(1,0,1,0),(1,0,-1,0),(0,1,0,1)\}$
17. $Y = (4,3,2,1)$, $W = span\{(0,2,0,2),(1,0,1,0)\}$ ans. $(3,2,3,2),(1,1,-1,-1)$

Part E. Given the inner product $<f, g> = \int_{0}^{1} f \, g \, dx$ on C[0,1];

18. show that $S = \{1,x\}$ is NOT orthogonal; use Gram-Schmidt to show that an orthogonal basis formed from S is $S' = \{1, -1/2 + x\}$.
19. find $proj_W y$ if $y = x^2$ and $W = span(S) = P_1$. This is the polynomial in W closest to y.

ans. $-1/6 + x$
20. Find the "z" polynomial such that $proj_W y + z = y$. Show that z is orthogonal to W by showing it is orthogonal to the orthogonal basis $S' = \{1, -1/2 + x\}$ which you already found in #18.

21. Find an approximation for y(1/2) using the projection you found in #19.
ans. exact value = 1/4, approximation = 1/3

22. Find the coordinates of the function h = 3 + 4x wrt. the orthogonal basis S' ={1, -1/2 + x} for W. Hint: recall how to find coordinates wrt. an orthogonal basis!

23. Make the orthogonal basis S' = {1, -1/2 + x} orthonormal   ans. {1, $-\sqrt{3} + 2\sqrt{3}\,x$ }

Part F. Given the inner product $<f, g> = \int_{-1}^{1} f\,g\,d\,x$ on C[-1,1];

24. show that $B = \{\ 1,\ x\ ,\ x^2 - \dfrac{1}{3}\ \}$ is orthogonal but not normal and serves as a basis for P_2 (a subspace of C[-1,1]).

25. find $proj_W\ y$ of $y = x^3$ onto the subspace W = span(B) = P_2. This is the polynomial in P_2 closest to y. ans. 3/5x

26. Use the projection of y onto W to approximate y(1/2); compare it to the actual value.

27. Write down how you would calculate the error in the projection from #25, viewed as an *appromation* **in W** to the function y(which is NOT in W).

ans. $\left\| \ y - proj_W\ y\ \right\| = \sqrt{\int_{-1}^{1} (x^3 - \dfrac{3}{5} x\ )^2\,d\,x}$

Part G. Given the set of vectors B = {(1,0,1,0),(0,2,2,0),(1,2,3,4)} in 4-space where W = span(B);

28. Show B is NOT orthogonal.

29 Use the Gram-Schmidt process to produce a basis B' from B which is orthogonal.
ans. B' = {(1,0,1,0),(-1,2,1,0),(0,0,0,4)}

30. Given the vector Y = (2,3,1,4); show that Y is NOT in W.

31. Find the projection of Y onto W. Hint: which basis should you use to project onto???
ans. (2/3,5/3,7/3,4)

32. Find the vector Z orthogonal to W such that $proj_W\ Y + Z = Y$. Verify that Z is orthogonal to W by showing it is orthogonal to the basis B' found in # 29.

33. What is the magnitude of the error vector Z, if we view $proj_W\ Y$ as the vector in W closest to Y ? ans. $\dfrac{4\sqrt{3}}{3}$

34. Find a basis for the subspace of $\mathfrak{R}^4$ which is orthogonal to W (this should be a line through the origin, since W is a hyperplane in 4-space). Show that Z lies on this line.

35. Find an equation for W. ans. x + y - z = 0

Part H. Given set of vectors S = {(1,0,-1,1),(0,1,0,1)} in 4-space with the usual dot product; let W = span(S).

36. Verify that S is NOT orthogonal; then show that {(1,0,-1,1), (-1/3,1,1/3,2/3)} is an orthogonal basis which results from using Gram-Schmidt.

37. Find $W^{\perp}$ by letting X = (a,b,c,d) be an arbitrary vector in W perp and solving the corresponding system created by dotting X with the basis vectors in S. Get a basis for W perp also.

ans. X = (c-d,-d,c,d), basis for W perp = {(1,0,1,0),(-1,-1,0,1)}

38. If Y = (6,3,4,4) show that (1,3,-1,4) is the vector in W closest to Y. Hint: think projection

39. Using Y and the answer to #38, show that the vector z = (5,0,5,0) is the vector in W perp which satisfies $proj_W\ y + z = y$.

Part I.

Given the matrix $A = \begin{bmatrix} 1 & 2 & 1 \\ 2 & 4 & 2 \\ 3 & 6 & 3 \end{bmatrix}$;

40. Find a basis for the null space of A and the row space of A; if W = row(A), show that $null(\ A\ ) = W^\perp$.

41. Use the Gram-Schmidt process on the basis of null(A) to produce a set of (two) vectors which is both independent and orthogonal. ans. (-2,1,0),(-1/5,-2/5,1)

42. Using the basis for null(A) found in #41, now add the (one) vector which forms a basis for row(A) and show you have a set of (three) independent & orthogonal vectors.

43. Normalize this set so you now have a basi for 3-space which is *orthonormal*.

ans. $\{\ (\ \frac{1}{\sqrt{6}}, \frac{2}{\sqrt{6}}, \frac{1}{\sqrt{6}}\ ),(\frac{-2}{\sqrt{5}}, \frac{1}{\sqrt{5}}, 0\ ),(\ \frac{-1}{\sqrt{30}}, \frac{-2}{\sqrt{30}}, \frac{5}{\sqrt{30}}\ )\ \}$

Part J. Given the matrix $A = \frac{1}{2} \begin{bmatrix} 1 & 0 & 1 \\ 0 & 2 & 0 \\ 1 & 0 & 1 \end{bmatrix}$;

44. Show that A is both idempotent and symmetric so that it represents a *projection* operator on 3-space.

45. Find the corresponding linear operator represented by A.

ans. T(x,y,y) = 1/2(x + z, 2y, x + z)

46. If W = rng(T); find W.

47. Take any vector in W, say X = (4,5,4); what is T(X)? Why is T(X) = X ?

ans. T(4,5,4) = (4,5,4), (4,5,4) is already in rng(T) and T is a projection

48. Find ker(T); show by doing dot products that ker(T) = $W^\perp$.

49. Find T(X) if X = 5(1,0,1) + 6(0,1,0) + 7(-1,0,1) ; don't add first - enjoy using the fact that T is LINEAR! ans. (5,6,5)

50. Now use $A = \frac{1}{7} \begin{bmatrix} 3 & 0 & 4 \\ 0 & 7 & 0 \\ 3 & 0 & 4 \end{bmatrix}$; show that this matrix is idempotent but NOT symmetric.

51. Let T be the linear operator represented by A from #50; find rng(T) and ker(T). Show that these two subspaces are NOT orthogonal to each other.

ans. rng(T) = {(1,0,1),(0,1,0)}, ker(T) = {(-4,0,3)}, dot products not all zero

Part K. Economical Gram-Schmidt; given B = {(1,0,1,0),(0,1,1,1)} and W = span(B);

52. Use the fact that the null space of the matrix A with the basis vectors in B as the first two rows has a null space orthogonal to W to find $W^\perp$.

53. Use Gram-Schmidt on the two basis vectors for null(A) to get an orthogonal basis for $W^{\perp}$.

ans. N = {(-1,-1,1,0),(1/3,-2/3,-1/3,1)}

54. Now use Gram-Schmidt on B to get an orthogonal basis for W.

55. You should now have FOUR vectors (2 for W and 2 for W perp) which should be orthogonal and thus form an orthogonal basis for $\mathfrak{R}^4$. Check this out by doing the dot products.

ans. {(1,0,1,0),(-1/2,1,1/2,1),(-1,-1,1,0),(1/3,-2/3,-1/3,1)}

56. Normalize the set from #55 to produce an orthonormal basis for $\mathfrak{R}^4$.

Part L. Given the set B = { $1, x, x^2, x^3$ }, V = span(B) so V = P_3;

57. Use Gram-Schmidt on this set B with the inner product $<f, g> = \int_{-1}^{1} f\, g\, d\,x$ and find an orthogonal set of functions which spans P_3 also. NB. Keeping in mind even vs. odd functions may save you some integration time. ans. $\{ 1, x, x^2 - \dfrac{1}{3}, x^3 - \dfrac{3}{5}x \}$

58. Let W = span$\{ 1, x^2 - \dfrac{1}{3} \}$ and U = span$\{ x, x^3 - \dfrac{3}{5}x \}$; show in fact that U = $W^{\perp}$

59. If $y = x^4$, find $proj_W\ y$. ans. $\dfrac{6}{7}x^2 - \dfrac{3}{35}$

60. find g and h such that y = g + h, where g is in W and h in $W^{\perp}$ using y from #59

61. If $y = x^5$, find $proj_U\ y$. ans. $\dfrac{10}{9}x^3 - \dfrac{5}{21}x$

5.5 Least Squares Approximation

Solving inconsistent systems

In this section we wish to find the "best solution" to an inconsistent system; part of the problem is what do we mean by the "best solution". For a simple problem to investigate, let us deal with the system

$$A X = \begin{bmatrix} 1 & 0 \\ 0 & 1 \\ 1 & 0 \end{bmatrix} \begin{bmatrix} x \\ y \end{bmatrix} = \begin{bmatrix} 1 \\ 1 \\ 2 \end{bmatrix} = B$$

It is clear that this system is **inconsistent**, since B does not lie in col(A), since
col(A) = span{(1,0,1),(0,1,0)}. If we think geometrically, what we need to do is to project B onto the column space of A; for convenience let W = col(A). Then we can solve the related system

$$A \overline{X} = proj_W B$$

where $\overline{X}$ is the "best solution" to this inconsistent system.

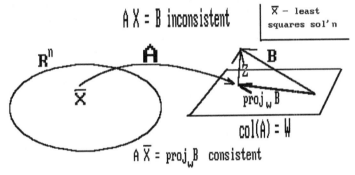

From our picture, we need to have

$$Z = B - proj_W B \perp W$$

Keep in mind that Z must be a vector in 3-space; if it is indeed perpendicular to col(A), the dot product of Z with each column of A must be zero:

$$C_1 \cdot (B - A \overline{X}) = 0 \qquad C_2 \cdot (B - A \overline{X}) = 0$$

This can be conveniently expressed in one matrix equation; since we are in essence multiplying columns of A times one vector (and matrix multiplication is ROW times COLUMN) we need to transpose A:

$$\begin{bmatrix} 1 & 0 & 1 \\ 0 & 1 & 0 \end{bmatrix} (B - A \overline{X}) = \begin{bmatrix} 0 \\ 0 \end{bmatrix}$$

From the point of view of chapter 4, we want the vector Z to be in the null space of A transpose. If we just realize that the matrix on the left is A transpose, we have:

$$A^t (B - A \overline{X}) = \theta$$

so by properties of matrix algebra, we can write

$$A^t B - A^t A \overline{X} = \theta$$

which can be better written as:

$$A^t A \overline{X} = A^t B \tag{1}$$

This last equation is the matrix version of the *normal equations* of the original (inconsistent) system. Look at how simple it is to produce- we have just **multiplied both sides of the original system by A transpose**! If we carry this out in our example, we get:

$$\begin{bmatrix} 1 & 0 & 1 \\ 0 & 1 & 0 \end{bmatrix} \begin{bmatrix} 1 & 0 \\ 0 & 1 \\ 1 & 0 \end{bmatrix} \begin{bmatrix} \overline{x} \\ \overline{y} \end{bmatrix} = \begin{bmatrix} 1 & 0 & 1 \\ 0 & 1 & 0 \end{bmatrix} \begin{bmatrix} 1 \\ 1 \\ 2 \end{bmatrix}$$

$$\begin{bmatrix} 2 & 0 \\ 0 & 1 \end{bmatrix} \begin{bmatrix} \overline{x} \\ \overline{y} \end{bmatrix} = \begin{bmatrix} 3 \\ 1 \end{bmatrix}$$

The best approximation to the solution of the original system $\overline{X}$ is clearly

$$\overline{x} = \frac{3}{2}, \quad \overline{y} = 1$$

Here we got a unique solution to the normal equations- this is not always the case.
Since the size of Z represents the error in our approximate solution, let's calculate the magnitude of Z:

$$\| \ Z \ \| = \| \ B - A \overline{X} \ \| = \| \ B - C \ \| = \left\| \begin{bmatrix} 1 \\ 1 \\ 2 \end{bmatrix} - \begin{bmatrix} \frac{3}{2} \\ 1 \\ \frac{3}{2} \end{bmatrix} \right\| = \sqrt{(-\frac{1}{2})^2 + 0^2 + (\frac{1}{2})^2} = \frac{\sqrt{2}}{2}$$

One can show that this error is *minimized* by using the approximate solution found from equ.(1) (ie. the normal equations). If we actually write out $\| \ B - A \overline{X} \ \|$ for an arbitrary system, we would have

$$\| \ B - A \overline{X} \ \| = \sqrt{(b_1 - c_1)^2 + \dots + (b_n - c_n)^2} \qquad \text{if } A \overline{X} = C$$

Squaring both sides implies

$$\| \ B - A \overline{X} \ \|^2 = (b_1 - c_1)^2 + \dots + (b_n - c_n)^2 \qquad \text{if } A \overline{X} = C$$

which explains the name "least squares", since "solving" an inconsistent system as we did implies that the *sum on the right is minimized* (which is the error squared).

Uniqueness of the least squares solution

We know from before that if A is an m by n matrix and rank(A) = n, then the columns of A are independent. One can prove that this means that $A^t A$ has rank n also and therefore is <u>invertible</u>. Thus the system $(A^t A) \overline{X} = A^t B$ has a **unique solution**. If rank(A) < n, then a unique solution does not exist.

EXAMPLE 1 Determine if the least squares solution to

$$x_1 + 2 x_2 = 0$$

$$- x_1 + 2 x_2 = 5$$

$$4 x_2 = 8$$

is unique.
Writing in matrix form:

$$\begin{bmatrix} 1 & 2 \\ -1 & 2 \\ 0 & 4 \end{bmatrix} \begin{bmatrix} x \\ y \end{bmatrix} = \begin{bmatrix} 0 \\ 5 \\ 8 \end{bmatrix} \text{ where } A = \begin{bmatrix} 1 & 2 \\ -1 & 2 \\ 0 & 4 \end{bmatrix}$$

Clearly rank(A) = 2 so the columns are *independent*. Thus $A^t A$ is invertible and the least squares solution will be unique. To continue with this example, we multiply both sides by A^t:

$$\begin{bmatrix} 1 & -1 & 0 \\ 2 & 2 & 4 \end{bmatrix} \begin{bmatrix} 1 & 2 \\ -1 & 2 \\ 0 & 4 \end{bmatrix} \begin{bmatrix} \bar{x} \\ \bar{y} \end{bmatrix} = \begin{bmatrix} 1 & -1 & 0 \\ 2 & 2 & 4 \end{bmatrix} \begin{bmatrix} 0 \\ 5 \\ 8 \end{bmatrix}$$

which produces the 2 by 2 system (the normal equations):

$$\begin{bmatrix} 2 & 0 \\ 0 & 24 \end{bmatrix} \begin{bmatrix} \bar{x} \\ \bar{y} \end{bmatrix} = \begin{bmatrix} -5 \\ 42 \end{bmatrix}$$

Note that since rank(A) = 2, $A^t A$ is invertible and the normal equations have a unique solution. We see easily that $\bar{y} = \dfrac{42}{24} = \dfrac{7}{4}$ and $\bar{x} = -\dfrac{5}{2}$. To see how good this approximate solution is, we can left-multiply $\bar{X}$ by A; remember that if the system had a solution, AX would turn out to be (0,5,8):

$$\begin{bmatrix} 1 & 2 \\ -1 & 2 \\ 0 & 4 \end{bmatrix} \begin{bmatrix} -\dfrac{5}{2} \\ \dfrac{7}{4} \end{bmatrix} = \begin{bmatrix} 1 \\ 6 \\ 7 \end{bmatrix}$$

The error in this approximation is measured by $\| A\bar{X} - B \|$; here we get

$$\| (1,6,7) - (0,5,8) \| = \| (1,1,-1) \| = \sqrt{3}$$

You should be able to show that $A\bar{X} = (1,6,7)$ is the projection of B = (0,5,8) onto col(A). (1,6,7) is the vector in col(A) *closest* to B.

EXAMPLE 2 Determine if the least squares method for the solution of the inconsistent system

$$x + 2y = 1$$

$$x + 2y = 2$$

$$x + 2y = 3$$

has a unique solution .

Writing the system in matrix form:

$$\begin{bmatrix} 1 & 2 \\ 1 & 2 \\ 1 & 2 \end{bmatrix} \begin{bmatrix} x \\ y \end{bmatrix} = \begin{bmatrix} 1 \\ 2 \\ 3 \end{bmatrix} \text{where } A = \begin{bmatrix} 1 & 2 \\ 1 & 2 \\ 1 & 2 \end{bmatrix}$$

Clearly rank(A) = 1 so the columns are *dependent*. Thus $A^t A$ will be *singular* and the least squares solution will NOT be unique.

EXERCISES 5.5

Part A. Given the matrices $A = \begin{bmatrix} 1 & 0 \\ 0 & 2 \\ 1 & 0 \end{bmatrix}$, $B = \begin{bmatrix} 1 \\ 2 \\ 4 \end{bmatrix}$;

1. Show that AX = B is inconsistent.
2. Find an equation of the plane which is col(A); show B does not lie on this plane.
3. Project B onto col(A) by projecting B onto each column separately and then adding the projections. NB. This is possible since the *columns of A are orthogonal*. ans. (5/2,2,5/2)
4. Solve $A \overline{X} = \beta$ where $\beta = proj_{col(A)} B$.
5. Show that the answer to #4 is the same as that obtained by pre-multiplying both sides of AX = B by A transpose. ans. (5/2,1)

Part B. Given the matrices $A = \begin{bmatrix} 1 & -1 \\ 1 & -1 \\ 1 & 2 \end{bmatrix}$, $B = \begin{bmatrix} 1 \\ 0 \\ 3 \end{bmatrix}$;

6. Show that AX = B is inconsistent.
7. Project B onto col(A) by projecting B onto each column separately and then adding the projections. NB. This is possible since the *columns of A are orthogonal*. ans. (1/2,1/2,3)
8. Solve $A \overline{X} = \beta$ where $\beta = proj_{col(A)} B$.
9. Show that the answer to #8 is the same as that obtained by pre-multiplying both sides of AX = B by A^t. ans. (4/3/,5/6)

Part C. Given the matrices $A = \begin{bmatrix} 1 & -1 & 0 \\ 0 & 2 & 0 \\ 1 & 1 & 0 \\ 0 & 0 & 4 \end{bmatrix}$, $B = \begin{bmatrix} 2 \\ 3 \\ 1 \\ 4 \end{bmatrix}$;

10. Show that AX = B is inconsistent.
11. Find an equation of the hyperplane which is col(A) and show B does not lie on this plane. ans. x + y - z = 0
12. Project B onto col(A) by projecting B onto each column separately and then adding. NB. This is possible since the *columns of A are orthogonal*. ans. (2/3,5/3,7/3,4)

13. Solve $A\overline{X} = \beta$ where $\beta = proj_{col(A)}B$. ans. (3/2,5/6,1)

14. Show that the answer to #13 is the same as that obtained by pre-multiplying both sides of $AX = B$ by A transpose.

Part D. The following points are supposed to lie on the line $y = mx + b$:
P(1,1), Q(2,5) and R(3,7).

15. Substitute the coordinates from each point in the equation to get 3 equations in m and b.
ans. $b + m = 1$, $b + 2m = 5$, $b + 3m = 7$

16. Write down the augmented matrix and show that the system is inconsistent.

17. Use least squares to get the best possible solution to $AX = B$. ans. $y = 3x - 5/3$

18. Plot the points and the line and see how well you did.

Part E. Show each system given is inconsistent; write the system in the form $AX = B$; then find the least squares solution to each inconsistent system by multiplying both sides by A^t if possible. If not possible, state why.

19.
$$x_1 + 2x_2 = 0$$
$$-x_1 + 2x_2 = 5$$ ans. (-5/2,7/4)
$$4x_2 = 8$$

20.
$$x_1 + x_2 + x_3 = 1$$
$$2x_1 + 2x_2 + 2x_3 = 3$$

21.
$$x_1 + x_3 = 1$$
$$x_2 + x_3 = 2$$
ans. (1,3/2,1/2)
$$x_1 + x_2 + x_3 = 3$$
$$x_1 + x_3 = 2$$

$$x_1 + 2\,x_2 = 3$$

22. $$2\,x_1 + 4\,x_2 = 7$$

$$2\,x_1 + 2\,x_2 = 4$$

F. The following points are supposed to lie on the parabola $y = a\,x^2 + b\,x + c$:
P(-1,2), Q(2,5), and R(3,8) and S(4,14).
23. Substitute these points into the equation and get a system in a, b and c.
$$a - b + c = 2$$

$$4\,a + 2\,b + c = 5$$
ans.

$$9\,a + 3\,b + c = 8$$

$$16\,a + 4\,b + c = 14$$

24. Write down the augmented matrix and show the system in inconsistent.
25. Using least-squares, find the "best parabola" that fits the given points.
ans. a = 267/352, b = 51/362, c = 260/181

G. Find $C = A^t\,A$ and determine whether C is invertible or not. Observe that C is singular when the columns of A are dependent.

26. $A = \begin{bmatrix} 1 & 1 \\ 2 & 1 \\ 3 & 1 \end{bmatrix}$

27. $A = \begin{bmatrix} 1 & 0 \\ 0 & 2 \\ 1 & 0 \end{bmatrix}$ ans. C invertible

28. Using A from #26, find the "best solution" to $A\,X = \begin{bmatrix} 1 \\ 3 \\ 4 \end{bmatrix}$ using least squares.

29. Using A from #27, solve the inconsistent system $A\,X = \begin{bmatrix} 1 \\ 2 \\ 3 \end{bmatrix}$ using least squares.

ans. $\bar{x} = 2,\ \bar{y} = 1$
30. Using A from #26, show that rank(A) = rank(C).

31. Using A from #27, show that rank(A) = rank(C).

32. Show that null(A) = null($A^t A$) using A from #26.

33. $A = \begin{bmatrix} 1 & 2 \\ 1 & 2 \\ 2 & 4 \end{bmatrix}$ ans. C singular

34. Using C from #33, if B = $\begin{bmatrix} 7 \\ 8 \\ 9 \end{bmatrix}$, show that AX = B is *inconsistent*. Then attempt the least

squares method for solving AX = B; show it will NOT produce a unique approximate solution.

5.6 Application of Orthogonality - Fourier Series

Most of the applications of orthogonality in function spaces (ie. inner product spaces whose elements are functions) involve an *integral* inner product. The idea is, given a particular subspace W in the IPS and a function NOT in W, to find the "best approximation" to the function which DOES lie in W. From Section 5.4, we know our method of attack- we need to *project* the function that we wish to approximate onto the subspace W. This projection will then be the "best approximation" which we seek. Of course having an orthogonal basis for the subspace is a key element of this process. Producing an orthogonal basis from any basis can always be accomplished by using the *Gram-Schmidt* process.

Introduction to Fourier Series

Consider the vector space of continuous functions $V = C [-\pi, \pi]$. A natural inner product to use on V is the integral inner product which we have already used:

$$< f, g > = \int_{-\pi}^{\pi} f \ g \, d x$$

We need to have a collection of subspaces of V which get successively larger (ie. their dimensions increase). Let's let T_1 be the following collection of trigonometric polynomials:

$$T_1 = \{ \ 1, \cos (\ x \ ), \sin (\ x \ ) \}$$

Verify that this set is orthogonal, hence independent so that $W_1 = span (\ T_1 \ )$ is a 3-dimensional subspace of the form

$$W_1 = \{ \ f (\ x \ ): f (\ x \ ) = k_1 \cdot 1 + k_2 \cos (\ x \ ) + k_3 \sin (\ x \ ), k_i \in \Re \ \}$$

Suppose we pick some function y which is not in W_1 and we wish to project this function y onto the subspace W_1. We recall from Section 5.4 that we must project onto an *orthogonal basis* for the subspace. Thus we can project y onto each basis function in T_1 and simply add the projections together. Remember the difference between integrating an even/odd function on a symmetric interval. This will save you some time in this section.

Let's pick a simple function such as y = x. Clearly this function is in $V = C [-\pi, \pi]$. The formula for the projection is thus:

$$proj_{W_1} \ y = proj_1 \ y + proj_{\cos (x)} \ y + proj_{\sin (x)} \ y$$

$$proj_{W_1} y = \frac{<x, 1>}{<1, 1>} \ 1 + \frac{<x, \cos (\ x \ )>}{<\cos (\ x \ ), \cos (\ x \ )>} \cos (\ x \ ) + \frac{<x, \sin (\ x \ )>}{<\sin (\ x \ ), \sin (\ x \ )>} \sin (\ x \ )$$

Keep in mind that all of these inner products that we need to calculate are *definite integrals*:

$$proj_{W_1} \; y = \frac{\int\limits_{-\pi}^{\pi} x \cdot 1 \, dx}{\int\limits_{-\pi}^{\pi} 1 \cdot 1 \, dx} \, 1 + \frac{\int\limits_{-\pi}^{\pi} x \cos(x) \, dx}{\int\limits_{-\pi}^{\pi} \cos(x) \cos(x) \, dx} \cos(x) + \frac{\int\limits_{-\pi}^{\pi} x \sin(x) \, dx}{\int\limits_{-\pi}^{\pi} \sin(x) \sin(x) \, dx} \sin(x)$$

Now the first two numerators involve integrals of functions which are *odd* and from calculus we know that these integrals are zero- we need to evaluate only the last integral (since odd x odd = even). In the denominator, it is nice to remember that

$$\int\limits_{-\pi}^{\pi} [\sin^2(x) + \cos^2(x)] \, dx = 2\pi \Rightarrow \int\limits_{-\pi}^{\pi} \cos^2(x) \, dx = \int\limits_{-\pi}^{\pi} \sin^2(x) \, dx = \pi$$

Hence we have

$$proj_{W_1} \; y = 0 \cdot 1 + 0 \cdot \cos(x) + \frac{2\pi}{\pi} \sin(x) = 2\sin(x)$$

Thus the projection of y = x on W_1 is the function 2 sin(x). Note that since f(x) = x is ODD, it has only odd functions in the projection (ie. 1 and cos(x) are even functions). Now suppose we let $T_2 = \{1, \cos(x), \sin(x), \cos(2x), \sin(2x)\}$ and consider the 5-dimensional space

$$W_2 = span(T_2) = \{ k_1 \cdot 1 + k_2 \cos(x) + k_3 \sin(x) + k_4 \cos(2x) + k_5 \sin(2x) \}$$

If we want to project y = x onto W_2, we again project it onto a basis for W_2; we've already done the first three functions. The first integral must be zero since y = x is *odd* and cosine is <u>even</u>:

$$proj_{\cos(2x)} \; y = \frac{<x, \cos(2x)>}{<\cos(2x), \cos(2x)>} \cos(2x) = \frac{\int\limits_{-\pi}^{\pi} x \cos(2x) \, dx}{\int\limits_{-\pi}^{\pi} \cos^2(2x) \, dx} \cos(2x) = 0$$

So the only work necessary is the inner product of y with sin(2x), since odd times odd is even:

$$proj_{\sin(2x)}\ y = \frac{<x,\sin(2x)>}{<\sin(2x),\sin(2x)>}\sin(2x) = \frac{\int_{-\pi}^{\pi} x\sin(2x)\,dx}{\int_{-\pi}^{\pi}\sin^2(2x)\,dx}\sin(2x) = \frac{-\pi}{\pi}\sin(2x) = -\sin(2x)$$

so that the projection of y = x onto W_2 is:

$$proj_{W_2}\ y = 2\sin(x) - \sin(2x)$$

Now we can repeat this process - let T_3 = {1, cos(x),sin(x), cos(2x), sin(2x), cos(3x), sin(3x)} and W_3 = span(T_3) which will be a 7-dimensional space. We would need to calculate the projection of y = x onto cos(3x) (we know that will be zero) and also the projection of y = x onto sin(3x) (which will not be zero):

$$proj_{\sin(3x)}\ y = \frac{<x,\sin(3x)>}{<\sin(3x),\sin(3x)>}\sin(3x) = \frac{\int_{-\pi}^{\pi} x\sin(3x)\,dx}{\int_{-\pi}^{\pi}\sin^2(3x)\,dx}\sin(3x) = \frac{2}{3}\sin(3x)$$

Thus we have $\quad proj_W\ y = 2\sin(x) - \sin(2x) + \frac{2}{3}\sin(3x)$

projectionofyontoW_3

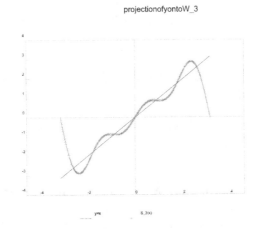

This is the function in W_3 closest to y = x. If we keep adding sinusoids of higher frequency, we will get better and better approximations of the function y = x in terms of cosines and sines- if we include all possible sines and cosines, we have an *infinite series* of cosines and sines. This series is called the *Fourier*[4] *series* for the function y = x.

Intuitively, we are thinking that as we add more and more sinusoids with higher frequencies to W_n, the projection of y onto W_n will get better and better.

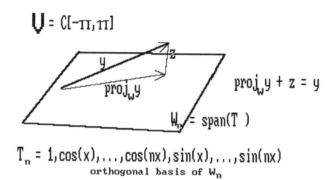

This means that in general, as n gets larger, the *error function* $\| Z_n \| = \| y - proj_{W_n} y \|$ gets smaller and each partial sum $S_n(x)$ does a better job of approximating the function y. If we let

$$S(x)= a_0 + \sum_{n=1}^{\infty} [a_n \cos (n x)+ b_n \sin (n x)]$$

then the coefficients $a_0, a_n, \& b_n$ are found by projecting y onto an infinite orthogonal basis

$$T = \{ 1, \cos (x), \sin (x),..., \cos (n x), \sin (n x),... \} \subseteq W$$

Here S(x) is the **Fourier series** for y, where W = span(T) is an infinite-dimensional subspace of $C [-\pi, \pi]$, consisting of linear combinations of cosines and sines with integral frequencies. The n-th partial sum would serve as an approximation to y; it is of the form

$$S_n(x)= a_0 + [a_1 \cos(x)+...+ a_n \cos (n\ x)] + [b_1 \sin (x)+...+ b_n \sin (n x)]$$

Remember that each term in the series is the projection of y onto that particular term; if a particular term is missing, then y was *orthogonal* to that term in the series. In general one wants a *formula* for these coefficients in terms of n. For example, from the previous problem we would be able to show that

---

[4] Fourier submitted his ideas on trigonometric series in conjunction with the solution to the problem of conduction of heat in a monograph to Institut de France in 1807.

$$a_0 = \frac{1}{2\pi} \int_{-\pi}^{\pi} y \, dx = 0 \quad a_n = \frac{1}{\pi} \int_{-\pi}^{\pi} y \cos(nx) \, dx = 0 \quad b_n = \frac{1}{\pi} \int_{-\pi}^{\pi} y \sin(nx) \, dx = \frac{-2\cos(n\pi)}{n}$$

and our Fourier series for y = x is of the form:

$$S(x) = 2\sin(x) - \frac{2\sin(2x)}{2} + \frac{2\sin(3x)}{3} - \frac{2\sin(4x)}{4} + \dots$$

Note that since y was odd, the series has only sine terms. Rewriting gives:

$$S(x) = 2[\sin(x) - \frac{\sin(2x)}{2} + \frac{\sin(3x)}{3} - \frac{\sin(4x)}{4} + \dots]$$

The coefficients of a Fourier series usually have a fairly simple pattern- and they decrease as n gets larger. When evaluating the integrals, a nice fact to remember is that an *even* function has only *cosines* (which includes the constant term) in its expansion and an odd function has only sines in its expansion. This will save you some time when doing all the integration. Of course the question of convergence[5] comes up naturally here. Suffice it to say that if

$$S_n(x) = a_0 + \sum_{k=1}^{n} [a_k \cos(kx) + b_k \sin(kx)]$$

then the error in using the n-th partial sum $S_n(x)$ is given by

$$\| Z_n \| = \| y - S_n \| = \sqrt{\int_{-\pi}^{\pi} (y - S_n(x))^2} = \| y \|^2 - \| S_n(x) \|^2$$

Looking at the norm of the error function, one sees that as $\| S_n(x) \|$ increases, the norm of y remains constant so that the error function is *shrinking*. The projection "child" is growing but it can never get larger than the "parent" function y.

The convergence of the series is "convergence in the mean" which implies that

$$\lim_{n \to \infty} \| y - S_n \| = \lim_{n \to \infty} \sqrt{\int_{-\pi}^{\pi} (y - S_n(x))^2} = 0$$

if $\int_{-\pi}^{\pi} y^2 \, dx$ is finite.

WARNING- the series may not converge *pointwise* for all x = c on $[-\pi, \pi]$.

---

[5] An excellent reference is Churchill's book *Fourier Series and Boundary Value Problems*

S(x) converges pointwise at x = c if

$$\lim_{n \to \infty} S(x)= y(c)$$

However, **if** the Fourier series for y DOES converge at x = c, it will converge to y(c).

EXAMPLE 2 Find formulae for the terms in the Fourier series of the function y where $y = x^2$ using

$$< f, g > = \int_{-\pi}^{\pi} f\ g\ d x$$

Since y is an EVEN function, we don't need to do the sine integrals. Thus there are only two integrations which must be done:

$$a_0 = \frac{1}{2\pi} \int_{-\pi}^{\pi} f(x)d x, \quad a_n = \frac{1}{\pi} \int_{-\pi}^{\pi} f(x)\cos(n x)d x$$

The factors in front of the integrals involving π arise from the fact that $\| 1 \| = \sqrt{2\pi}$ and that the norm of any cosine or sine is $\sqrt{\pi}$; but in the projection formulae, the denominators contain the magnitude *squared* of the basis function.
Let's first project y onto $f_0 = 1$;

$$a_0 = \frac{1}{2\pi} \int_{-\pi}^{\pi} y\ d x = \frac{1}{2\pi} \int_{-\pi}^{\pi} x^2\ d x \quad = \frac{\left[\dfrac{x^3}{3}\right]_{-\pi}^{\pi}}{2\pi} = \frac{\pi^2}{3}$$

From calculus we know that this is the *average value* of the function y; a good way to start!
Now for the projection onto the cosine functions:

$$a_n = \frac{1}{\pi} \int_{-\pi}^{\pi} x^2 \cos(n x)d x$$

$$a_n = \frac{1}{\pi} [\ \frac{n^2 x^2 \sin(n x) - 2 \sin(n x) + 2 n x \cos(n x)}{n^3}\ ]_{-\pi}^{\pi} a_n = \frac{4 \cos(n \pi)}{n^2}$$

Thus the series is of the form

$$S(x) = \frac{\pi^2}{3} - 4 [\ \frac{\cos(x)}{1^2} - \frac{\cos(2 x)}{2^2} + \frac{\cos(3 x)}{3^2} - ...\ ]$$

Since y is an *even* function, no sine terms appear in the series since they are odd. Thus the series has no sine terms, since the corresponding coefficients are *zero*.

Note that the *farther out* in the series you go, the smaller the coefficients get. This means as frequency increases, the *amplitudes* of the sinusoids *decrease*.

projofyontoW=span(T_3)

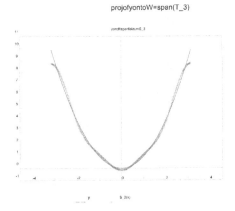

EXAMPLE 3 Find formulae for the terms in the Fourier series of the function y where $y = \sin^3(x)$ using

$$< f, g > = \int_{-\pi}^{\pi} f \ g \ d \ x$$

Since y is an ODD function, we need to do only the sine integrals.

$$b_n = \frac{1}{\pi} \int_{-\pi}^{\pi} \sin^3(x) y \sin(nx) dx \Rightarrow b_1 = \frac{3}{4}, b_2 = 0, b_3 = -\frac{1}{4}, b_4 = 0, b_5 = 0, \dots$$

Thus the Fourier "series" here is actually just a trigonometric polynomial; we can write

$$S(x) = \frac{3}{4} \sin(x) - \frac{1}{4} \sin(3x)$$

and in fact y = S(x) since y can be written as a linear combination of sin(x) and sin(3x). We could write y in this form using old-fashioned trig identities.

Discontinuous functions and Fourier series

One can find a Fourier series even for functions which are *not continuous* as long as they are piecewise continuous; specifically we must work in the IPS V of **square-integrable** functions:

$$V = \{ y : \int_{-\pi}^{\pi} y^2 d \ x \text{ exists} \}$$

Of course $C[-\pi, \pi]$ is a proper subspace of this IPS, since a continuous function is automatically

277

square-integrable.

EXAMPLE 4 Find a Fourier series for the discontinuous function

$$y = \begin{cases} 0 & -\pi \le x < 0 \\ 1 & 0 \le x \le \pi \end{cases}$$

This function has a jump discontinuity at x = 0 but it will still have a Fourier series. The first term is the *average value* of the function:

$$a_0 = \frac{1}{2\pi} \int_{-\pi}^{\pi} y \, dx = \frac{\pi}{2\pi} = \frac{1}{2}$$

All the cosine coefficients come from the integral

$$a_n = \frac{1}{\pi} \int_{-\pi}^{\pi} y \, \cos(nx) \, dx$$

It turns out that all the cosine coefficients are zero; thus we need to get the sine coefficients:

$$b_n = \frac{1}{\pi} \int_{-\pi}^{\pi} y \, \sin(nx) \, dx$$

Since the value of the function is zero on the first half of the interval, we can just integrate over the second half:

$$b_n = \frac{1}{\pi} \int_{0}^{\pi} 1 \, \sin(nx) \, dx = \frac{[-\cos(nx)]_0^{\pi}}{n\pi} = \frac{1 - \cos(n\pi)}{n\pi}$$

Thus we can write out a few terms in the Fourier series for y:

$$S(x) = \frac{1}{2} + \frac{2}{\pi} [\sin(x) + \frac{\sin(3x)}{3} + \frac{\sin(5x)}{5} + \dots]$$

Note that the coefficients *decrease* as you farther out in the series; also the even harmonics are missing (ie. no even frequencies) for this function y. Below is a graph for you to enjoy.

Fourierseriesofadisc.function

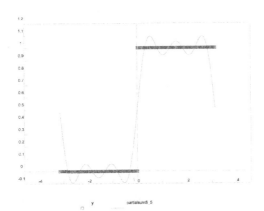

y   partialsum5 5

As a vindication of the power of the Fourier series, if you let $x = \frac{\pi}{2}$, y = 1; using all terms up to

n = 5 gives an approximation $S_5(\frac{\pi}{2}) = 1.052$ which is rather close to the true value. Using more

terms in the series would increast the accuracy of this approximation.

Interesting numerical series can sometimes be summed using particular values of x where the series

converges pointwise; for example, the reader should be able to show that by substituting x = $\frac{\pi}{2}$, we

get

$$1 = \frac{1}{2} + \frac{2}{\pi}(1 - \frac{1}{3} + \frac{1}{5} - \frac{1}{7} + \dots) \Rightarrow \frac{\pi}{4} = 1 - \frac{1}{3} + \frac{1}{5} - \frac{1}{7} + \dots$$

EXERCISES 5.6

Part A. Use the inner product $< f, g > = \int_{-\pi}^{\pi} f \, g \, d \, x$ for $C [-\pi, \pi]$ where $W = span\{B\} =$ if

B = {1, cos(x), sin(x)} and y = x + 1;

1. Find $proj_W y$. ans. 1 + 2 sin(x)

2. Find the function z satisfying $proj_W y + z = y$.

3. Find an approximation for y(1) in W using the answer to #1. ans. 1 + 2sin(1)

4. Now add the functions {cos(2x),sin(2x)} to B. Show that the new larger basis B is orthogonal.

5. Project y onto W using the basis B from #4.   ans. 1 + 2sin(x) - sin(2x)

6. Find the function z satisfying $proj_W y + z = y$.

7. Find an approximation for y(1) in W using the answer to #5. Compare to the exact value y(1). Is this a better approximation? ans. 1 + 2sin(1) - sin(2) = 1.7736, y(1) = 2, YES

8. Conjecturing that all cosine coefficients are zero, find a formula for the sine coefficients.

9. Use the formula from #8 to write FIVE non-zero terms in the Fourier series for y.
ans. y = 1 + 2[sin(x) - sin(2x)/2 + sin(3x)/3 - sin(4x)/4 + ...]

Part B. Use the inner product $< f, g >= \int_{-1}^{1} f\, g\, d\,x$ for $V = C[-1,1]$, where $W = \text{span}\{B\} = $ if

$$B = \{\, \frac{1}{\sqrt{2}}, \cos(\pi x), \sin(\pi x)\,\} \text{ and } y = x^2.$$

10. Show that the set B is actually ortho*normal*.

11. Find $proj_W\, y$. ans. $\dfrac{1}{3} - \dfrac{4\cos(\pi x)}{\pi^2}$

12. Find the function z satisfying $proj_W\, y + z = y$.

13. Find an approximation for y(1/2) in W using the answer to #11. ans. 1/3

14. Now add the functions $\cos(2\pi x), \sin(2\pi x)$ to B. Project y onto W using this larger basis B.

15. Find the function z satisfying $proj_W\, y + z = y$.

ans. $x^2 - [\, \dfrac{1}{3} - \dfrac{4\cos(\pi x)}{\pi^2} + \dfrac{4\cos(2\pi x)}{4\pi^2}\,]$

16. Conjecturing that all sine coefficients are zero, find a formula for the cosine coefficients.

17. Use the formula from #16 to write FOUR non-zero terms in the Fourier series for y.

ans. y = $\dfrac{\sqrt{2}}{3} - \dfrac{4}{\pi^2}[\, \cos(\pi x) - \dfrac{\cos(2\pi x)}{2^2} + \dfrac{\cos(3\pi x)}{3^2} - \dfrac{\cos(4\pi x)}{4^2} + \dots\,]$

Part C: Use the integral inner product $< f, g >= \int_{-\pi}^{\pi} f\, g\, d\,x$ for $V = \{\, f : \int_{-\pi}^{\pi} f\, d\,x\ exists\ \}$

where $y = \begin{array}{ll} -1 & -\pi \le x < 0 \\ 1 & 0 \le x \le \pi \end{array}$

18. Let B = {1, cos(x),sin(x)} and find the projection of y onto W, where W = span(B).

19. Find a function z, orthogonal to W such that $z = y - \dfrac{4}{\pi}\sin(x)$.

20. Use the projection from #18 to find an approximation for y(1).

21. Add {cos(2x),sin(2x),cos(3x),sin(3x)} to the basis B; project y onto the new, larger subspace W. ans. $\dfrac{4\sin(x)}{\pi} + \dfrac{4\sin(3x)}{3\pi}$ NB. In this problem, the *even harmonics* are missing

22. Find a formula for the sine coefficients in the Fourier series for y.

23. Write out four terms in the Fourier series for y.

ans. $\dfrac{4}{\pi}[\, \sin(x) + \dfrac{\sin(3x)}{3} + \dfrac{\sin(5x)}{5} + \dfrac{\sin(7x)}{7} + \dots\,]$

NB. Since y is ODD, we observe only sine terms in the Fourier series; all is well.

Part D. Use the interval [-1,1] with V = C[-1,1] and inner product $< f, g >= \int_{-1}^{1} f\ g\ d\ x$; let

$B = \{\ 1, \cos(\ \pi\ x\ ), \sin(\ \pi\ x\ )\ \}$ and W = span(B).

24. Show that B is an orthogonal set wrt. the inner product given.

25. Let y = x; project y onto W. ans. $2\dfrac{\sin(\ \pi\ x\ )}{\pi}$

26. Find a function in V which is orthogonal to W and satisfies $proj_W\ y + z = y$.

27. Use your answer to #25 to approximate y(1/2). ans. 0.6366

28. Find a formula for the Fourier coefficients of y.

29. Write out at least four terms in the Fourier series for y.

ans. $\dfrac{2}{\pi}[\ \sin(\ \pi\ x\ ) + \dfrac{\sin(\ 3\ \pi\ x\ )}{3} + \dfrac{\sin(5\ \pi\ x)}{5} + \dfrac{\sin(\ 7\ \pi\ x\ )}{7} + ...\ ]$

30. Now enlarge B so that $B = \{\ 1, \cos(\ \pi\ x\ ), \sin(\ \pi\ x\ ), ..., \cos(3\ \pi\ x\ ), \sin(3\ \pi\ x\ )\ \}$ and let

$$y = \begin{cases} 1 & -1 \le x < 0 \\ 3 & 0 \le x \le 1 \end{cases}$$

Find $proj_W\ y$ if W = span(B).

Part E: Use V = C[-L,L] with the inner product $< f, g >= \int_{-L}^{L} f\ g\ d\ x$ and the orthogonal basis

$B = \{\ 1, \cos(\ \omega\ x\ ), \sin(\ \omega\ x\ )\ \}$

31. Find the frequency omega so that the period T = 2L. ans. $\omega = \dfrac{\pi}{L}$

32. Show that B is an *orthogonal* basis wrt. the given inner product if $\omega = \dfrac{\pi}{L}$.

33. Find the projection of the function

$$y = \begin{cases} 0 & -L \le x < 0 \\ 1 & 0 \le x \le L \end{cases}$$

onto W = span(B). ans. $\dfrac{1}{2} + \dfrac{2}{\pi}\sin(\dfrac{\pi\ x}{L})$

34. Get a Fourier series for y, writing out at least four terms.

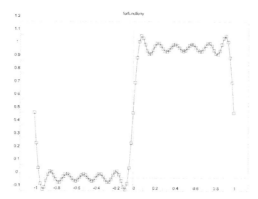

partial sum of Fourier series with n = 11

5.7 Chapter Five Outline and Review

5.1. Scalar Product

The scalar (dot) product in $\Re^n$ satisfies the following properties:

1) $X \cdot Y = Y \cdot X$ (symmetry)
2) $X \cdot (Y + Z) = X \cdot Y + X \cdot Z$ (additivity)
3) $(kX \cdot Y) = k(X \cdot Y)$ (homogeneity)
4) $X \cdot X \geq 0$ and $X \cdot X = 0$ iff $X = \theta$ (positivity)

Norm (magnitude or length) of a vector: $\|X\| = \sqrt{X \cdot X}$

Distance: $d(X, Y) = \|X - Y\|$

Angle: $\cos(\phi) = \dfrac{X \cdot Y}{\|X\| \|Y\|}$ $0 \leq \phi \leq \pi$

Orthogonality: nonzero vectors X and Y are orthogonal (perpendicular) if $X \cdot Y = 0$

Projection: to project Y onto X -

$$proj_X Y = (\frac{Y \cdot X}{X \cdot X}) X = (\frac{Y \cdot X}{\|X\|^2}) X$$

where the "error" vector

$$Z = Y - proj_X Y$$

is always orthogonal to X
(its magnitude measures the distance between Y and its projection onto X)

Orthogonal complement- if W is a subspace of V, then $W^\perp$ (W perp) consists of all vectors in V orthogonal to a *basis* for W

Important example - for any matrix A, null(A) is the *orthogonal complement* of row(A).

5.2 Inner Product Spaces

The function $\| \cdot \|$ is a **norm** on the vector space V if

N1) $\|X\| \geq 0$ & $\|X\| = 0$ iff $X = \theta$

N2) $\|kX\| = |k| \|X\|$

N3) $\|X + Y\| \leq \|X\| + \|Y\|$ (triangle inequality)

Normed vector space(NVS)- vector space plus a norm; model is Euclidean n-space with the norm

$$\|X\| = \sqrt{x_1^2 + x_2^2 + ... + x_n^2}$$

Inner product axioms:
IP1) $< X, Y > = < Y, X >$ (symmetry)
IP2) $< X + Y, Z > = < X, Z > + < Y, Z >$ (additivity)

IP3) $< k\, X, Y > = k < X, Y >$ (homogeneity)

IP4) $< X, X > \geq 0$ and $< X, X > = 0$ iff $X = \theta$ (positivity)

These axioms reflect the properties of the scalar product on $\Re^n$.

Inner product space (IPS)- vector space V plus inner product (model is scalar product on $\Re^n$)

Important example- C[a,b] is an IPS if $< f , g > = \int_a^b f\ g\ d\ x$

(this is the *standard* inner product for C[a,b])

THEOREM Every inner product space(IPS) is a normed vector space(NVS).

Just let

$$\| X \| = \sqrt{< X, X >}$$

then the IPS becomes a NVS also .

In any IPS:

1) Cauchy-Schwarz $| < X, Y > | \leq \| X \| \| Y \|$

2) Orthogonality $X \perp Y$ iff $<X, Y> = 0$

3) Pythagoras $X \perp Y \Rightarrow \| X + Y \|^2 = \| X \|^2 + \| Y \|^2$

4) Angle $\cos(\phi) = \dfrac{< X, Y >}{\| X \| \| Y \|}$ where $0 \leq \phi \leq \pi$

5.3 Orthonormal Bases

Orthogonal basis- vectors are pairwise orthogonal (eg. **i-j-k** in 3-space)

Normal vectors- magnitude(norm) is one

Orthonormal set- orthogonal and normal

If B = $\{ X_1, X_2, ..., X_n \}$ is an *orthogonal* basis for V(an IPS) then

$$X = \frac{< X, X_1 >}{< X_1, X_1 >} X_1 + \frac{< X, X_2 >}{< X_2, X_2 >} X_2 + ... + \frac{< X, X_n >}{< X_n, X_n >} X_n$$

If B is ortho*normal*, then this simplifies to

$$X = < X, X_1 > X_1 + < X, X_2 > X_2 + ... + < X, X_n > X_n$$

5.4 Projection and Gram-Schmidt

Projecting Y onto X in IPS V: $proj_X Y = \dfrac{< Y, X >}{< X, X >} X = \dfrac{< Y, X >}{\| X \|^2} X$

Gram-Schmidt- making an arbitrary basis into an orthonormal basis; key is *subtracting away a projection* to get an orthogonal vector

Given B = $\{ X_1, X_2, X_3 \}$ in IPS V;

let $Y_1 = X_1$ then

$$Y_2 = X_2 - proj_{Y_1} X_2 = X_2 - \frac{<X_2,Y_1>}{<Y_1,Y_1>} Y_1$$

and

$$Y_3 = X_3 - proj_{Y_1} X_3 - proj_{Y_2} X_3 = X_3 - \frac{<X_3,Y_1>}{<Y_1,Y_1>} Y_1 - \frac{<X_3,Y_2>}{<Y_2,Y_2>} Y_2$$

where B' = { Y_1, Y_2, Y_3 } is orthogonal; now normalize by dividing each vector by its norm

Projecting onto a subspace: Given an *orthogonal* set B where W = span(B) and Y does NOT lie in W; then to project Y onto W, simply *add all projections of Y onto each basis vector* in B:

$$proj_W Y = proj_{X_1} Y + proj_{X_2} Y + ...$$

where

$$proj_W Y + Z = Y \Rightarrow Z = Y - proj_W Y$$

where Z - a vector orthogonal to W

5.5 Least Squares Approximation

To solve the inconsistent system AX = B:

1) multiply both sides by A^t

2) solve the corresponding system ($A^t A$) $\overline{X} = A^t B$ to get the *least squares* solution $\overline{X}$

Geometrically- you are projecting B onto col(A), thus finding the vector IN col(A) which is closest to B and solving the resulting system ($\overline{X}$ minimizes the error in the approx. solution)

Unique solution- if A(m x n) has independent columns (so rank(A) = n)

5.6 Application of Orthogonality - Fourier Series

Given B = {1, cos(x), sin(x)} in the IPS $V = C [-\pi, \pi]$, where W = span(B), using the inner

product $< f, g > = \int_{-\pi}^{\pi} f \, g \, d \, x$; projecting a function $y \notin W$ onto W is done by

1) projecting y onto each basis vector in B and adding:

$$proj_W y = proj_{f_1} y + proj_{f_2} y + proj_{f_3} y$$

2) this is possible since B is an orthogonal basis

3) the *error vector* z (orthogonal to W) is found using

$$proj_W y + z = y \Rightarrow z = y - proj_W y$$

$\| z \|$ is *minimized* by this projection, since the projection is the vector closest to y in W

Fourier series: if T = {1,cos(x),sin(x),...,cos(nx),sin(nx),...}, W = span(T), and $V = C [-\pi, \pi]$ then projecting a function y onto W creates the *Fourier series* for y, where

$$a_0 = \frac{1}{2\pi} \int_{-\pi}^{\pi} y\, dx, \quad a_n = \frac{1}{\pi} \int_{-\pi}^{\pi} y \cos(nx)\, dx, \quad b_n = \frac{1}{\pi} \int_{-\pi}^{\pi} y \sin(nx)\, dx$$

The partial sums

$$S_n = a_0 + [\, a_1 \cos(x) + ... + a_n \cos(nx)\,] + [\, b_1 \sin(x) + ,,, + b_n \sin(nx)\,]$$

create better and better approximations of the function y as n increases; in fact

$$\lim_{n \to \infty} \| y - S_n \| = 0$$

and the Fourier series S(x) is given by:

$$S(x) = a_0 + \sum_{n=1}^{\infty} [\, a_n \cos(nx) + b_n \sin(nx)\,]$$

for any function y which has $\int_{-\pi}^{\pi} y^2\, dx < \infty$ where S(x) converges in the mean to y.

Review Problems

Part A: Given X = (1,2) and Y = (8,-4) using the ordinary dot product find:
1. norm of X and norm of Y ans. $\sqrt{5}, 4\sqrt{5}$
2. norm of X + Y ans. $\sqrt{85}$
3. d(X,Y) ans. $\sqrt{85}$
4. angle ϕ between X and Y ans. $\dfrac{\pi}{2}$
5. projection of Y onto X ans. (0,0)

Part B: Given $\| f \| = \max_{x \in [0,4]} | f(x)|$ is a norm on V = C[0,4] and f = 2x and g = x - 3;
6. Find $\| f \|$, $\| g \|$. ans. 8, 3
7. Show $\| -4g \| = |-4| \| g \|$. ans. both equal 12
8. Find $\| f + g \|$. ans. 9
9. Show $\| f + g \| \le \| f \| + \| g \|$.

Part C: Given the function $< p, q > = a_1 a_2 + b_1 b_2$ on P_1 (for $p = a + bx$):
10. Show the function is an inner product on P_1
11. Find $\| 3 - 4x \|$. ans. 5
12. Find d(p,q) if p = 3 - 4x and q = -2 + 5x. ans. $\sqrt{106}$

13. Find $proj_q\, p$ using p & q from #12. ans. 1/29(52 - 130x)

14. Find the form of any linear polynomial orthogonal to p.ans. b(4/3 + x)

Part D: Given W = span$\{(1,1,0,2),(1,1,0,-1)\}$ in $\mathcal{R}^4$;

15. Find $W^\perp$. ans. span($\{(-1,1,0,0),(0,0,1,0)\}$)

16. Given the vector X = (3,4,5,6) find a vector X_w in W and X_p in W perp such that
X = X_w + X_p. ans.(7/2,7/2,0,6) + (-1/2,1/2,5,0)

Part E: Let W = span$\{(1,1,0,2),(0,1,1,1),(1,2,1,4)\}$;

17. Use these vectors to form the rows of matrix A and find a basis for row(A).
ans. $\{(1,1,0,2),(0,1,1,1),(0,0,0,1)\}$

18. Using null(A), find $W^\perp$. ans. span($\{(1,-1,1,0)\}$)

19. Write the vector X = (1,2,3,4) as the sum of a vector in W and a vector in $W^\perp$.
ans. (1/3,8/3,7/3,4) + (2/3,-2/3,2/3,0)

20. Show that col(A) is the orthogonal complement of the *left null space* of A (ie. $null(\,A^t\,)$) by finding null(A^t). ans.$\{(0,0,0)\}$

Part F: Given W = span$\{(-1,-7,1,1)\}$ (ie. a line in 4-space);

21. find a basis for $W^\perp$ ans. S = $\{(1,0,0,1),(0,1,0,7),(0,0,1,-1)\}$

22. find an equation for $W^\perp$ by letting X = (x,y,z,w) be an arbitrary vector in $W^\perp$
ans. x + 7y - z - w = 0

23. show that the 3 vectors given in the set S above are the rows from the reduced echelon form of a matrix A whose null space is spanned by (-1,-7,1,1)

Part G: Use the Gram-Schmidt process to find an orthogonal set using the given set of vectors with the inner product given.

24. $B = \{\, 1, x, x^2 \,\}$ in C[0,1] with inner product $< f, g >= \int_0^1 f\, g\, d\, x$

ans. $\{\, 1, x - \dfrac{1}{2}, x^2 - x - 6 \,\}$

25. $B = \{\, 1, x, x^2 \,\}$ in C[-1,1] with inner product $< f, g >= \int_{-1}^1 f\, g\, d\, x$

ans. $\{\, 1, x, x^2 - \dfrac{1}{3} \,\}$

26. normalize your answer to #25 ans. $\dfrac{1}{\sqrt{2}}$, $\dfrac{\sqrt{6}}{2}\, x$, $\dfrac{3\sqrt{10}}{4}\,(\, x^2 - \dfrac{1}{3} \,)$

27. B = $\{(1,0,1,0),(0,1,0,1),(0,1,1,1)\}$ (use ordinary dot product)
ans.$\{(1,0,1,0),(0,1,0,1,),(-1/2,0,1/2,0)\}$

28. normalize your answer to #27

ans.$\{ (\frac{1}{\sqrt{2}}, 0, \frac{1}{\sqrt{2}}, 0), (0, \frac{1}{\sqrt{2}}, 0, \frac{1}{\sqrt{2}}), (\frac{-1}{\sqrt{2}}, 0, \frac{1}{\sqrt{2}}, 0) \}$

29. $B = \{ 1+x, 1-x, x^2 \}$ where $< p, q > = a_0 a_1 + b_0 b_1 + c_0 c_1$

ans. B is already orthogonal

30. normalize your answer to # 29 ans. $B = \{ \frac{1+x}{\sqrt{2}}, \frac{1-x}{\sqrt{2}}, x^2 \}$

Part H: Given the hyperplane x - z - w = 0 in $\Re^4$;

31. By solving for w and thinking that if X = (x,y,z,w) lies in the hyperplane, find a basis B for this hyperplane consisting of THREE vectors.
ans.B = {(1,0,0,1),(0,1,0,0),(0,0,1,-1)}

32. Show that B is NOT orthogonal; make it orthogonal using Gram-Schmidt.
ans. {(1,0,0,1),(0,1,0,0),(1/2,0,1,-1/2)}

33. Using the 3 basis vectors from #32 as the ROWS of the matrix A, get a basis for null(A) and show it is orthogonal to the hyperplane. Thus if W is the hyperplane, null(A) is $W^{\perp}$.ans.{(-1,0,1,1)}

34. You now have FOUR vectors; show this is a basis for $\Re^4$. ans. set is independent

Part I: Orthogonal complements: Let W = span(S) where S = {(1,1,0),(0,1,1)} and Y = (4,-1,4);

35. Find $W^{\perp}$. Hint: let W = row(A) for some 2 by 3 matrix A. ans. $W^{\perp} = \{ t (1, -1, 1) \}$

36. Find $proj_W Y$. Be careful, S is NOT orthogonal (call Gram-Schmidt)!
ans. $proj_W Y = (1, 2, 1)$

37. Find a vector Z in $W^{\perp}$ such that $proj_W Y + Z = Y$; show it actually IS orthogonal to W.
ans. Z = (3,-3,3)

38. Show $proj_W Y \neq proj_{(1,1,0)} + proj_{(0,1,1)}$; why does this not work???

ans. S was not an orthogonal set

39. Switch the roles of W and its orthogonal projection- ie. let W = span({(1,-1,1)}), project Y onto W and find a vector Z in the new $W^{\perp}$ such that $proj_W Y + Z = Y$.
ans. $proj_W Y = (3, -3, 3)$, Z = (1,2,1)

Part J: Given S = {(1,1,0,1,1),(1,-1,0,-1,1)} where W = span(S);

40. Find $W^{\perp}$. Hint: Use the fact that if row(A) = span(W), then null(A) = W perp
ans. W perp = span({(0,-1,0,1,0),(0,0,1,0,0),(-1,0,0,0,1)})

41. Use Gram-Schmidt (if necessary) to make the basis for W perp orthogonal.
ans. given spanning set is already orthogonal

42. Write down a basis for $\Re^5$ using W and $W^{\perp}$.
ans. B = {(1,1,0,1,1),(1,-1,0,-1,1),(0,-1,0,1,0),(0,0,1,0,0),(-1,0,0,0,1)}

43. Is it necessary to use Gram-Schmidt on your answer to #42?
ans. NO; set is already orthogonal

45. If X = (1,2,3,4,5), find $(X)_B$. ans. $(X)_B = (3,0,1,3,2)$

Part K: Let T be a linear operator on $\mathfrak{R}^3$ and L = span({(1,2,-1)});

46. Let X = (x,y,z) and find $proj_L X$; then T(X) = $proj_L X$.

ans. T(x,y,z) = 1/6(x+2y-z,2x+4y-2z,-x-2y+z)

47. Find the standard matrix A of this transformation so that [T(X)] = A[X].

NB. A check on your matrix is that it must be *idempotent* (ie. $A^2 = A$) and symmetric.

ans. $A = \dfrac{1}{6}\begin{bmatrix} 1 & 2 & -1 \\ 2 & 4 & -2 \\ -1 & -2 & 1 \end{bmatrix}$

48. Find rank(T) and nullity(T); determine if T is 1-1 and/or onto. ans. 1, 2, neither

49. Show that if X lies on L, then T(X) = X (ie. the subspace L is the set of *fixed points* of this transformation). ans. L = {t(1,2,-1)}= rng(T)

50. Find the orthogonal complement of L = rng(T).

ans. span(S) where S = {(-2,0,1),(1,0,1)}

Part L: Show each system given is inconsistent; then find the least squares solution to each inconsistent system given.

$$x_1 + 2\,x_2 = 0$$

51. $-x_1 + 2\,x_2 = 5$ ans. (-5/2,7/4)

$$4\,x_2 = 8$$

$$x_1 + x_3 = 1$$

$$x_2 + x_3 = 2$$

52.   ans. (1,3/2,1/2)

$$x_1 + x_2 + x_3 = 3$$

$$x_1 + x_3 = 2$$

$$-x_1 + 2\,x_2 = 1$$

53. $2\,x_1 + 2\,x_2 = 4$ ans. (11/13,12/13)

$$3\,x_1 + 3\,x_2 = 5$$

Part M: Given $A = \begin{bmatrix} 1 & -1 \\ 1 & -1 \\ 2 & 1 \end{bmatrix}$, $B = \begin{bmatrix} 0 \\ 1 \\ 3 \end{bmatrix}$;

54. Show AX = B is inconsistent.

55. Find $proj_{col(A)}$ B. NB. The columns of A are orthogonal. ans. (1/2,1/2,3)

56. Solve $A X = proj_{col(A)}$ B .ans. (7/6,2/3)

57. Now multiply both sides of AX = B by A^t and find the least squares solution. Observe that your answers to #56 and #57 are the same.

Part N: Given the set $S = \{ 1, x, x^2 - \frac{1}{2} \}$ for W = span(S) with inner product $<f, g> = \int_{-1}^{1} fg \, dx$ in V = C[-1,1];

58. find $proj_W y$ if $y = x^3$ ans. 3/5 x

59. find the function z in $W^\perp$ such that $proj_W$ $y + z = y$. ans. $x^3 - \frac{3}{5} x$

60. If $proj_W$ y is viewed as an approximation to the function y, how would you calculate the error in this approximation? ans. $\sqrt{\int_{-1}^{1} (x^3 - \frac{3}{5} x)^2 \, dx}$

Part O: Let W = span{1, cos(x), sin(x)}; using the inner product $<f, g> = \int_{-\pi}^{\pi} f \, g \, dx$:

61. find the projection of y = x onto W ans. 2sin(x)

62. find a function z orthogonal to W such that $proj_W$ $y + z = y$. ans. x - 2 sin(x)

63. find the projection of $y = x^2$ onto W. ans. $\frac{\pi^2}{3} - 4 \cos(x)$

64. find a function z orthogonal to W such that $proj_W$ $y + z = y$.

ans. $x^2 - \frac{\pi^2}{3} + 4 \cos(x)$

65. find the projection of $y = \begin{cases} 0, & -\pi \le x < 0 \\ 4, & 0 \le x \le \pi \end{cases}$ onto W.

ans. $2 + \frac{8}{\pi} \sin(x)$

66. find a function z orthogonal to W such that $proj_W$ $y + z = y$.

ans. $y - [2 + \frac{8}{\pi} \sin(x)]$

67. Find the Fourier series for y from #65; write out at least 4 non-zero terms

ans. $2 + \frac{8}{\pi} [\sin(x) + \frac{\sin(3 x)}{3} + \frac{\sin(5 x)}{5} + ...]$

Part P: Given $V = M_{2,2}$ with function $<,>|M_{2,2} \to \Re$ where $<A, B> = tr(\ A B^t\ )$;

68. Show that $<A, A> \geq 0$ and $<A,A> = 0$ iff $A = \begin{bmatrix} 0 & 0 \\ 0 & 0 \end{bmatrix}$.

69. Show that $<A,B> = <B,A>$. Hint: recall $tr(AB) = tr(BA)$.

70. Show $<A + B,C> = <A,C> + <B,C>$.

71. Show $<kA,B> = k<A,B>$. Thus you have shown that this function $tr(\ A B^t\ )$ forms an inner product on V. Recall $tr(kA) = k\ tr(A)$.

72. If $S = \{ \begin{bmatrix} 1 & 0 \\ 0 & 2 \end{bmatrix}, \begin{bmatrix} 0 & 1 \\ 1 & 0 \end{bmatrix} \}$ $W = span(S)$ find $W^{\perp}$ using the trace inner product.

Hint: let $C = \begin{bmatrix} a & b \\ c & d \end{bmatrix}$ and force it to be orthogonal to W. ans. $C = \begin{bmatrix} c & d \\ -2d & -c \end{bmatrix}$

Part Q: Given the function $y = |x|$; using the inner product $<f, g> = \int_{-\pi}^{\pi} f\ g\ d\ x$ find:

73. Projection of y onto $W = span(\{1,\cos(x),\sin(x)\})$. ans. $\dfrac{\pi}{2} - \dfrac{4}{\pi} \cos(\ x\ )$

74. Projection of y onto $W = span(\{1,\cos(x),\sin(x),\cos(2x),\sin(2x)\})$. ans. same as #73

75. How would you calculate the error in your answer to #73? Do not evaluate the integral!

ans. $\left\| |x| - (\dfrac{\pi}{2} - \dfrac{4}{\pi}\cos(\ x\ )) \right\| = \sqrt{\int_{-\pi}^{\pi}(|x| - \dfrac{\pi}{2} + \dfrac{4}{\pi}\cos(\ x\ ))^2 d\ x}$

Part R: given the function $y = \begin{cases} 1 & -\pi \leq x < 0 \\ 3 & 0 \leq x \leq \pi \end{cases}$; using the inner product $<f, g> = \int_{-\pi}^{\pi} f\ g\ d\ x$ find:

76. the Fourier series for the function y.

ans. $S(\ x\ ) = 2 + \dfrac{4}{\pi}[\ \sin(\ x\ ) + \dfrac{\sin(\ 3x\ )}{3} + \dfrac{\sin(\ 5x\ )}{5} + ...\ ]$

77. Find an approximation for $y(\dfrac{\pi}{2})$ using $S_3(\ x\ )$; to what should the series converge at this point? ans. $2 + \dfrac{8}{3\pi}, 3$

CHAPTER 6

EIGENVECTORS

CHAPTER OVERVIEW

In this chapter we consider the problem of finding "special vectors" associated with a square matrix. This problem is best thought of by remembering that a square matrix represents a *linear operator* T on $\Re^n$; [T(X)] = A[X]. We want the image T(X) to be a scalar multiple of the input X:

$$[T (X)]= A X = \lambda X$$

Any non-zero vector X for which this equation holds is called an **eigenvector** of A; the scalar λ is called an **eigenvalue**. Since this equation always holds true if X is the zero vector, we exclude $X = \theta$. In section 1 we determine how to find the eigenvalues and eigenvectors for a given matrix A.

In section 2, we show that the key to dealing with many applications is to factor the matrix A in the important form $A = P D P^{-1}$, where D is a **diagonal** matrix. Calculating the *powers* of a diagonal matrix D is easy, so this allows us to calculate powers of A efficiently.

In section 3, we use this process of diagonalization to "uncouple" a linear **system of differential equations**. Along the way, we discover how to calculate the *matrix exponential* exp(At).

In section 4, the eigenvalue/eigenvector problem is expanded to matrices with imaginary eigenvalues. Nothing really changes except that we need to be able to do arithmetic in the complex number field.

6.1 Eigenvalues and Eigenvectors

Consider the matrix $A = \begin{bmatrix} 1 & 3 \\ 3 & 1 \end{bmatrix}$. We know from Chapter 4 that this matrix represents the *linear operator* $T(x,y) = (x + 3y, 3x + y)$ on $\mathfrak{R}^2$ with respect to the standard basis, where $[T(X)] = A[X]$. Here however, we are concerned with certain pre-images, which have a very simple type of image. For example, if we pick $X = (1,1)$, $T(1,1) = (4,4) = 4(1,1) = 4\,X$. Using the standard matrix A, we'd have:

$$[\,T(\,1,1\,)\,] = \begin{bmatrix} 1 & 3 \\ 3 & 1 \end{bmatrix} \begin{bmatrix} 1 \\ 1 \end{bmatrix} = \begin{bmatrix} 4 \\ 4 \end{bmatrix} = 4 \begin{bmatrix} 1 \\ 1 \end{bmatrix}$$

In other words, if $X = (1,1)$ then the **image** $T(X) = (4,4)$ is simply a **SCALAR MULTIPLE** of the pre-image $X = (1,1)$. Since multiplying a vector by a scalar is the simplest thing we can do to a vector, we are interested in how the number "4" turned out to be the scalar multiplier. Let's generalize this problem. Thinking of X as a coordinate matrix (ie. an n x 1 column matrix), we wish to solve

$$A\,X \;=\; \lambda\,X$$

where $X \neq \theta$ and $\lambda \in \mathfrak{R}$. We outlaw the zero vector, since this would trivially satisfy the equation for any λ. In linear algebra, this is a classical problem called the **eigenvalue**[1] problem. It can be solved by writing it in the form:

$$A\,X - \lambda\,X = \theta$$

or rewriting carefully so we can "factor out" the EIGENVECTOR X:

$$(\,A - \lambda\,I_n\,)\,X = \theta$$

For our matrix A we would have:

$$\left(\begin{bmatrix} 1 & 3 \\ 3 & 1 \end{bmatrix} - \lambda \begin{bmatrix} 1 & 0 \\ 0 & 1 \end{bmatrix} \right) \begin{bmatrix} x \\ y \end{bmatrix} = \begin{bmatrix} 0 \\ 0 \end{bmatrix}$$

So the system in matrix form is:

$$\begin{bmatrix} 1-\lambda & 3 \\ 3 & 1-\lambda \end{bmatrix} \begin{bmatrix} x \\ y \end{bmatrix} = \begin{bmatrix} 0 \\ 0 \end{bmatrix}$$

This 2 x 2 homogeneous linear system has **nontrivial** solutions only if

---

[1] A. Cauchy was first to use "characteristic value"; D. Hilbert was first to use "eigenvector".

$$\begin{vmatrix} 1-\lambda & 3 \\ 3 & 1-\lambda \end{vmatrix} = (\,1-\lambda\,)(\,1-\lambda\,)-9=0$$

This equation is called the **characteristic equation**; we are forcing the coefficient matrix $(A-\lambda\mathbf{I})$ to be *singular* so that $(A-\lambda\mathbf{I})\,X=\theta$ will have infinitely-many solutions (the EIGENVECTORS). The left side is a degree two polynomial (the *characteristic polynomial*) in λ corresponding to the fact that we have a 2 x 2 coefficient matrix A. For our example we calculate

$$p(\,\lambda\,)=\lambda^2-2\,\lambda-8=0$$

We can factor the characteristic polynomial to obtain:

$$p(\,\lambda\,)=(\,\lambda-4\,)(\,\lambda+2\,)=0 \Rightarrow \lambda=-2,4$$

The solutions $\lambda=-2$ and $\lambda=4$ are the **eigenvalues** of the matrix A; in this case they are real numbers. Now we need to find the pre-images (eigenvectors in dmn(T)) that go along with the eigenvalues. We return to the defining system of equations. For $\lambda=4$ we get:

$$(\,A-\lambda I\,)\,X = \begin{bmatrix} 1-4 & 3 \\ 3 & 1-4 \end{bmatrix} \begin{bmatrix} x \\ y \end{bmatrix} = \begin{bmatrix} 0 \\ 0 \end{bmatrix}$$

whose augmented matrix is:

$$\begin{bmatrix} -3 & 3 & 0 \\ 3 & -3 & 0 \end{bmatrix} \sim \begin{bmatrix} 1 & -1 & 0 \\ 0 & 0 & 0 \end{bmatrix}$$

and from the echelon form we have x - y = 0 or x = y. Let y = t; then x = t and the eigenvectors of $\lambda=4$ are then of the form

$$X = \begin{bmatrix} t \\ t \end{bmatrix} = t \begin{bmatrix} 1 \\ 1 \end{bmatrix} \quad t \neq 0$$

We can (and should!) verify whether these eigenvectors work by picking one at random and seeing if the image is the correct scalar multiple of the pre-image; for example if we choose X = (5,5) the image should be Y = 4(5,5) = (20,20) = 4 X:

$$A\,X = \begin{bmatrix} 1 & 3 \\ 3 & 1 \end{bmatrix} \begin{bmatrix} 5 \\ 5 \end{bmatrix} = \begin{bmatrix} 20 \\ 20 \end{bmatrix} = 4 \begin{bmatrix} 5 \\ 5 \end{bmatrix} = \lambda\,X$$

This check can be performed on any eigenvalue/eigenvector combination.

DEFINITION

For any eigenvalue λ, the set $E_\lambda=\{\ X:A\,X=\lambda\,X\ \}\cup\{\ \theta\ \}$ is called the **eigenspace** of the eigenvalue λ.

This set of eigenvectors (plus the zero vector) is actually a *subspace* of $\mathfrak{R}^2$, since each vector in this eigenspace is a multiple of (1,1). Geometrically $E_{\lambda=4}$, the eigenspace of $\lambda = 4$, is a *line* through the origin whose dimension is one (since the one vector (1,1) spans this space).

Now to continue with the other eigenvalue $\lambda = -2$, we must solve the homogeneous system

$$(A - \lambda I) X = \begin{bmatrix} 1-(-2) & 3 \\ 3 & 1-(-2) \end{bmatrix} \begin{bmatrix} x \\ y \end{bmatrix} = \begin{bmatrix} 0 \\ 0 \end{bmatrix}$$

The augmented matrix is:

$$\begin{bmatrix} 3 & 3 & 0 \\ 3 & 3 & 0 \end{bmatrix} \sim \begin{bmatrix} 1 & 1 & 0 \\ 0 & 0 & 0 \end{bmatrix}$$

and from the row-echelon form we have x + y = 0 or x = -y. Let y = s; then x = -s and the eigenvectors of $\lambda = -2$ are then of the form

$$X = \begin{bmatrix} -s \\ s \end{bmatrix} = s \begin{bmatrix} -1 \\ 1 \end{bmatrix} \quad s \neq 0$$

We get a *subspace* of $\mathfrak{R}^2$ once again (if we throw in the zero vector), since each vector in this eignespace is a multiple of (-1,1). Geometrically the eigenspace of $\lambda = -2$ is also a *line* through the origin whose dimension is one (since the one vector (-1,1) spans this space). Let's pick a particular eigenvector in the eigenspace and make sure it works; let X = (-7,7); then

$$A X = \begin{bmatrix} 1 & 3 \\ 3 & 1 \end{bmatrix} \begin{bmatrix} -7 \\ 7 \end{bmatrix} = \begin{bmatrix} 14 \\ -14 \end{bmatrix} = -2 \begin{bmatrix} -7 \\ 7 \end{bmatrix} = \lambda X$$

so all is right with the world.

Note that if we marry the two basis vectors we found, we produce an *independent* set of vectors S = {(1,1),(-1,1)}, which could serve as a basis for $\mathfrak{R}^2$. Here A is symmetric so its eigenvalues are *real* and the corresponding eigenvectors are *orthogonal*.

The eigenvalues

The set of all eigenvalues of a matrix A is called its **spectrum**; we write $\sigma(A)$ to indicate this set of real numbers. From our example

$$\sigma(A) = \{ -2, 4 \}$$

The *spectral radius*[2] of A is defined by $\rho(A) = \max\{ |\lambda| : \lambda \in \sigma(A) \}$; it is a non-negative real number. From our example, $\rho(A) = 4$.

EXAMPLE 1 Find the spectral radius of $A = \begin{bmatrix} -5 & 7 \\ 0 & 3 \end{bmatrix}$.

---

[2] In the complex plane, all eigenvalues lie on a disk of radius ρ centered at the origin.

The characteristic equation is $p(\lambda) = (-5-\lambda)(3-\lambda) = 0$ so the set of eigenvalues is $\sigma(A) = \{-5,3\}$. Taking absolute values, it is clear that $\lambda = -5$ has the largest absolute value. Thus $\rho(A) = 5$.

WARNING- A real matrix may have NO REAL EIGENVALUES!

For example, the matrix $A = \begin{bmatrix} 0 & -1 \\ 1 & 0 \end{bmatrix}$ represents a counter-clockwise rotation of 90 degrees in the plane; clearly such a matrix CANNOT have a real eigenvalue! We deal only with real eigenvalues until Section 6.4.

THEOREM The eigenspace corresponding to a particular eigenvalue λ forms a subspace of $\Re^n$.
Proof: Suppose X_1, X_2 are two eigenvectors corresponding to the eigenvector λ of the matrix A. Then we know that

$$A X_1 = \lambda X_1 \quad \& \quad A X_2 = \lambda X_2$$

and this means that

$$A X_1 + A X_2 = A(X_1 + X_2) = \lambda X_1 + \lambda X_2 = \lambda(X_1 + X_2)$$

Hence $X_1 + X_2$ is also an eigenvector of A so the eigenspace is *closed* under vector addition. Now for scalar multiplication- if X is an eigenvector of A corresponding to the eigenvalue λ then

$$A(kX) = k(AX) = k(\lambda X) = \lambda(kX)$$

since λ and k are just scalars and scalars commute. This equation says that kX is also an eigenvector with eigenvalue λ so the eigenspace is closed under scalar multiplication. Thus the eigenspace is actually a subspace of $\Re^n$.

The dimension of the eigenspace is called the GEOMETRIC MULTIPLICITY of λ. A *simple root* of the characteristic equation always has geometric multiplicity ONE; its eigenspace is then a *line* through the origin.

General procedure for finding eigenvalues/eigenvectors:
1) solve the characteristic equation

$$\det(A - \lambda I_n) = 0$$

2) for each eigenvalue λ, solve the homogeneous system

$$(A - \lambda I_n)X = \theta$$

which must have *infinitely-many* solutions (at least one row of your augmented matrix MUST DISAPPEAR - if not, you've made a mistake someplace!).

EXAMPLE 2 Find the eigenvalues/eigenvectors for $A = \begin{bmatrix} 0 & 0 & 3 \\ 0 & 3 & 0 \\ 3 & 0 & 0 \end{bmatrix}$.

$$p(\lambda) = \det(A - \lambda I) = \begin{vmatrix} 0-\lambda & 0 & 3 \\ 0 & 3-\lambda & 0 \\ 3 & 0 & 0-\lambda \end{vmatrix} = \lambda^2(3-\lambda) - 9(3-\lambda) = (3-\lambda)(\lambda^2-9)$$

Factoring the characteristic polynomial and setting each factor equal to zero gives us the eigenvalues: $p(\lambda) = (3-\lambda)(\lambda-3)(\lambda+3) = 0 \Rightarrow \lambda = -3, 3$. Thus we have the spectrum of the matrix A: $\sigma(A) = \{-3, 3\}$.

Let's get the eigenvectors corresponding to $\lambda = -3$:

$$(A-(-3)I)X = \begin{bmatrix} 3 & 0 & 3 \\ 0 & 6 & 0 \\ 3 & 0 & 3 \end{bmatrix}\begin{bmatrix} x_1 \\ x_2 \\ x_3 \end{bmatrix} = \begin{bmatrix} 0 \\ 0 \\ 0 \end{bmatrix} \Rightarrow \begin{bmatrix} 3 & 0 & 3 & 0 \\ 0 & 6 & 0 & 0 \\ 3 & 0 & 3 & 0 \end{bmatrix} \sim \begin{bmatrix} 1 & 0 & 1 & 0 \\ 0 & 1 & 0 & 0 \\ 0 & 0 & 0 & 0 \end{bmatrix}$$

so y = 0 and x + z = 0 thus x = -z. If z = r then x = -r so the eigenvectors are of the form:

$$X = \begin{bmatrix} -r \\ 0 \\ r \end{bmatrix} = r\begin{bmatrix} -1 \\ 0 \\ 1 \end{bmatrix} \quad r \neq 0$$

Thus here the eigenspace is spanned by the vector (-1,0,1). Keep in mind we could use (1,0,-1) or (-2,0,2) etc.; any non-zero multiple of (-1,0,1) would do. The eigenspace is one-dimensional (a line through the origin) and thus its geometric multiplicity is one.

Now for the second eigenvalue $\lambda = 3$:

$$(A-3I)X = \begin{bmatrix} -3 & 0 & 3 \\ 0 & 0 & 0 \\ 3 & 0 & -3 \end{bmatrix}\begin{bmatrix} x_1 \\ x_2 \\ x_3 \end{bmatrix} = \begin{bmatrix} 0 \\ 0 \\ 0 \end{bmatrix} \Rightarrow \begin{bmatrix} -3 & 0 & 3 & 0 \\ 0 & 0 & 0 & 0 \\ 3 & 0 & -3 & 0 \end{bmatrix} \sim \begin{bmatrix} 1 & 0 & -1 & 0 \\ 0 & 0 & 0 & 0 \\ 0 & 0 & 0 & 0 \end{bmatrix}$$

so x - z = 0 thus x = z. Letting y = s and z = t means x = t. So the eigenvectors are of the form:

$$X = \begin{bmatrix} t \\ s \\ t \end{bmatrix} = s\begin{bmatrix} 0 \\ 1 \\ 0 \end{bmatrix} + t\begin{bmatrix} 1 \\ 0 \\ 1 \end{bmatrix} \quad s, t \neq 0$$

Thus the geometric multiplicity of the eigenvalue $\lambda = 3$ is two. Note that here the geometric multiplicity of $\lambda = 3$ equals its algebraic multiplicity (since $\lambda = 3$ was a *double* root of the characteristic equation).

The eigenspace of $\lambda = 3$ is spanned by the two vectors $\{(0,1,0),(1,0,1)\}$; geometrically it is a plane through the origin. A keen observation is that the set of independent eigenvectors $S=\{(-1,0,1),(0,1,0),(1,0,1)\}$ forms an orthogonal set; this is true since A here is *symmetric*.

Defective Matrices and the Rank + Nullity Theorem

Using the matrix A from EX2, we could form a basis for $\Re^3$ by using the three vectors we found as bases for the eigenspaces; hence this matrix is called NON-DEFECTIVE. It has enough independent eigenvectors to span $\Re^n$ (here since A is 3 x 3, n = 3). If a matrix does NOT have enough eigenvectors to span $\Re^n$, it is called *defective*. If each eigenvalue has algebraic multiplicity ONE, the matrix cannot be defective. A defective matrix can only arise when the geometric multiplicity is LESS THAN the algebraic.

The *eigenspace* is simply the **null space** of A - λ I. This means we can use the "rank plus nullity" theorem nicely. If we find the rank of A - λ I, the dimension of the corresponding null space satisfies

$$rank\ (\ A - \lambda I\ ) + nullity\ (\ A - \lambda I\ ) = n \Rightarrow dim(\ E_\lambda\ ) = n - rank\ (\ A - \lambda I\ )$$

From the previous 3 x 3 example, for $\lambda = -3$, rank(A - λ I) = 2 so the nullity is 1; the eigenspace is the **null space** of A - λ I so this eigenspace must be $3 - 2 = 1$-dimensional.
On the other hand for $\lambda = 3$, rank(A - λ I) = 1 so the nullity is 2; the eigenspace is $3 - 1 = 2$-dimensional. The geometric multiplicity matched the algebraic for this eigenvalue; this need not be the case when an eigenvalue is a repeated root. The geometric multiplicity could be *less than* the algebraic; this would cause an "eigenvector shortage". And the matrix would be defective.

A classic case of a defective matrix is a *Jordan block* matrix like $A = \begin{bmatrix} k & 1 \\ 0 & k \end{bmatrix}$; it has only one eigenvalue $\lambda = k$ (multiplicity TWO) and its eigenspace is spanned by X = (1,0). You can't find two independent eigenvectors for this matrix.

THEOREM The geometric multiplicity of an eigenvalue is less than or equal to its algebraic multiplicity.

Of course if an n x n matrix has n *distinct* eigenvalues, the geometric multiplicity matches the algebraic (they are both ONE) and we all go home happy. Then every eigenspace is a line through the origin. Such a matrix CANNOT be defective.

Now for some facts about eigenvalue and matrices; given a square matrix A:
1. A is singular iff $\lambda = 0$ is an eigenvalue of A

2. If A is invertible and λ is an eigenvalue of A, then $\frac{1}{\lambda}$ is an eigenvalue of A^{-1}

3. Eigenvectors corresponding to different eigenvalues are independent.
4. ***The eigenvalues of a triangular matrix sit on its main diagonal.

Checks for eigenvalue calculations:
1) product of eigenvalues = det(A)
2) sum of eigenvalues = trace(A)
3) Cayley-Hamilton theorem: a matrix satisfies its own characteristic equation.

EXAMPLE 3 Show that the matrix $A = \begin{bmatrix} 1 & 3 \\ 3 & 1 \end{bmatrix}$ satisfies all three of these.

1) $\lambda_1 \lambda_2 = (-2)(4) = \det(A) = 1 - 9 = -8$

2) $\lambda_1 + \lambda_2 = (-2) + (4) = \text{tr}(A) = 1 + 1 = 2$

3) Recall that $p(\lambda) = \lambda^2 - 2\lambda - 8$. First we need to get the proper "matrix version" of the characteristic equation: $p(A) = A^2 - 2A - 8I = \theta$

Now substitute A in the equation:

$$\begin{bmatrix} 1 & 1 \\ 4 & 1 \end{bmatrix}^2 - 2\begin{bmatrix} 1 & 1 \\ 4 & 1 \end{bmatrix} - 3\begin{bmatrix} 1 & 0 \\ 0 & 1 \end{bmatrix} = \begin{bmatrix} 5 & 2 \\ 8 & 5 \end{bmatrix} - \begin{bmatrix} 2 & 2 \\ 8 & 2 \end{bmatrix} - \begin{bmatrix} 3 & 0 \\ 0 & 3 \end{bmatrix} = \begin{bmatrix} 0 & 0 \\ 0 & 0 \end{bmatrix}$$

This idea can be used to find the inverse of A;

$$A^2 - 2A - 8I = \theta \Rightarrow 8I = A^2 - 2A \Rightarrow I = \frac{1}{8}(A^2 - 2A)$$

Now just multiply both sides by the inverse of A:

$$A^{-1} = \frac{1}{8}(A - 2I)$$

Eigenvectors and linear transformations

Remember that every square matrix represents a linear transformation (operator) $T \mid \Re^n \to \Re^n$. If we pick a particular **non-zero** eigenvalue λ and find its eigenspace, we can show that when the transformation is restricted to the eigenspace E_λ, it must be one-to-one and onto.

Proof: Suppose λ is an eigenvector of T with eigenspace E_λ. Let $X_1, X_2 \in E_\lambda$; then clearly

$$T(X_1) = \lambda X_1 \text{ and } T(X_2) = \lambda X_2$$

Therefore

$$T(X_1) = T(X_2) \Rightarrow \lambda X_1 = \lambda X_2 \Rightarrow \lambda X_1 - \lambda X_2 = \theta \Rightarrow \lambda(X_1 - X_2) = \theta$$

There are two possibilities here; either $\lambda = 0$ or $X_1 - X_2 = \theta$. But by hypothesis, $\lambda \neq 0$ so therefore $X_1 = X_2$ and T is one-to-one. The onto part of the proof is left as an exercise.

Thus, when restricted to this eigenspace E_λ with non-zero eigenvalue, a linear operator T is one-to-one and onto (the eigenspace is then an example of an *invariant subspace*[3]).

---

[3] A subspace W is invariant under T if whenever X is in W, so is T(X).

Vectors in the eigenspace

Looking back at the introductory example, if we pick a particular eigenvector corresponding to $\lambda = 4$, say $X = \begin{bmatrix} 10 \\ 10 \end{bmatrix}$ and find its image using A:

$$AX = \begin{bmatrix} 1 & 3 \\ 3 & 1 \end{bmatrix}\begin{bmatrix} 10 \\ 10 \end{bmatrix} = \begin{bmatrix} 40 \\ 40 \end{bmatrix} = 4\begin{bmatrix} 10 \\ 10 \end{bmatrix} = 4X = Y$$

we get another vector in the eigenspace. If we would now left-multiply Y by A (ie. output becomes input), we would get 4Y, which again lies in the eigenspace. Moral - if a vector is IN a particular eigenspace, it can *never* get out (ie. under multiplication by A). On the other hand, any vector NOT in this eigenspace can never get in!

6.1 EXERCISES

Part A. Given the matrix A and an eigenvector X, find the corresponding eigenvalue.

1. $A = \begin{bmatrix} 1 & -2 \\ 3 & 6 \end{bmatrix}$ $X = \begin{bmatrix} -1 \\ 1 \end{bmatrix}$ ans. $\lambda = 3$

2. $A = \begin{bmatrix} 1 & -2 \\ 3 & 6 \end{bmatrix}$ $X = \begin{bmatrix} -2 \\ 3 \end{bmatrix}$

3. $A = \begin{bmatrix} 1 & 3 \\ -2 & 6 \end{bmatrix}$ $X = \begin{bmatrix} 6 \\ 4 \end{bmatrix}$ ans. $\lambda = 3$

4. $A = \begin{bmatrix} 1 & 3 \\ -2 & 6 \end{bmatrix}$ $X = \begin{bmatrix} 3 \\ 2 \end{bmatrix}$

5. $A = \begin{bmatrix} 1 & 3 \\ -2 & 6 \end{bmatrix}$ $X = \begin{bmatrix} 5 \\ 5 \end{bmatrix}$ ans. $\lambda = 4$

6. $A = \begin{bmatrix} \dfrac{1}{2} & \dfrac{1}{6} \\ \dfrac{-1}{4} & \dfrac{1}{12} \end{bmatrix}$ $X = \begin{bmatrix} -12 \\ 12 \end{bmatrix}$

Part B. Given the linear operator T(x,y) = (x + 2y, 8x + y);
7. Show that X= (1,2) is an eigenvector of T; find the corresponding eigenvalue. ans. $\lambda = 5$
8. Show that X=(1,-2) angenvector of T; find the corresponding eigenvalue.
9. Explain why X = (3,5) is NOT an eigenvector of T. ans. T(X) NOT a scalar multiple of X
10. Show that X = 6(1,2) + 7(1,-2) is NOT an eigenvector even though it is the sum of two eigenvectors; explain why.

11. Find A, the standard matrix of the linear operator T. Then find the null space of A − λI, if λ = 5. Show that this null space is the eigenspace of λ = 5. ans. $A = \begin{bmatrix} 1 & 2 \\ 8 & 1 \end{bmatrix}$ $X = s \begin{bmatrix} 1 \\ 2 \end{bmatrix}$

12. Using A from #11, show that $X = t \begin{bmatrix} 1 \\ -2 \end{bmatrix}$ is the null space of A − λI, if λ = -3. Verify that this null space is the eigenspace of λ = -3.

13. Find S(x,y) if its standard matrix is A − 5**I**; show that S is singular.

ans. S(x,y) = (-4x+2y,8x-4y)

14. Find S(x,y) if its standard matrix is A − (-3)**I**; show that S is neither one-to-one nor onto. Why is this so?

15. Find the inverse of A − λ**I**. ans. $\begin{bmatrix} \dfrac{1-\lambda}{\lambda^2 - 2\lambda - 15} & \dfrac{-2}{\lambda^2 - 2\lambda - 15} \\[3mm] \dfrac{-8}{\lambda^2 - 2\lambda - 15} & \dfrac{1-\lambda}{\lambda^2 - 2\lambda - 15} \end{bmatrix}$

16. Find the values of λ for which the inverse of A − λ**I** does not exist. Are they familiar?

Part C. For the matrix A, find the characteristic polynomial and the spectrum of A. Then find the corresponding eigenspace for each eigenvalue.

17. $A = \begin{bmatrix} 1 & 2 \\ 0 & 3 \end{bmatrix}$ ans. $p(\lambda) = \lambda^2 - 4\lambda + 3$, $\sigma(A) = \{1,3\}$

$\lambda = 1, E_{\lambda=1} = \{ s \begin{bmatrix} 1 \\ 0 \end{bmatrix} \}$ $\lambda = 3, E_{\lambda=3} = \{ t \begin{bmatrix} 1 \\ 1 \end{bmatrix} \}$

18. $A = \begin{bmatrix} 1 & 2 \\ 2 & 1 \end{bmatrix}$

19. $A = \begin{bmatrix} 1 & 2 & 3 \\ 0 & 2 & 2 \\ 0 & 0 & 3 \end{bmatrix}$ ans. $p(\lambda) = -\lambda^3 + 6\lambda^2 - 11\lambda + 6$, $\sigma(A) = \{1,2,3\}$

$\lambda = 1, E_{\lambda=1} = \{ r \begin{bmatrix} 1 \\ 0 \\ 0 \end{bmatrix} \}$ $\lambda = 2, E_{\lambda=2} = \{ s \begin{bmatrix} 2 \\ 1 \\ 0 \end{bmatrix} \}$, $\lambda = 3, E_{\lambda=3} = \{ t \begin{bmatrix} 7 \\ 4 \\ 2 \end{bmatrix} \}$

20. $A = \begin{bmatrix} 3 & -1 & -1 \\ -1 & 3 & -1 \\ -1 & -1 & 3 \end{bmatrix}$

21. $A = \begin{bmatrix} 2 & 1 & 0 \\ 0 & 2 & 0 \\ 0 & 0 & 2 \end{bmatrix}$　　ans. $p(\lambda) = (2 - \lambda)^3$, $\sigma(A) = \{2\}$　$E_{\lambda=2} = \{ s \begin{bmatrix} 1 \\ 0 \\ 0 \end{bmatrix} + t \begin{bmatrix} 0 \\ 0 \\ 1 \end{bmatrix} \}$

22. $A = \begin{bmatrix} 0 & 1 & 0 \\ 1 & 0 & 0 \\ 0 & 0 & 1 \end{bmatrix}$

.

Part D. Given the matrix $A = \begin{bmatrix} 1 & 0 \\ 2 & -3 \end{bmatrix}$;

23. Find the eigenvalues and corresponding eigenspaces.

ans. $\lambda = -3, E_{\lambda=-3} = \{ s \begin{bmatrix} 0 \\ 1 \end{bmatrix} \}$　$\lambda = 1, E_{\lambda=1} = \{ t \begin{bmatrix} 2 \\ 1 \end{bmatrix} \}$

24. Show that $B = P^{-1} A P$, where P is a matrix whose columns are formed using $X_1 = (0,1)$ and $X_2 = (2,1)$, has the *same eigenvalues* as A.

25. Show that the eigenvectors of B are NOT the same as the corresponding eigenvectors of A. Conclusion: Similar matrices in general share eigenvalues but not eigenvectors.

26. Prove that if (λ, X) is an eigenvalue/eigenvector pair for A, and B is similar to A, then (λ, Y) is an eigenvalue/eigenvector pair for B if $Y = P^{-1} X$.

27. Let $C = A^t$; show that C has the same eigenvalues as A but different eigenvectors.

ans. $\lambda = -3, E_{\lambda=-3} = \{ s \begin{bmatrix} 1 \\ -2 \end{bmatrix} \}$　$\lambda = 1, E_{\lambda=1} = \{ t \begin{bmatrix} 1 \\ 0 \end{bmatrix} \}$

28. Find the eigenvalues and eigenvectors of A^2. Show that the eigenvalues are different but the eigenvectors are the same as those of A. What is the relationship between the eigenvalues of A and its square?

29. Find the eigenvalues/vectors for A^{-1}. What is the eigenvalue relationship between A and its inverse?? What about the eigenvectors??

ans. $\lambda = -\dfrac{1}{3}, E_{\lambda=-\frac{1}{3}} = \{ s \begin{bmatrix} 0 \\ 1 \end{bmatrix} \}$　$\lambda = 1, E_{\lambda=1} = \{ t \begin{bmatrix} 2 \\ 1 \end{bmatrix} \}$

eigenvalues of A^{-1} are reciprocals of the eigenvalues of A, same eigenvectors
NB. This reminds you that if A is singular, one of its eigenvalues is zero.

Part E. Given the matrix $A = \begin{bmatrix} \frac{1}{2} & 0 & \frac{1}{2} \\ 0 & 1 & 0 \\ \frac{1}{2} & 0 & \frac{1}{2} \end{bmatrix}$;

30. Show that A is a *projection* matrix, since it is idempotent ($A^2 = A$) and symmetric.

31. Find the characteristic polynomial and the eigenvalues; then find a basis for each eigenspace.

ans. $\lambda = 0$, $\left\{ \begin{bmatrix} -1 \\ 0 \\ 1 \end{bmatrix} \right\}$ $\lambda = 1$, $\left\{ \begin{bmatrix} 1 \\ 0 \\ 1 \end{bmatrix}, \begin{bmatrix} 0 \\ 1 \\ 0 \end{bmatrix} \right\}$

32. Show that for the corresponding operator T represented by A that:
a) rng(T) = span({(1,0,1),(0,1,0)})
b) ker(T) = span({(-1,0,1)})
33. Show that if W = rng(T) then $W^{\perp} = ker(T)$.

Part F.
34. Show that for a 2 x 2 matrix, the characteristic polynomial is $p(\lambda) = \lambda^2 - tr(A)\lambda + \det(A)$.

35. Use the result from #34 to find a matrix A whose characteristic polynomial is $p(\lambda) = \lambda^2 - 5\lambda + 6$. ans. $A = \begin{bmatrix} 2 & b \\ 0 & 3 \end{bmatrix}$ (or any matrix similar to this)

36. The matrix $A = \begin{bmatrix} 0 & 0 & c \\ 1 & 0 & b \\ 0 & 1 & a \end{bmatrix}$ is called a (3 x 3) *companion matrix*. Show that the characteristic

polynomial of such a matrix is $p(\lambda) = -\lambda^3 + a\lambda^2 + b\lambda + c$.

37. Use the idea from # 36 to construct a companion matrix A whose eigenvalues are 2,3 and 4.

Hint: what should the factors of the characteristic polynomial be? ans. $A = \begin{bmatrix} 0 & 0 & 24 \\ 1 & 0 & -26 \\ 0 & 1 & 9 \end{bmatrix}$

38. Find the eigenvectors from # 37.

39. Show that the matrix $J = \begin{bmatrix} 3 & 1 & 0 \\ 0 & 3 & 1 \\ 0 & 0 & 3 \end{bmatrix}$ has only one eigenvalue; get the only lonely eigenvector.

J is a *Jordan block* matrix. Such a matrix is *defective* since it has only one independent eigenvector (and not three).

ans. $\lambda = 3$, $t\begin{bmatrix} 1 \\ 0 \\ 0 \end{bmatrix}, t \neq 0$

40. A matrix A is *idempotent* if $A^2 = A$; prove that such a matrix has eigenvalues $\{0,1\}$.

41. Prove that if $B = P^{-1} A P$ (ie. B is similar to A) then the characteristic polynomials of A and B are the same.

42. The celebrated Cayley-Hamilton theorem says that a matrix *satisfies its own characteristic equation.* Using the matrix $A = \begin{bmatrix} 1 & 2 \\ 0 & 3 \end{bmatrix}$, the characteristic polynomial is $p(\lambda) = \lambda^2 - 4\lambda + 3 = 0$ and therefore the matrix version is $A^2 - 4A + 3I = \begin{bmatrix} 0 & 0 \\ 0 & 0 \end{bmatrix}$. Show that the given A satisfies this equations; then show that any matrix *similar* to A (ie. a matrix of the form $B = P^{-1} A P$) satisfies this equation also.

43. Use the Cayley-Hamilton theorem to find a formula for A^{-1} in terms of powers of A less than 2 and verify that it works. Hint: isolate I and multiply by A^{-1}.

ans. $A^{-1} = \dfrac{1}{3}(4I - A)$

44. Prove that a linear operator $T \mid \Re^n \to \Re^n$ such that [T(X)] = A[X], has a *fixed point* (ie. some X such that T(X) = X) if λ = 1 is an eigenvalue of T with eigenvector X.

Part G. Given the Markov matrix $M = \begin{bmatrix} 0.2 & 0.5 \\ 0.8 & 0.5 \end{bmatrix}$;

45. Show that $\sigma(M) = \{-0.3, 1\}$. Verify that a basis for the eigenspace corresponding to λ = 1 is $\begin{bmatrix} 5 \\ 8 \end{bmatrix}$. Find a basis for the eigenspace of $\lambda = -0.3$ ans. $\begin{bmatrix} -1 \\ 1 \end{bmatrix}$

46. Show that if $X_0 = t \begin{bmatrix} 5 \\ 8 \end{bmatrix}$, $t > 0$ is some initial population and $X_{n+1} = MX_n$, then the population never changes (ie. X_0 is a fixed point).

47. Show that in this Markov process, the eigenvalue λ = 1 DOMINATES by choosing an initial population, say $X_0 = \begin{bmatrix} 80 \\ 180 \end{bmatrix} = 20 \begin{bmatrix} -1 \\ 1 \end{bmatrix} + 20 \begin{bmatrix} 5 \\ 8 \end{bmatrix}$, called an eigenvector decomposition (see # 45).

Let $X_1 = M X_0$, $X_2 = M X_1$, $X_3 = M X_2$, keeping the result in the form of a **sum** of eigenvectors as you do this. Can you see what is happening to the component in the direction of (-1,1)? Why?

ans. $X_3 = \begin{bmatrix} 100.54 \\ 159.46 \end{bmatrix}$ The component in the direction of X = (-1,1) is decreasing since it keeps getting multiplied by λ = -0.3.

Part H. Given the linear operator T on $\Re^3$ whose standard matrix is $A = \begin{bmatrix} 1 & 0 & 1 \\ 0 & 1 & 3 \\ 0 & 0 & 2 \end{bmatrix}$;

clearly the only eigenvalues are $\lambda = 1$ (algebraic multiplicity two) and $\lambda = 2$ (algebraic multiplicity one);

48. Find the corresponding eigenspaces, call them W_1 and W_2. Show that W_1 is the x-y plane and W_2 is a line in the direction of (1,3,1).
49. Show that W_1 and W_2 share only the zero vector (an eigenvector cannot serve two masters).
ans. $X \in W_1 \cap W_2 \Rightarrow X = (0,0,0)$
50. Show that $T|W_1 \to W_1$ is one-to-one and onto. Show that the same holds for W_2.
51. Write the vector $X = (8,19,5)$ as a linear combination of a vector in W_1 (call it X_1) and a vector in W_2 (call it X_2); apply T to this vector X and show that the image of X is again a linear combination of X_1 and X_2. Why is this so?
ans. $X_1 = (3,4,0)$, $X_2 = (5,15,5)$, since X_1 & X_2 are eigenvectors and T is linear

Part I.
52. Complete the proof of the theorem about linear transformations and eigenspaces by showing that if $\lambda \neq 0$ is an eigenvector of $T(X) = AX$ with corresponding eigenspace E_λ, then T, when restricted to this eigenspace, is onto. Remember $\lambda \neq 0$!

53. Using $A = \begin{bmatrix} 2 & 1 & 0 \\ 0 & 2 & 0 \\ 0 & 0 & 2 \end{bmatrix}$, show that the corresponding linear operator T is one-to-one on E, using

the eigenvalue $\lambda = 2$ and the eigenspace $E = \text{span}(\{(1,0,0), (0,0,1)\})$
54. Show using the set-up from #53 that T is also onto E.

Part J. Suppose the population dynamics of a certain beetle[4] (a bug not a car) is modeled by matrix

equation $X_{n+1} = L X_n$ where $L = \begin{bmatrix} 0 & 0 & 6 \\ \dfrac{1}{2} & 0 & 0 \\ 0 & \dfrac{1}{3} & 0 \end{bmatrix}$ & $X = \begin{bmatrix} x \\ y \\ z \end{bmatrix}$;

here n = number of months, x = juveniles, y = young adults, and z = mature beetles.
55. Find the eigenvalues of L and show that only one is real; find its eigenspace.

ans. $\lambda = 1$, $E_{\lambda=1} = \{ t \begin{bmatrix} 6 \\ 3 \\ 1 \end{bmatrix} \}$

---

[4] P.H. Leslie came up with this matrix model in 1945, studying rats.

56. If X is the initial population distribution at t = 0 and the total population (ie. x + y + z) is 100 when t = 0, find X which guarantees LX = X.

57. Show that the population distribution of the beetles is cyclic by finding X_1, X_2, X_3 if $X_0 = \begin{bmatrix} 30 \\ 20 \\ 10 \end{bmatrix}$.

Why is this so? ans. $X_3 = \begin{bmatrix} 30 \\ 20 \\ 10 \end{bmatrix}$; since $L^3 = I$.

58. What initial population distribution would guarantee no change in the population?

ans. $X_0 = k \begin{bmatrix} 6 \\ 3 \\ 1 \end{bmatrix}$

Part K. Given the symmetric matrix $A = \begin{bmatrix} 1 & 2 \\ 2 & 1 \end{bmatrix}$;

59. Show that A does NOT have all positive eigenvalues. ans. eigenvalues = {-1,3}

60. Show that the matrix $B = A^t A$ does have all positive eigenvalues. Find the corresponding eigenvectors for each eigenvalue found. A SYMMETRIC matrix with all positive eigenvalues is called *positive definite*. So B is positive definite.

61. If B is positive definite, $X^t BX > 0$ for all non-zero X. Verify this is true for X = $\begin{bmatrix} -8 \\ 5 \end{bmatrix}$.

62. Show that $X^t AX < 0$ if X = $\begin{bmatrix} -8 \\ 5 \end{bmatrix}$; this means that A can NOT be positive definite.

63. Since all entries of A are *positive*, A is a *positive* matrix and we write A > 0. Verify that A has a positive eigenvalue r = $\rho(A)$ and that $\lambda = r$ has a positive eigenvector. This is true for any positive matrix[5]. So A is positive but NOT positive definite.

64. Let $A = \begin{bmatrix} a & b \\ c & d \end{bmatrix}$ be a positive matrix; using Perron's Theorem, prove that A cannot have imaginary eigenvalues.

_____

[5] This is Perron's Theorem; see *Matrix Analysis*, pg. 500.

6.2 Diagonalization

Introduction

The process of diagonalization of a matrix is intimately connected with the eigenvalue problem. Finding the eigenvalue/eigenvector combinations for a particular matrix allows one in general to diagonalize a matrix. Certainly we wonder why we would wish to do this. The reason is that if we think of a square matrix A as a linear operator T on $V = \Re^n$, then there are infinitely-many different matrix representatives of this linear operator- all we need do is change the basis for V and we have another matrix which represents this same linear operator. Finding the simplest representative means finding the right basis B', which will make the matrix representative of T a *diagonal* matrix. The reason is that *multiplication by a diagonal matrix* is the simplest kind of matrix multiplication. And there are important applications in which we need to have the simplest representative of the operator T; in general this is the diagonal matrix, which represents T.

Let's consider the matrix $A = \begin{bmatrix} 1 & 3 \\ 3 & 1 \end{bmatrix}$; our goal is to find two matrices P and D such that

AP = PD, where $D = \begin{bmatrix} d_1 & 0 \\ 0 & d_2 \end{bmatrix}$ & $P = \begin{bmatrix} p_1 & q_1 \\ p_2 & q_2 \end{bmatrix}$.

If we just write out this equation using the given matrix A, we have:

$$\begin{bmatrix} 1 & 3 \\ 3 & 1 \end{bmatrix}\begin{bmatrix} p_1 & q_1 \\ p_2 & q_2 \end{bmatrix} = \begin{bmatrix} p_1 & q_1 \\ p_2 & q_2 \end{bmatrix}\begin{bmatrix} d_1 & 0 \\ 0 & d_2 \end{bmatrix} \Rightarrow \begin{bmatrix} p_1+3p_2 & q_1+3q_2 \\ 3p_1+p_2 & 3q_1+q_2 \end{bmatrix} = \begin{bmatrix} d_1p_1 & d_2q_1 \\ d_1p_2 & d_2q_2 \end{bmatrix}$$

The first column on the left (of AP) must equal the first column (of PD) on the right:

$$\begin{bmatrix} 1 & 3 \\ 3 & 1 \end{bmatrix}\begin{bmatrix} p_1 \\ p_2 \end{bmatrix} = d_1 \begin{bmatrix} p_1 \\ p_2 \end{bmatrix}$$

This shows that d_1 must be the **eigenvalue** of A corresponding to the *eigenvector* $C_1 = \begin{bmatrix} p_1 \\ p_2 \end{bmatrix}$ and

similarly for d_2 and its *eigenvector* $C_2 = \begin{bmatrix} q_1 \\ q_2 \end{bmatrix}$. Thus P is the MATRIX OF EIGENVECTORS of A and

d_1 and d_2 form the *diagonal matrix D* of its eigenvalues. Hence to diagonalize the matrix A, it must have enough eigenvectors; if A is defective, diagonalization is NOT possible.

Continuing with the given example, we showed in Section 6.1 that A has two eigenvalues $\lambda = -2$ and $\lambda = 4$ with corresponding eigenvectors $X_1 = s \begin{bmatrix} 1 \\ -1 \end{bmatrix}, s \neq 0 \quad X_2 = t \begin{bmatrix} 1 \\ 1 \end{bmatrix}, t \neq 0$.

P is the matrix of eigenvectors whose *first column* is the basis of the eigenspace corresponding to $\lambda = -2$ and whose *second column* is the basis of the eigenspace corresponding to $\lambda = 4$. We now have all the ingredients to complete our goal:

$$AP = \begin{bmatrix} 1 & 3 \\ 3 & 1 \end{bmatrix} \begin{bmatrix} 1 & 1 \\ -1 & 1 \end{bmatrix} = \begin{bmatrix} 1 & 1 \\ -1 & 1 \end{bmatrix} \begin{bmatrix} -2 & 0 \\ 0 & 4 \end{bmatrix} = PD$$

The reader can verify for himself that the above equation is true; this is a good check on your calculations. Note that *the order of the eigenvalues in D must follow the order of the eigenvectors in P* (ie. column ONE of P corresponds to $\lambda = -2$, which is the FIRST entry on the diagonal of D). By the way, P is not *unique*; for example if I used eigenvectors (5,-5) and (7,7) in P then we'd get :

$$AP = \begin{bmatrix} 1 & 3 \\ 3 & 1 \end{bmatrix} \begin{bmatrix} 5 & 7 \\ -5 & 7 \end{bmatrix} = \begin{bmatrix} 5 & 7 \\ -5 & 7 \end{bmatrix} \begin{bmatrix} -2 & 0 \\ 0 & 4 \end{bmatrix} = PD$$

You can verify that this equation is still true.

A clear consequence of the equation AP = PD is $A = P D P^{-1}$ which means that A is *similar* to D. The behavior of A is linked to the behavior of D, since they share the same eigenvalues; for example, what happens if we raise A to successively higher powers?

$$A^2 = (P D P^{-1})(P D P^{-1}) = (P D)(P^{-1} P)(D P^{-1}) = (P D) I (D P) = P D^2 P^{-1}$$

Inductively, it is easy to see that $A^n = P D^n P^{-1}$. In our example, we have

$$A^n = \begin{bmatrix} 1 & 1 \\ -1 & 1 \end{bmatrix} \begin{bmatrix} (-2)^n & 0 \\ 0 & (4)^n \end{bmatrix} \begin{bmatrix} \frac{1}{2} & -\frac{1}{2} \\ \frac{1}{2} & \frac{1}{2} \end{bmatrix}$$

We can observe that as n increases, successive powers of A will *get larger in size* due to the diagonal entries in D (ie. the eigenvalues control this behavior!)
.
THEOREM If a matrix A has n **distinct** eigenvalues, then it has n *linearly independent* eigenvectors.

This theorem implies that a matrix cannot be defective if its characteristic equation has *distinct* roots (ie. no repeated roots). If a matrix has repeated roots in the characteristic equation, then there is a danger that the geometric multiplicity of a repeated root may be LESS than the algebraic multiplicity- in that case there will be a shortage of eigenvectors and the matrix will NOT be diagonalizable. Fortunately most matrices are not defective!

EXAMPLE 1 Find the matrix P of eigenvectors and the diagonal matrix D for the matrix A, if

$$A = \begin{bmatrix} 1 & 0 & 0 \\ 0 & -5 & 3 \\ 0 & -6 & 4 \end{bmatrix}$$

The characteristic equation is $p(\lambda) = -(\lambda + 2)(\lambda - 1)^2 = 0$ so $\sigma(A) = \{-2, 1\}$ where $\lambda = 1$ is a double root.

For the first eigenvalue, we solve the linear system $(A-(-2)I)X = \theta$ whose augmented matrix is:

$$\begin{bmatrix} 3 & 0 & 0 & 0 \\ 0 & -3 & 3 & 0 \\ 0 & -6 & 6 & 0 \end{bmatrix} \sim \begin{bmatrix} 1 & 0 & 0 & 0 \\ 0 & 1 & -1 & 0 \\ 0 & 0 & 0 & 0 \end{bmatrix} \Rightarrow x_2 = x_3 = r, x_1 = 0$$

This means the eigenvectors are of the form $X = \begin{bmatrix} 0 \\ r \\ r \end{bmatrix} = r \begin{bmatrix} 0 \\ 1 \\ 1 \end{bmatrix}$ $r \neq 0$; this is a line through the origin.

For the second eigenvalue $\lambda = 1$ we have the system:

$$\begin{bmatrix} 0 & 0 & 0 & 0 \\ 0 & -6 & 3 & 0 \\ 0 & -6 & 3 & 0 \end{bmatrix} \sim \begin{bmatrix} 0 & 1 & -\dfrac{1}{2} & 0 \\ 0 & 0 & 0 & 0 \\ 0 & 0 & 0 & 0 \end{bmatrix} \Rightarrow x_2 - \frac{1}{2}x_3 = 0 \Rightarrow x_2 = \frac{1}{2}x_3 \Rightarrow x_2 = t, x_3 = 2t, \ x_1 = s$$

The eigenvectors are of the form $X = \begin{bmatrix} s \\ t \\ 2t \end{bmatrix} = s\begin{bmatrix} 1 \\ 0 \\ 0 \end{bmatrix} + t\begin{bmatrix} 0 \\ 1 \\ 2 \end{bmatrix}$ $s, t \neq 0$. This represents a plane through the origin. Hence we have THREE independent eigenvectors from the two eigenspaces: S = {(0,1,1), (1,0,0), (0,1,2} so A is not defective. We can write down the eigenvector matrix P and the diagonal matrix D:

$$P = \begin{bmatrix} 0 & 1 & 0 \\ 1 & 0 & 1 \\ 1 & 0 & 2 \end{bmatrix} \quad D = \begin{bmatrix} -2 & 0 & 0 \\ 0 & 1 & 0 \\ 0 & 0 & 1 \end{bmatrix}$$

Note that the **columns** of P are formed by the eigenvectors (in order as we found them). To verify our calculations, we can evaluate AP and compare it to PD.

$$AP = \begin{bmatrix} 1 & 0 & 0 \\ 0 & -5 & 3 \\ 0 & -6 & 4 \end{bmatrix}\begin{bmatrix} 0 & 1 & 0 \\ 1 & 0 & 1 \\ 1 & 0 & 2 \end{bmatrix} = \begin{bmatrix} 0 & 1 & 0 \\ 1 & 0 & 1 \\ 1 & 0 & 2 \end{bmatrix}\begin{bmatrix} -2 & 0 & 0 \\ 0 & 1 & 0 \\ 0 & 0 & 1 \end{bmatrix} = PD$$

If they are the same, we have done everything correctly. Keep in mind that A and D are *similar*; they share the *same eigenvalues* but have different eigenvectors.

Application- Markov Processes

Suppose the Boston Store cafeteria always has the same number of customers per week, split between coffee drinkers and tea drinkers. Someone determines that in a week, 50 % of the coffee drinkers switch to tea while 80 % of the tea drinkers switch to coffee.
This leads to a system of equations of the form

$$\begin{aligned}.5\,c_0 + .8\,t_0 = c_1\\[4pt].5\,c_0 + .2\,t_0 = t_1\end{aligned} \Rightarrow \begin{bmatrix}.5 & .8\\ .5 & .2\end{bmatrix}\begin{bmatrix}c_0\\ t_0\end{bmatrix}=\begin{bmatrix}c_1\\ t_1\end{bmatrix}\; where\; M =\begin{bmatrix}.5 & .8\\ .5 & .2\end{bmatrix}$$

which we can write as $X_1 = M\,X_0$. Here c = number of coffee drinkers and t = number of tea drinkers.

If we think of this as a *process*, where X_1 depends on X_0, then we can find X_2 using X_1: $X_2 = M\,X_1$ but we have X_1 in terms of X_0 so this means that $X_2 = M\,(\,M\,X_0\,) = M^2\,X_0$ and so forth so that $X_n = M^n\,X_0$. If

1) the present distribution X_n depends only on the previous distribution X_{n-1},

2) the probabilities of switching from tea to coffee or vice-versa do not change and

3) the population in the process does not change

then we have a **Markov process**. Here X is called the *state vector* and M is the *transition matrix*. Note that the matrix M has only non-negative entries where the *sum of the entries in each column* is ONE. A practical goal would be to determine the long-term behavior of this process, ie. find $\lim_{t \to \infty}(\,X_n\,)$. In order to find this limit, we need successively higher powers of M; if we can diagonalize M, it will be easy to raise M to higher powers.

By solving the characteristic equation $\det(\,M - \lambda I\,) = 0$ we easily see that the spectrum of M is given by $\sigma(\,M\,) = \{\,-\dfrac{3}{10}, 1\,\}$. The corresponding eigenspaces are $E_{\lambda=-\frac{3}{10}} = r\begin{bmatrix}-1\\ 1\end{bmatrix}, E_{\lambda=1} = s\begin{bmatrix}8\\ 5\end{bmatrix}$.

This means that the matrix of eigenvectors is $P = \begin{bmatrix}-1 & 8\\ 1 & 5\end{bmatrix}$ and $D = \begin{bmatrix}-\dfrac{3}{10} & 0\\ 0 & 1\end{bmatrix}$; hence MP = PD

and M can be written in the form $M = P D P^{-1}$. By raising M to higher powers, we have $M^2 = (\,P D P^{-1}\,)(P D P^{-1}\,) = P D^2 P^{-1}$ etc. which means that D is sandwiched between P and its inverse- D does all the work!

We can observe easily that $D^n = \begin{bmatrix}(-\dfrac{3}{10})^n & 0\\ 0 & 1\end{bmatrix} \to \begin{bmatrix}0 & 0\\ 0 & 1\end{bmatrix}$ which means

$$M^n X_0 = P D^n P^{-1} X_0 \to \begin{bmatrix}-1 & 8\\ 1 & 5\end{bmatrix}\begin{bmatrix}0 & 0\\ 0 & 1\end{bmatrix}\begin{bmatrix}\dfrac{5}{13} & \dfrac{8}{13}\\[6pt] \dfrac{1}{13} & \dfrac{1}{13}\end{bmatrix}X_0 = \begin{bmatrix}\dfrac{8}{13} & \dfrac{8}{13}\\[6pt] \dfrac{5}{13} & \dfrac{5}{13}\end{bmatrix}X_0$$

For example if initially there were $c_0 = 65$ coffee drinkers and $t_0 = 65$ tea drinkers, eventually there would be c = 80 coffee drinkers and t = 50 tea drinkers. What is interesting here is that if $\dfrac{c_0}{t_0} = \dfrac{8}{5}$,

then nothing would change. Let's verify this; suppose initially c = 80 and t = 50; then

$$X_1 = \begin{bmatrix}c_1\\ t_1\end{bmatrix} = \begin{bmatrix}.5 & .8\\ .5 & .2\end{bmatrix}\begin{bmatrix}80\\ 50\end{bmatrix} = \begin{bmatrix}80\\ 50\end{bmatrix} = X_0$$

Thus every Markov process has a steady state (output = input) if the initial state vector is an *eigenvector* corresponding to the eigenvalue $\lambda = 1$.

Linear Operators and Changing Bases

Going back to the introductory example $A = \begin{bmatrix} 1 & 3 \\ 3 & 1 \end{bmatrix}$, we may think of A as the standard matrix of the linear operator $[T(X)] = A[X]$. Can we find a basis for $V = \Re^2$ such that the matrix of T with respect to this basis is as simple as possible (ie. diagonal)? Using a basis of eigenvectors of A, we let B' = $\{(1,-1),(1,1)\}$; then we can consider the matrix P as the *transition matrix* from the new basis B' to the old (standard) basis B = $\{(1,0),(0,1)\}$. The inverse of this matrix is the transition matrix Q from the standard basis B to the new basis B'.

We can now verify our "goal" equation AP = PD, rewritten in the form $D = P^{-1} A P$.

If we now find the new matrix A', which represents the same linear operator as A does, then we know from Chapter 4 that we calculate it thus:

$$A' = P^{-1} A P = \begin{bmatrix} \frac{1}{2} & -\frac{1}{2} \\ \frac{1}{2} & \frac{1}{2} \end{bmatrix} \begin{bmatrix} 1 & 3 \\ 3 & 1 \end{bmatrix} \begin{bmatrix} 1 & 1 \\ -1 & 1 \end{bmatrix} = \begin{bmatrix} -2 & 0 \\ 0 & 4 \end{bmatrix} = D \Rightarrow [T(X)]_{B'} = D[X]_{B'}$$

The result of this calculation is the main point of this section- the matrix we get by sandwiching A between P (the matrix of eigenvectors) and its inverse Q, is a **diagonal** matrix. The elegant simplicity of the diagonal matrix D justifies all the work that we did to find it. Remember that if you wish to use the matrix D to represent T, all calculations must be done with respect to the non-standard basis B' = $\{(1,-1),(1,1)\}$.

EXERCISES 6.2

Part A. Given the matrix $A = \begin{bmatrix} 1 & 4 \\ 4 & 7 \end{bmatrix}$;

1. Find the P and D matrices for A. ans. $P = \begin{bmatrix} -2 & 1 \\ 1 & 2 \end{bmatrix}$ $D = \begin{bmatrix} -1 & 0 \\ 0 & 9 \end{bmatrix}$

2. Find the eigenvalues and corresponding eigenvectors for D.

3. Show that $G = \begin{bmatrix} -11 & 10 \\ -20 & 19 \end{bmatrix}$ is *similar* to D by finding the P and D matrices for G.

ans. $P = \begin{bmatrix} 1 & 1 \\ 1 & 2 \end{bmatrix}$ $D = \begin{bmatrix} -1 & 0 \\ 0 & 9 \end{bmatrix}$

4. Show that A is *similar* to G.

5. Find the determinant and trace of A, D and G. ans. -9, 8

6. A > 0 (ie. A is a POSITIVE matrix) ; show that A has a positive eigenvalue r = ρ(A). Verify that $\lambda = r$ has a positive eigenvector X.

Part B. For each matrix A (if possible) find the eigenvector matrix P and the diagonal matrix D; in each case check that AP = PD. If this is NOT possible, state why.

7. $A = \begin{bmatrix} 2 & 3 \\ 3 & 2 \end{bmatrix}$ ans. $P = \begin{bmatrix} -1 & 1 \\ 1 & 1 \end{bmatrix}$ $D = \begin{bmatrix} -1 & 0 \\ 0 & 5 \end{bmatrix}$

8. $A = \begin{bmatrix} 0 & 1 \\ 1 & 0 \end{bmatrix}$

9. $A = \begin{bmatrix} 3 & 4 \\ 4 & 3 \end{bmatrix}$ ans. $P = \begin{bmatrix} -1 & 1 \\ 1 & 1 \end{bmatrix}$ $D = \begin{bmatrix} -1 & 0 \\ 0 & 7 \end{bmatrix}$

10. $A = \begin{bmatrix} 1 & 0 & 2 \\ 0 & -3 & 3 \\ 0 & 0 & 6 \end{bmatrix}$

11. $A = \begin{bmatrix} 2 & 2 & 2 \\ 0 & 3 & -3 \\ 0 & 0 & 4 \end{bmatrix}$ ans. $P = \begin{bmatrix} 1 & 2 & -2 \\ 0 & 1 & -3 \\ 0 & 0 & 1 \end{bmatrix}$ $D = \begin{bmatrix} 2 & 0 & 0 \\ 0 & 3 & 0 \\ 0 & 0 & 4 \end{bmatrix}$

12. $A = \begin{bmatrix} -1 & 0 & 1 \\ 1 & -1 & 1 \\ 0 & 0 & 2 \end{bmatrix}$

13. $A = \begin{bmatrix} 0 & 1 & 0 \\ 0 & 0 & 1 \\ 0 & 0 & 2 \end{bmatrix}$ ans. not possible, A is defective

14. $A = \begin{bmatrix} 1 & 0 & 0 \\ 0 & -3 & 0 \\ 2 & 3 & 6 \end{bmatrix}$

15. $A = \begin{bmatrix} 1 & 2 & 3 \\ 2 & 1 & 4 \\ 3 & 3 & 7 \end{bmatrix}$ ans. $P = \begin{bmatrix} -1 & 5 & 5 \\ 1 & 2 & 6 \\ 0 & -3 & 11 \end{bmatrix}$ $D = \begin{bmatrix} -1 & 0 & 0 \\ 0 & 0 & 0 \\ 0 & 0 & 10 \end{bmatrix}$

Part C. Given the matrix M, where $M = \begin{bmatrix} 0.25 & 0.6 \\ 0.75 & 0.4 \end{bmatrix}$;

16. Find the eigenvalues and corresponding eigenvectors of M.

312

17. If $X_{n+1} = M \, X_n$, find X_1, then X_2 and X_3 by left-multiplying by M;

let $X_0 = U_0 + V_0 = \begin{bmatrix} -20 \\ 20 \end{bmatrix} + \begin{bmatrix} 40 \\ 50 \end{bmatrix}$ and use the fact that U and V are *eigenvectors* of M. Don't add

U and V until you're done! $\qquad$ ans. $X_3 = \begin{bmatrix} 40.8575 \\ 49.1425 \end{bmatrix}$

18. Using MP = PD, solve for M in terms of P and D (ie. diagonalize M).

19. Let X = PY and rewrite $X_{n+1} = M \, X_n$ in terms of Y. Find Y_3 in terms of D if $X_0 = \begin{bmatrix} 20 \\ 70 \end{bmatrix}$.

Don't forget to find Y_0 first! NB. Observe how much easier it is to see what's happening to the

components of Y, since you are multiplying by D, not M. $\qquad$ ans. $Y_3 = D^3 \begin{bmatrix} 20 \\ 10 \end{bmatrix}$

20. If the total population is 180, what initial state X_0 would result in no change in passing from
$\qquad$ X_0 to X_1?? NB. You are finding the *fixed point* of the corresponding linear operator T
(ie. the *steady state* of the Markov process).

21. If $X_0 = kU_0 + jV_0 = k \begin{bmatrix} -1 \\ 1 \end{bmatrix} + j \begin{bmatrix} 4 \\ 5 \end{bmatrix}$ find X_n in terms of U and V. ans. $X_n = k(\dfrac{-7}{20})^n U_0 + jV_0$

Part D. Given the matrix $A = \begin{bmatrix} 1 & 2 \\ 2 & 1 \end{bmatrix}$;

22. Find the matrices P and D such that AP = PD.

23. If $[\, T(\,X\,)\,]_B = A\,[X\,]_B$ and B = {(1,0),(0,1)}, find $[\, T(\,X\,)\,]_{B'} = A'\,[\,X\,]_{B'}$ if

B' = {(-1,1),(1,1)}. $\qquad$ ans. $A' = \begin{bmatrix} -1 & 0 \\ 0 & 3 \end{bmatrix} = P^{-1} A P$

24. Find $[\, T(\,1,7\,)\,]_{B'}$ using A' from #23.

25. Find T(1,7) by using the answer from #24 ; then just calculate T(1,7) directly using A.

ans. $[\, T(\,X\,)\,]_{B'} = D \begin{bmatrix} 3 \\ 4 \end{bmatrix} = \begin{bmatrix} -3 \\ 12 \end{bmatrix} \Rightarrow [\, T(\,X\,)\,]_B = P \begin{bmatrix} -3 \\ 12 \end{bmatrix} = \begin{bmatrix} 15 \\ 9 \end{bmatrix}$

Part E. Suppose the robin population is modeled by the matrix equation

$X_{n+1} = \begin{bmatrix} 0 & 10 \\ \dfrac{3}{10} & \dfrac{1}{2} \end{bmatrix} \begin{bmatrix} j \\ a \end{bmatrix} = A \, X_n \; where \; X = \begin{bmatrix} j \\ a \end{bmatrix}$ where j = juveniles and a = adults (n = number of

years).

26. Find the eigenvalues and corresponding eigenvectors for A.

27. Using AP = PD find A in terms of P and D. ans.
$$A = \begin{bmatrix} -20 & 5 \\ 3 & 1 \end{bmatrix} \begin{bmatrix} -3 & 0 \\ 2 & \\ 0 & 2 \end{bmatrix} \begin{bmatrix} -\dfrac{1}{35} & \dfrac{1}{7} \\ \dfrac{3}{35} & \dfrac{4}{7} \end{bmatrix} = PDP^{-1}$$

28. Let X = PY and rewrite $X_{n+1} = A X_n$ in terms of Y; find Y_{n+1} .

If $X_0 = \begin{bmatrix} 30 \\ 20 \end{bmatrix}$ find Y_3 in terms of Yo. What is X3 ?

29. Show that an initial population of the form $X_0 = k \begin{bmatrix} 5 \\ 1 \end{bmatrix}$ will result in a doubling of the population every year.

Part F. Given the linear operator T(x,y) = (y,6x+y);
30. Find A, the standard matrix of T.

31. Find the matrices P and D satisfying AP = PD.   ans. $P = \begin{bmatrix} 1 & 1 \\ -2 & 3 \end{bmatrix}$, $D = \begin{bmatrix} -2 & 0 \\ 0 & 3 \end{bmatrix}$

32. Find the simplest matrix representative for T (ie. find a diagonal representative for T).
33. What basis must one use for $\Re^2$ if one wishes to use the matrix D from #32?
ans. B' = {(1,-2),(1,3)}
34. Find the eigenvalues/eigenvectors for D. Are they the same as the eigenvalues/eigenvectors of the matrix A?
35. Let G = HDH^{-1} where H is any 2 x 2 invertible matrix (your choice); show that the eigenvalues of G are the same as those of D and A but the eigenvectors are different.

Part G. Given
$$\frac{dx}{dt} = x + 2y$$
$$;$$
$$\frac{dy}{dt} = 8x + y$$

36. Write the system in the form dX/dt = AX if $X = \begin{bmatrix} x \\ y \end{bmatrix}$.

37. Find the matrices P and D which satisfy AP = PD. ans. $P = \begin{bmatrix} 1 & 1 \\ -2 & 2 \end{bmatrix}$, $D = \begin{bmatrix} -3 & 0 \\ 0 & 5 \end{bmatrix}$

38. Let X = PY (you are now changing to a basis of eigenvectors B'); rewrite the system dX/dt = AX in terms of Y and P. Hint: if X = PY then d/dt(X) = P(dY/dt) since P is *constant*.

39. Now solve for dY/dt; you should get an uncoupled system. ans. dY/dt = DY.
40. If $Y = \begin{bmatrix} u \\ v \end{bmatrix}$, write out the original system in terms of u and v. NB. It should look much simpler than the original!

41. If the solution of dY/dt = DY is $Y(t) = \begin{bmatrix} e^{-3t} & 0 \\ 0 & e^{5t} \end{bmatrix} \begin{bmatrix} c_1 \\ c_2 \end{bmatrix}$, find X(t). Hint: see #38.

ans. $X(t) = \begin{bmatrix} x(t) \\ y(t) \end{bmatrix} = \begin{bmatrix} e^{-3t} & e^{5t} \\ -2e^{-3t} & 2e^{5t} \end{bmatrix} \begin{bmatrix} c_1 \\ c_2 \end{bmatrix}$

42. Actually verify that x = x(t) and y = y(t) from # 41 actually satisfy the *original system*.

Part H.

43. Given the Jordan block matrix $J = \begin{bmatrix} k & 1 \\ 0 & k \end{bmatrix}$; show that J has only one eigenvalue. Describe the

corresponding eigenspace. Thus A here is defective. ans. $\lambda = k$, $E_{\lambda=k} = \{t \begin{bmatrix} 1 \\ 0 \end{bmatrix}\}$

44. Given the Jordan block matrix $J = \begin{bmatrix} k & 1 & 0 \\ 0 & k & 1 \\ 0 & 0 & k \end{bmatrix}$; find all eigenvalues/eigenvectors. Explain why

J is defective.

45. Let $A = \begin{bmatrix} a & 1 & 0 \\ 0 & a & 0 \\ 0 & 0 & b \end{bmatrix}$; A has a 2 x 2 Jordan block in the upper-left corner. Show that A has only

two independent eigenvectors and is defective. The culprit of course is the eigenvalue $\lambda = a$.

46. Show that $A = \begin{bmatrix} a & 0 & 0 \\ 0 & a & 0 \\ 0 & 0 & b \end{bmatrix}$ is NOT defective!

6.3 Systems of Linear Differential Equations

Consider the system of linear differential equations $\dfrac{d\,x}{d\,t}=x$.

$$\dfrac{d\,y}{d\,t}=\ 3\,y$$

This can be written in matrix form as: $\begin{bmatrix} \dfrac{d\,x}{d\,t} \\[2mm] \dfrac{d\,y}{d\,t} \end{bmatrix} = \begin{bmatrix} x \\ 3\,y \end{bmatrix}$

If we define the derivative of a matrix as the derivative of each entry:

$$X=\begin{bmatrix} x \\ y \end{bmatrix} \Rightarrow \dfrac{d}{d\,t}X=\dfrac{d}{d\,t}\begin{bmatrix} x \\ y \end{bmatrix}=\begin{bmatrix} \dfrac{d\,x}{d\,t} \\[2mm] \dfrac{d\,y}{d\,t} \end{bmatrix}$$

then this system can be rewritten more conveniently as:

$$\begin{bmatrix} \dfrac{d\,x}{d\,t} \\[2mm] \dfrac{d\,y}{d\,t} \end{bmatrix}=\begin{bmatrix} 1 & 0 \\ 0 & 3 \end{bmatrix}\begin{bmatrix} x \\ y \end{bmatrix}=A\,X$$

It is really a *linear system* of two homogeneous differential equations; the reason is we can rewrite it as

$$\dfrac{d\,X}{d\,t}-A\,X=\begin{bmatrix} 0 \\ 0 \end{bmatrix}=\theta$$

We can compare it easily with the *scalar equation* dx/dt - ax = 0. Now the solution of the scalar equation is $x=e^{a\,t}c$ which the reader can easily check. If there is any justice, the solution of the matrix equation dX/dt = AX should be of the same form: $X=e^{A\,t}C$. In our case, the original system was *uncoupled* - the equation with dx/dt had only x in it and the equation with dy/dt had only y in it. We could solve these equations separately and obtain:

$$x=e^{t}c_{1} \qquad\qquad y=e^{3\,t}c_{2}$$

In order to write this solution in matrix form, we easily show that:

$$\begin{bmatrix} x \\ y \end{bmatrix}=\begin{bmatrix} e^{t} & 0 \\ 0 & e^{3\,t} \end{bmatrix}\begin{bmatrix} c_{1} \\ c_{2} \end{bmatrix}$$

If indeed the matrix differential equation behaves like the scalar one, then we must admit that the matrix exponential exp(At) here must be of the form

$$\exp(\,A\,t\,)=e^{A\,t}=\begin{bmatrix} e^{t} & 0 \\ 0 & e^{3\,t} \end{bmatrix}$$

so the solution can be written as:

$$X(t) = \begin{bmatrix} e^t & 0 \\ 0 & e^{3t} \end{bmatrix} \begin{bmatrix} c_1 \\ c_2 \end{bmatrix} = \exp(At)C$$

It was easy to calculate the matrix exponential here because A was a **diagonal** matrix. What if A is NOT diagonal?

EXAMPLE 1 Find the solution of the linear system of differential equations

$$\begin{matrix} \dfrac{dx}{dt} = x + 2y \\ \\ \dfrac{dy}{dt} = 2x + y \end{matrix} \Rightarrow \dfrac{dX}{dt} = \begin{bmatrix} \dfrac{dx}{dt} \\ \dfrac{dy}{dt} \end{bmatrix} = \begin{bmatrix} 1 & 2 \\ 2 & 1 \end{bmatrix} \begin{bmatrix} x \\ y \end{bmatrix} = AX$$

First we need to find the diagonal representative of the coefficient matrix A, which is $A = \begin{bmatrix} 1 & 2 \\ 2 & 1 \end{bmatrix}$.

If we find the eigenvalues and eigenvectors, we have:

$$\lambda_1 = -1 \quad X_1 = \begin{bmatrix} -1 \\ 1 \end{bmatrix} \qquad \lambda_2 = 3 \quad X_2 = \begin{bmatrix} 1 \\ 1 \end{bmatrix}$$

The matrix equation AP = PD then is satisfied by:

$$P = \begin{bmatrix} -1 & 1 \\ 1 & 1 \end{bmatrix} \qquad D = \begin{bmatrix} -1 & 0 \\ 0 & 3 \end{bmatrix}$$

At this point, we make a change of variable: let X = PY. Here P is the *transition matrix,* which takes us from the NEW basis of eigenvectors of A to the OLD standard basis for $\mathfrak{R}^2$. With this change of coordinates, we have dX/dt = d/dt(PY) = P dY/dt since P is constant:

$$\frac{dX}{dt} = AX \Rightarrow P\frac{dY}{dt} = A(PY)$$

If we solve for dY/dt we get

$$\frac{dY}{dt} = (P^{-1}AP)Y = DY$$

since if AP = PD then $P^{-1}AP = D$. With this change of coordinates, we have *uncoupled* the system, since now we have a diagonal matrix on the right, and we can proceed as before. If we use a new coordinate system (ie. the "Y system") whose basis vectors consist of the eigenvectors from A (which span $\mathfrak{R}^2$) then we UNCOUPLE the original system (written in terms of the old standard basis ie. the "X system").

So the solution for Y is as before:

$$Y(t) = e^{Dt} C = \begin{bmatrix} e^{-t} & 0 \\ 0 & e^{3t} \end{bmatrix} \begin{bmatrix} c_1 \\ c_2 \end{bmatrix}$$

But we don't really want Y- we want X! So we need to translate back to the original "X system" using the matrix P; X = PY so

$$X(t) = P Y(t) = P e^{Dt} C = \begin{bmatrix} -1 & 1 \\ 1 & 1 \end{bmatrix} \begin{bmatrix} e^{-t} & 0 \\ 0 & e^{3t} \end{bmatrix} \begin{bmatrix} c_1 \\ c_2 \end{bmatrix} = \begin{bmatrix} -e^{-t} & e^{3t} \\ e^{-t} & e^{3t} \end{bmatrix} \begin{bmatrix} c_1 \\ c_2 \end{bmatrix} = \begin{bmatrix} -c_1 e^{-t} + c_2 e^{3t} \\ c_1 e^{-t} + c_2 e^{3t} \end{bmatrix}$$

Note that if this last matrix is broken up into the *sum* of two column matrices, the importance of the eigenvectors shines forth:

$$X(t) = \begin{bmatrix} -c_1 e^{-t} + c_2 e^{3t} \\ c_1 e^{-t} + c_2 e^{3t} \end{bmatrix} = e^{-t} \begin{bmatrix} -1 \\ 1 \end{bmatrix} c_1 + e^{3t} \begin{bmatrix} 1 \\ 1 \end{bmatrix} c_2 = e^{\lambda_1 t} X_1 c_1 + e^{\lambda_2 t} X_2 c_2 = E_1(t)c_1 + E_2(t)c_2$$

One can verify rather quickly that

$$E_1(t) = e^{-t} \begin{bmatrix} -1 \\ 1 \end{bmatrix}, \qquad E_2(t) = e^{3t} \begin{bmatrix} 1 \\ 1 \end{bmatrix}$$

both separately satisfy dX/dt = AX. So X(t) is a linear combination of E1(t) and E2(t).

The solution X(t) = P exp(Dt)C can be checked by direct substitution. We must verify that dX/dt = AX. Well,

dX/dt = P D **exp(Dt)C** and A X = A(P exp(Dt)C) = A P **exp(Dt)C**.

Clearly if PD = AP we have the correct solution to dX/dt = AX.

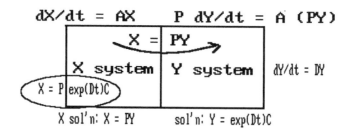

To summarize- the general solution of **dX/dt = AX** is X (t) = P exp(Dt) C
where :
P = matrix of *eigenvectors*, exp(Dt) is the *matrix exponential* of D, and
C is a column matrix of constants.

318

Calculus on the left matches matrix algebra on the right - and the key is to find the eigenvalues of A in order to uncouple the system and find exp(Dt) and P (the matrix of eigenvectors).

Initial value problems

If we are given *initial conditions*, then we must find C. Back in EX1 suppose that $x(0) = 3$ and $y(0) = 5$; recall that

$$X(t) = P\,Y(t) = P\,e^{Dt}\,C = \begin{bmatrix} -1 & 1 \\ 1 & 1 \end{bmatrix}\begin{bmatrix} e^{-t} & 0 \\ 0 & e^{3t} \end{bmatrix}\begin{bmatrix} c_1 \\ c_2 \end{bmatrix}$$

then we have:

$$X(0) = \begin{bmatrix} 3 \\ 5 \end{bmatrix} = \begin{bmatrix} -1 & 1 \\ 1 & 1 \end{bmatrix}\begin{bmatrix} 1 & 0 \\ 0 & 1 \end{bmatrix}\begin{bmatrix} c_1 \\ c_2 \end{bmatrix} = P\,I\,C = P\,C$$

Now to solve for c_1 & c_2, we need to left-multiply by the inverse of P:

$$X(0) = \begin{bmatrix} 3 \\ 5 \end{bmatrix} = P\begin{bmatrix} c_1 \\ c_2 \end{bmatrix} \Rightarrow C = \begin{bmatrix} c_1 \\ c_2 \end{bmatrix} = P^{-1}\,X(0)$$

We can then find the matrix of constants C:

$$C = P^{-1}\,X(0) = -\frac{1}{2}\begin{bmatrix} 1 & -1 \\ -1 & -1 \end{bmatrix}\begin{bmatrix} 3 \\ 5 \end{bmatrix} = \begin{bmatrix} 1 \\ 4 \end{bmatrix}$$

Now get the solution of the system X(t):

$$X(t) = P\,Y(t) = \begin{bmatrix} -e^{-t} & e^{3t} \\ e^{-t} & e^{3t} \end{bmatrix}\begin{bmatrix} 1 \\ 4 \end{bmatrix} = \begin{bmatrix} -e^{-t} + 4\,e^{3t} \\ e^{-t} + 4\,e^{3t} \end{bmatrix} = e^{-t}\begin{bmatrix} -1 \\ 1 \end{bmatrix} + e^{3t}\begin{bmatrix} 4 \\ 4 \end{bmatrix}$$

which gives a *particular solution* (ie. no arbitrary constants) to the initial-value problem. The reader can easily see that the initial conditions $x(0) = 3$ and $y(0) = 5$ are satisfied. $(x(t),y(t))$ is plotted parametrically below; you can see that this trajectory heads towards the eigenspace of $\lambda = 3$ and grows rapidly. The solution is plotted on the t-interval $[0,1]$ only.

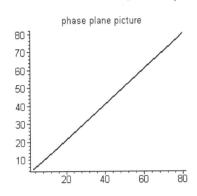

phase plane picture

The matrix exponential exp(At)

Using the solution X(t) from the above discussion, if A is diagonalizable, then the solution of the system can be written in the form:

$$X(t) = P\,e^{Dt}\,C = P\,e^{Dt}(\,P^{-1}\,X(\,0\,)\,) = (\,P\,e^{Dt}\,P^{-1}\,)\,K$$

where I have let X(0) = K (a column matrix of constants). The matrix $P\,e^{Dt}\,P^{-1}$ should be exp(At) - **the matrix exponential**. Mimicking the power series for exp(at):

$$e^{at} = 1 + (\,at\,) + \frac{(\,at\,)^2}{2!} + \dots \frac{(\,at\,)^n}{n!} + \dots$$

The matrix version of this series is:

$$e^{At} = I + (\,At\,) + \frac{1}{2!}(\,At\,)^2 + \frac{1}{3!}(\,At\,)^3 + \dots = I + At + A^2\frac{t^2}{2!} + A^3\frac{t^3}{3!} + \dots$$

Since this series converges for all A, this is a good way to *define* the exponential of a matrix. Actually *calculating* this matrix uses the diagonalization process; if $A = P\,D\,P^{-1}$ then

$$\exp(\,At) = I + At + \frac{A^2 t^2}{2!} + \dots = I + (\,P\,D\,P^{-1}\,)t + \frac{(\,P\,D\,P^{-1}\,)^2 t^2}{2!} + \dots$$

$$\exp(\,At) = P\,I\,P^{-1} + P\,D\,P^{-1}t + \frac{P\,D^2\,P^{-1}t^2}{2!} + \dots = P\,[\,I + Dt + \frac{D^2 t^2}{2!} + \dots\,]\,P^{-1} = P\exp(\,Dt\,)P^{-1}$$

Checking calculations: Since finding exp(At) involves some intense calculations, it is a good idea to verify your work. The most important criterion is to see if d/dt[**exp(At)**] = **A exp(At)**. A second (simpler) criterion on all this work is to see if exp(At) = *I when t = 0*; if not you made a mistake somewhere.

The chart below should make the relationships clear between the X system (standard basis) and the Y system (basis of eigenvectors).

X system (coupled)	Y system (uncoupled)
$X = \begin{bmatrix} x \\ y \end{bmatrix}$	$Y = \begin{bmatrix} u \\ v \end{bmatrix}$
$X(t) = P\,e^{Dt}\,C$	$Y(t) = e^{Dt}\,C$
$X(\,0\,) = P\,Y(\,0\,)$	$Y(\,0\,) = C$
$X(\,0\,) = K = P\,C$	$C = P^{-1}\,X(\,0\,) = P^{-1}\,K$

EXAMPLE 2 Find exp(At) for $A = \begin{bmatrix} 1 & 2 \\ 2 & 1 \end{bmatrix}$.

From before we have:

$$\lambda = -1 \Rightarrow E_{\lambda = -1} = \left\{ \begin{bmatrix} -1 \\ 1 \end{bmatrix} s \right\}, \quad \lambda = 3 \Rightarrow E_{\lambda = 3} = \left\{ \begin{bmatrix} 1 \\ 1 \end{bmatrix} t \right\}$$

This means that we can construct P and find its inverse:

$$P = \begin{bmatrix} -1 & 1 \\ 1 & 1 \end{bmatrix}, \quad P^{-1} = -\frac{1}{2} \begin{bmatrix} 1 & -1 \\ -1 & -1 \end{bmatrix}$$

We can then write down exp(At) using these matrices:

$$\exp(At) = \begin{bmatrix} -1 & 1 \\ 1 & 1 \end{bmatrix} \begin{bmatrix} e^{-t} & 0 \\ 0 & e^{3t} \end{bmatrix} \left(\frac{1}{2}\right) \begin{bmatrix} -1 & 1 \\ 1 & 1 \end{bmatrix} = \frac{1}{2} \begin{bmatrix} e^{-t} + e^{3t} & -e^{-t} + e^{3t} \\ -e^{-t} + e^{3t} & e^{-t} + e^{3t} \end{bmatrix}$$

$$\exp(At) = \frac{1}{2} \begin{bmatrix} e^{-t} + e^{3t} & e^{3t} - e^{-t} \\ e^{3t} - e^{-t} & e^{-t} + e^{3t} \end{bmatrix}$$

This requires a lot of work but the result is almost a work of art- in can be checked in several ways; the best way is to show it satisfies dX/dt = AX. It also must be true that if t = 0 then exp(At) = **I**. The general solution of dX/dt = AX can also be given as X(t) = exp(At) K, where K is a column matrix of constants. In an initial-value problem, K = X(0). The matrix exponential is usually called the FUNDAMENTAL MATRIX solution, usually represented by $\Phi(t)$ so we can write $X(t) = \Phi(t) K$.

Some important properties of the matrix exponential exp(At):
1) exp(A(0)) = **I**
2) d/dt(exp(At)) = A exp(At)
3) exp(At) is invertible and $(\exp(At))^{-1} = \exp(-At)$
4) exp(At)exp(Bt) = exp((A+B)t) iff AB = BA

EXERCISES 6.3

Part A. Given the matrix $A = \begin{bmatrix} 1 & 2 \\ 0 & 3 \end{bmatrix}$;

1. Write the system of differential equations dX/dt = AX in matrix form where $X = \begin{bmatrix} x \\ y \end{bmatrix}$.

ans. $\dfrac{dX}{dt} = AX = \begin{bmatrix} 1 & 2 \\ 0 & 3 \end{bmatrix} \begin{bmatrix} x \\ y \end{bmatrix}$

2. Write out the corresponding scalar equations for dx/dt and dy/dt.

3. Find the eigenvalues/eigenvectors for A.

$$\text{ans. } \lambda = 1, X = \{ s \begin{bmatrix} 1 \\ 0 \end{bmatrix} \} \quad \lambda = 3, X = \{ t \begin{bmatrix} 1 \\ 1 \end{bmatrix} \} \quad s, t \neq 0$$

4. Find the matrix of eigenvectors P; then let X = PY and rewrite dX/dt = AX in terms of Y.

5. Find the solution of the system dX/dt = AX using exp(Dt).

$$\text{ans. } X(t) = PY = P\,e^{Dt}\,C = \begin{bmatrix} 1 & 1 \\ 0 & 1 \end{bmatrix} \begin{bmatrix} e^t & 0 \\ 0 & e^{3t} \end{bmatrix} \begin{bmatrix} c_1 \\ c_2 \end{bmatrix}$$

6. Multiply out the solution X(t) in #5, writing X(t) as a 2 x 2 matrix times a 2 x 1 column matrix C.

7. Show that your answer to #6 satisfies dX/dt = AX by direct substitution.

8. Show that $E_1(t) = e^t \begin{bmatrix} 1 \\ 0 \end{bmatrix}, E_2(t) = e^{3t} \begin{bmatrix} 1 \\ 1 \end{bmatrix}$ each separately satisfies dX/dt = AX.

9. Now completely multiply out the answer to #5 to see the connection between the separate solutions in #8 and the factored version from #5.

$$\text{ans. } X(t) = \begin{bmatrix} c_1 e^t + c_2 e^{3t} \\ c_2 e^{3t} \end{bmatrix}$$

10. Find the matrix exponential $\Phi(t)$ for the matrix A; show that exp(At) = I if t = 0.

Part B. Find the solution X(t) = P exp(Dt) C for each linear system of differential equations. If initial conditions are given, find C to get a particular solution.

11.
$$\frac{d x}{d t} = 2 x$$
$$\frac{d y}{d t} = -2 y$$
$$\text{ans. } X(t) = \begin{bmatrix} e^{2t} & 0 \\ 0 & e^{-2t} \end{bmatrix} \begin{bmatrix} c_1 \\ c_2 \end{bmatrix}$$

12.
$$\frac{d x}{d t} = -2 x$$
$$\frac{d y}{d t} = 6 y$$
, IC's: x(0) = 20, y(0) = 50

13. $\frac{dx}{dt} = 2y$

$\frac{dy}{dt} = 2x$

, IC's: x(0) = 4, y(0) = 10 $\quad$ ans. $X(t) = \begin{bmatrix} e^{-2t} & e^{2t} \\ -e^{-2t} & e^{2t} \end{bmatrix} \begin{bmatrix} -3 \\ 7 \end{bmatrix}$

14. $\frac{dx}{dt} = -2x + 3y$

$\frac{dy}{dt} = 4y$

15. $\frac{dx}{dt} = x$

$\frac{dy}{dt} = -2x + 3y$

IC's: x(0) = 10, y(0) = -20 $\quad$ ans. $X(t) = \begin{bmatrix} e^t & 0 \\ e^t & e^{3t} \end{bmatrix} \begin{bmatrix} 10 \\ -30 \end{bmatrix}$

16. $\frac{dx}{dt} = 2x + y - z$

$\frac{dy}{dt} = -2y + 4z$

$\frac{dz}{dt} = 6z$

17. $\frac{dx}{dt} = x + z$

$\frac{dy}{dt} = 2y$

$\frac{dz}{dt} = x + z$

IC's: x(0) = 10, y(0) = 20, z(0) = 30 $\quad$ ans. $X(t) = \begin{bmatrix} -1 & 0 & e^{2t} \\ 0 & e^{2t} & 0 \\ 1 & 0 & e^{2t} \end{bmatrix} \begin{bmatrix} 10 \\ 20 \\ 20 \end{bmatrix}$

18.

$$\frac{dx}{dt} = x + 2y + 3z$$

$$\frac{dy}{dt} = 4y + 5z$$

$$\frac{dz}{dt} = 6z$$

Part C. Solve each system by find the fundamental matrix solution $\Phi(t) = \exp(At)$; if initial conditions are given, find a particular solution of the system.

19.

$$\frac{dx}{dt} = -2x + 4y$$

$$\frac{dy}{dt} = 3y$$

ans. $X(t) = \dfrac{1}{5}\begin{bmatrix} 5e^{-2t} & 4e^{3t} - 4e^{-2t} \\ 0 & 5e^{3t} \end{bmatrix}\begin{bmatrix} k_1 \\ k_2 \end{bmatrix}$

20.

$$\frac{dx}{dt} = 2x + 3y$$

$$\frac{dy}{dt} = 3x - 6y$$

IC's: $x(0) = 19$, $y(0) = -7$

21.

$$\frac{dx}{dt} = z$$

$$\frac{dy}{dt} = y$$

$$\frac{dz}{dt} = x$$

ans. $X(t) = \dfrac{1}{2}\begin{bmatrix} e^t + e^{-t} & 0 & e^t - e^{-t} \\ 0 & 2e^t & 0 \\ e^t - e^{-t} & 0 & e^t + e^{-t} \end{bmatrix}\begin{bmatrix} k_1 \\ k_2 \\ k_3 \end{bmatrix}$

Part D. Given the system

$$\frac{dx}{dt} = y$$

$$\frac{dy}{dt} = z \quad ;$$

$$\frac{dz}{dt} = 0$$

22. Write the system in the form dX/dt = AX; the matrix A should have an especially simple form- it is called a *nilpotent* matrix since $A^n = 0$ for some positive integer n.
23. Find the integer n mentioned in # 22. ans. n = 3
24. Find exp(At) using the series definition (it should truncate!).

25. Find the solution of the system using your answer to #24.

ans. $X(t) = \begin{bmatrix} 1 & t & \dfrac{t^2}{2} \\ 0 & 1 & t \\ 0 & 0 & 1 \end{bmatrix} \begin{bmatrix} k_1 \\ k_2 \\ k_3 \end{bmatrix}$

26. Prove that the solution works by substitution in the original system.

27. Solve the corresponding initial-value problem if x(0) = 1, y(0) = 2 and z(0) = 3.

ans. $X = \begin{bmatrix} 1 & t & \dfrac{t^2}{2} \\ 0 & 1 & t \\ 0 & 0 & 1 \end{bmatrix} \begin{bmatrix} 1 \\ 2 \\ 3 \end{bmatrix}$

Part E. Given the order two differential equation $y'' + 3\,y' + 2\,y = 0$;

28. Let dy/dt = v so that y" = dv/dt; then solve the given differential equation for dv/dt. Realize that this means the original differential equation can be written as a system in y and v:

$$\frac{d\,y}{d\,t} = v$$

$$\frac{d\,v}{d\,t} = -2\,y - 3\,v$$

29. Let $Y = \begin{bmatrix} y \\ v \end{bmatrix} \Rightarrow \dfrac{d\,Y}{d\,t} = \begin{bmatrix} \dfrac{d\,y}{d\,t} \\ \dfrac{d\,v}{d\,t} \end{bmatrix}$ so that the system can be written as dY/dt = AY.

Write the system in this matrix form.

ans. $\begin{matrix} \dfrac{d\,y}{d\,t} = v \\ \\ \dfrac{d\,v}{d\,t} = -2\,y - 3v \end{matrix} \Rightarrow \dfrac{d\,Y}{d\,t} = \begin{bmatrix} \dfrac{d\,y}{d\,t} \\ \dfrac{d\,v}{d\,t} \end{bmatrix} = \begin{bmatrix} 0 & 1 \\ -2 & -3 \end{bmatrix} \begin{bmatrix} y \\ v \end{bmatrix} = A\,Y$

30. Realize that the original order two equation is easily solved and its solution is given by:

$$y = c_1 e^{-2t} + c_2 e^{-t}$$

Thus find the solution $Y = \begin{bmatrix} y \\ v \end{bmatrix}$ of the system in matrix form, separating the different exponential

functions. Hint: v = dy/dt

31. Rewrite the solution you found as the **sum of two** solutions, each containing an exponential function times a column matrix. Amazingly you have found the eigenvectors of A w/o using the method of Section 6.1. What are the eigenvalues?

Hint: look at the exponential multipliers of the eigenvectors.

ans. eigenvalues: $\{-2,-1\}$, $\quad Y(t) = e^{-2t} \begin{bmatrix} 1 \\ -2 \end{bmatrix} c_1 + e^{-t} \begin{bmatrix} 1 \\ -1 \end{bmatrix} c_2$.

32. Now you should be able to write down P and D satisfying AP = PD. Verify that indeed this matrix equation is true.

Part F. Given the system of differential equations $\quad \dfrac{dx}{dt} = -3x + y$;

$$\dfrac{dy}{dt} = -3y$$

33. Show that A has only one eigenvalue (repeated) and is *defective* so it cannot be diagonalized. Write A in the form $-3\mathbf{I} + N$, where $N = \begin{bmatrix} 0 & 1 \\ 0 & 0 \end{bmatrix}$. Use the power series definition for exp(Nt) and the

fact that exp(At) = exp(-3$\mathbf{I}$ t) exp(Nt) to calculate exp(At). $\qquad$ ans. $\begin{bmatrix} e^{-3t} & t\,e^{-3t} \\ 0 & e^{-3t} \end{bmatrix}$

34. Verify that d/dt [exp(At)] = A exp(At) and that when t = 0, exp(At) = $\mathbf{I}$.

35. Solve the system with initial conditions x(0) = 5, y(0) = 6. ans. $X(t) = \begin{bmatrix} e^{-3t} & t\,e^{-3t} \\ 0 & e^{-3t} \end{bmatrix} \begin{bmatrix} 5 \\ 6 \end{bmatrix}$

Part G. Given $A = \begin{bmatrix} 1 & 2 \\ 0 & 3 \end{bmatrix}$;

36. Find exp(At).
37. Show that A commutes with exp(At).
38. For each eigenvalue/eigenvector pair of A, show that exp(λt) is an *eigenvalue* of exp(At) with the same eigenvector X of A.

39. Find exp(A). $\qquad$ ans. $\begin{bmatrix} e & e^3 - e \\ 0 & e^3 \end{bmatrix}$

40. Find exp(-A) and show that it is the inverse of exp(A).

41. If exp(2A) = $\begin{bmatrix} e^2 & e^6 - e^2 \\ 0 & e^6 \end{bmatrix}$ = B, what is log(B) ? ans. 2A = $\begin{bmatrix} 2 & 4 \\ 0 & 6 \end{bmatrix}$;

Part H. Solving dX/dt = AX numerically :

42. Mimicking Euler's method for dx/dt = f(x,t) which uses $x(t + h) \approx x(t) + h\dfrac{dx}{dt}$ where h = step

size, find the first approximation for x(0+h) and y(0+h) if x(0) = 3, y(0) = 4 in terms of h using

A = $\begin{bmatrix} 1 & 2 \\ 0 & 3 \end{bmatrix}$. $\qquad$ Hint: remember that dX/dt = $\begin{bmatrix} \dfrac{dx}{dt} \\ \dfrac{dy}{dt} \end{bmatrix}$ = AX.

326

43. Apply this method with h = 0.1 and approximate x(0.1) and y(0.1). ans. x = 5.4, y = 6.5
NB. This is easily done with a spreadsheet – use a smaller step size, like h = 0.01 for better accuracy!
44. Show that subsequent approximations are found by simply using $X_{n+1} = (I + hA)X_n$.

Part I. Prove:
45. d/dt [exp(At)] = A exp(At) Hint: use the power series expansion for exp(At)
46. the inverse of exp(At) is exp(-At).
47. X(t) = P exp(Dt)C is the solution of dX/dt = AX if AP = PD.

6.4 Complex Eigenvalues

Introduction

In this section we allow the matrix A to have imaginary eigenvalues and also consider some applications where the eigenvalues turn out to be imaginary.

Consider the simple matrix

$$A = \begin{bmatrix} 0 & 3 \\ -3 & 0 \end{bmatrix}$$

If we find the characteristic polynomial, we get:

$$A - \lambda I = \begin{bmatrix} 0 & 3 \\ -3 & 0 \end{bmatrix} - \lambda \begin{bmatrix} 1 & 0 \\ 0 & 1 \end{bmatrix} = \begin{bmatrix} -\lambda & 3 \\ -3 & -\lambda \end{bmatrix} \Rightarrow p(\lambda) = \begin{vmatrix} -\lambda & 3 \\ -3 & -\lambda \end{vmatrix} = \lambda^2 + 9$$

Solving the characteristic equation will give us the eigenvalues:

$$\lambda^2 + 9 = 0 \Rightarrow \lambda = \pm 3i$$

We have imaginary eigenvalues, which are CONJUGATES of each other- this is expected since for a polynomial equation with real coefficients, any imaginary roots come in *complex conjugate* pairs. We don't really change what we did in Section 6.1, we just need to be able to do complex arithmetic. We need to find the eigenvectors; let's deal with $\lambda = 3i$ first.

$$A - (3i)I = \begin{bmatrix} 0 & 3 \\ -3 & 0 \end{bmatrix} - 3i \begin{bmatrix} 1 & 0 \\ 0 & 1 \end{bmatrix} = \begin{bmatrix} -3i & 3 \\ -3 & -3i \end{bmatrix} \Rightarrow (A - \lambda I)X = \begin{bmatrix} -3i & 3 \\ -3 & -3i \end{bmatrix} \begin{bmatrix} x \\ y \end{bmatrix} = \begin{bmatrix} 0 \\ 0 \end{bmatrix}$$

The augmented matrix of the system is thus

$$\begin{bmatrix} -3i & 3 & 0 \\ -3 & -3i & 0 \end{bmatrix}$$

Putting this in echelon form (keep in mind at least one row must always disappear):

$$\begin{bmatrix} -3i & 3 & 0 \\ -3 & -3i & 0 \end{bmatrix} \underset{i\,row_1}{\sim} \begin{bmatrix} 3 & 3i & 0 \\ -3 & -3i & 0 \end{bmatrix} \underset{row_1 + row_2}{\sim} \begin{bmatrix} 3 & 3i & 0 \\ 0 & 0 & 0 \end{bmatrix}$$

$$\begin{bmatrix} 3 & 3i & 0 \\ 0 & 0 & 0 \end{bmatrix} \underset{\frac{1}{3}row_1}{\sim} \begin{bmatrix} 1 & i & 0 \\ 0 & 0 & 0 \end{bmatrix} \Rightarrow x + iy = 0 \Rightarrow x = -iy$$

Letting y = s, the corresponding eigenvectors are of the form:

$$X = \begin{bmatrix} x \\ y \end{bmatrix} = \begin{bmatrix} -is \\ s \end{bmatrix} = \begin{bmatrix} -i(it) \\ it \end{bmatrix} = t \begin{bmatrix} 1 \\ i \end{bmatrix}$$

We now deal with $\lambda = -3i$ (ie. the conjugate of the first eigenvalue):

$$A - (-3i)I = \begin{bmatrix} 0 & 3 \\ -3 & 0 \end{bmatrix} + 3i \begin{bmatrix} 1 & 0 \\ 0 & 1 \end{bmatrix} = \begin{bmatrix} 3i & 3 \\ -3 & 3i \end{bmatrix} \Rightarrow (A - \lambda I)X = \begin{bmatrix} 3i & 3 \\ -3 & 3i \end{bmatrix} \begin{bmatrix} x \\ y \end{bmatrix} = \begin{bmatrix} 0 \\ 0 \end{bmatrix}$$

The augmented matrix of the system is:

$$\begin{bmatrix} 3i & 3 & 0 \\ -3 & 3i & 0 \end{bmatrix}$$

Putting this in echelon form:

$$\begin{bmatrix} 3i & 3 & 0 \\ -3 & 3i & 0 \end{bmatrix} \underset{-i\,row_1}{\sim} \begin{bmatrix} 3 & -3i & 0 \\ -3 & 3i & 0 \end{bmatrix} \underset{row_1 + row_2}{\sim} \begin{bmatrix} 3 & -3i & 0 \\ 0 & 0 & 0 \end{bmatrix} \underset{\frac{1}{3}row_1}{\sim} \begin{bmatrix} 1 & -i & 0 \\ 0 & 0 & 0 \end{bmatrix} \Rightarrow x - iy = 0 \Rightarrow x = iy$$

Letting $y = s$ means that the corresponding eigenvectors are of the form:

$$X = \begin{bmatrix} x \\ y \end{bmatrix} = \begin{bmatrix} is \\ s \end{bmatrix} = \begin{bmatrix} i(-it) \\ -it \end{bmatrix} = t \begin{bmatrix} 1 \\ -i \end{bmatrix}, t \neq 0$$

In the future we can save time by noticing that the eigenvector of $\lambda = -3i$ is just the *conjugate* of the eigenvector corresponding to $\lambda = 3i$ (ie. everything comes out in conjugate pairs). We can check our calculations in the usual way; pick any eigenvector and left-multiply by A; the product must be of the form λX. So for example, if we let $X = \begin{bmatrix} 4 \\ 4i \end{bmatrix}$, the product should be 3i X:

$$AX = \begin{bmatrix} 0 & 3 \\ -3 & 0 \end{bmatrix} \begin{bmatrix} 4 \\ 4i \end{bmatrix} = \begin{bmatrix} 12i \\ -12 \end{bmatrix} \quad \lambda X = 3i X = 3i \begin{bmatrix} 4 \\ 4i \end{bmatrix} = \begin{bmatrix} 12i \\ 12i^2 \end{bmatrix} = \begin{bmatrix} 12i \\ -12 \end{bmatrix}$$

This makes us feel comfortable with the complex arithmetic we need, since the defining equation of an eigenvalue/eigenvector pair is $AX = \lambda X$.

The matrix exponential

Let's use the matrix A from the introduction to get exp(At). Recall that we need to diagonalize A, find the matrix of eigenvectors P which satisfies the matrix equation AP = PD where D is a diagonal matrix with the eigenvalues on its main diagonal. From before, we can write down P at once by using a basis for each eigenspace. For $\lambda = 3i$ we have a basis (1,i) and for $\lambda = -3i$ we have a basis (1,-i); thus we can write down P:

$$P = \begin{bmatrix} 1 & 1 \\ i & -i \end{bmatrix}$$

We also can write down the diagonal matrix of eigenvalues:

$$D = \begin{bmatrix} 3i & 0 \\ 0 & -3i \end{bmatrix}$$

Then we have the matrix equation:

$$A P = \begin{bmatrix} 0 & 3 \\ -3 & 0 \end{bmatrix} \begin{bmatrix} 1 & 1 \\ i & -i \end{bmatrix} = \begin{bmatrix} 1 & 1 \\ i & -i \end{bmatrix} \begin{bmatrix} 3i & 0 \\ 0 & -3i \end{bmatrix} = P D$$

You can observe that this equation is correct by doing the indicated multiplication yourself.
In order to calculate the matrix exponential exp(At), we get it by remembering that we just solve
AP = PD for A and let D do all the work:

$$A = P D P^{-1} \Rightarrow \exp(A t) = P \exp(D t) P^{-1} = \begin{bmatrix} 1 & 1 \\ i & -i \end{bmatrix} \begin{bmatrix} e^{3it} & 0 \\ 0 & e^{-3it} \end{bmatrix} \begin{bmatrix} \frac{1}{2} & \frac{-1}{2}i \\ \frac{1}{2} & \frac{1}{2}i \end{bmatrix}$$

After multiplying out and using Euler's identity

$$e^{3it} = \cos(3t) + i \sin(3t)$$

we get the beautiful matrix exponential of A:

$$\exp(A t) = \begin{bmatrix} \cos(3t) & \sin(3t) \\ -\sin(3t) & \cos(3t) \end{bmatrix}$$

Of course, if we think of the power series definition of exp(At), it clearly cannot have any *imaginary* numbers in it, since A itself only had real entries. Also if t = 0 you get the identity matrix **I**; and more importantly, if you differentiate exp(At), you should get A exp(At). This I leave for you to verify. As an application of exp(At), we could use the matrix exponential to write down immediately the solution of the corresponding system of differential equations dX/dt = AX. It would simply be

$$X(t) = \Phi(t) K = \exp(A t) K = \begin{bmatrix} \cos(3t) & \sin(3t) \\ -\sin(3t) & \cos(3t) \end{bmatrix} \begin{bmatrix} k_1 \\ k_2 \end{bmatrix}$$

EXAMPLE 1 Write the order two linear differential equation $\dfrac{d^2 y}{dt^2} + \dfrac{g}{L} y = 0$ as a system of differential equations in y and v and solve using the matrix exponential.

This differential equation comes from linearizing the pendulum equation $m L \dfrac{d^2 y}{dt^2} + m g \sin(y) = 0$, which describes the angular displacement y of a simple pendulum. If we let $v = \dfrac{d y}{d t}$ then as a system we get (for convenience letting g = 10 and L = 1/10) :

$$\frac{dy}{dt} = v$$

$$\frac{dv}{dt} = -100\,y$$

Rewriting in matrix form implies:

$$Y = \begin{bmatrix} y \\ v \end{bmatrix} \Rightarrow \frac{dY}{dt} = \begin{bmatrix} \dfrac{dy}{dt} \\[2mm] \dfrac{dv}{dt} \end{bmatrix} = \begin{bmatrix} 0 & 1 \\ -100 & 0 \end{bmatrix} \begin{bmatrix} y \\ v \end{bmatrix} = AY$$

where $A = \begin{bmatrix} 0 & 1 \\ -100 & 0 \end{bmatrix}$. Solving the characteristic equation will give us the eigenvalues:

$$\lambda^2 + 100 = 0 \Rightarrow \lambda = \pm 10i$$

We need to find the eigenvectors; let's deal with $\lambda = 10i$ first.

$$A - (10i)I = \begin{bmatrix} -10i & 1 \\ -100 & -10i \end{bmatrix} \Rightarrow (A - \lambda I)X = \begin{bmatrix} -10i & 1 \\ -100 & -10i \end{bmatrix} \begin{bmatrix} x \\ y \end{bmatrix} = \begin{bmatrix} 0 \\ 0 \end{bmatrix}$$

The augmented matrix of the system is:

$$\begin{bmatrix} -10i & 1 & 0 \\ -100 & -10i & 0 \end{bmatrix} \sim \begin{bmatrix} 1 & -\dfrac{1}{10i} & 0 \\ 0 & 0 & 0 \end{bmatrix} \Rightarrow x - \frac{1}{10i}y = 0 \Rightarrow x = \frac{1}{10i}y$$

Letting y = s, the corresponding eigenvectors are of the form

$$\begin{bmatrix} x \\ y \end{bmatrix} = \begin{bmatrix} \dfrac{s}{10i} \\[2mm] s \end{bmatrix} = t \begin{bmatrix} 1 \\ 10i \end{bmatrix}$$

if we let s = (10i)t.

For $\lambda = -10i$ we get $\begin{bmatrix} x \\ y \end{bmatrix} = t \begin{bmatrix} 1 \\ -10i \end{bmatrix}$ (ie. the *conjugate* of the first eigenvector).

Thus the matrix P of eigenvectors is of the form $P = \begin{bmatrix} 1 & 1 \\ 10i & -10i \end{bmatrix}$, which means $P^{-1} = \begin{bmatrix} \dfrac{1}{2} & \dfrac{-i}{20} \\[2mm] \dfrac{1}{2} & \dfrac{i}{20} \end{bmatrix}$.

Finally we can get exp(At):

$$Y = \begin{bmatrix} y \\ v \end{bmatrix} = \begin{bmatrix} 1 & 1 \\ 10i & -10i \end{bmatrix} \begin{bmatrix} e^{10it} & 0 \\ 0 & e^{-10it} \end{bmatrix} \begin{bmatrix} \dfrac{1}{2} & \dfrac{-i}{20} \\[2mm] \dfrac{1}{2} & \dfrac{i}{20} \end{bmatrix} \begin{bmatrix} k_1 \\ k_2 \end{bmatrix}$$

Multiplying things out on the right gives us the solution of the system:

$$Y(t) = \begin{bmatrix} y(t) \\ v(t) \end{bmatrix} = \begin{bmatrix} \cos(10t) & \frac{1}{10}\sin(10t) \\ -10\sin(10t) & \cos(10t) \end{bmatrix} \begin{bmatrix} k_1 \\ k_2 \end{bmatrix}$$

This is of the form $Y(t) = \Phi(t)K = \exp(At)\begin{bmatrix} k_1 \\ k_2 \end{bmatrix}$.

Thus $y(t) = k_1\cos(10t) + \frac{k_2}{10}\sin(10t)$ $v(t) = -10k_1\sin(10t) + k_2\cos(10t)$,

so y(t) and v(t) are is periodic (not unexpected) with period $T = \frac{2\pi}{10} = \frac{\pi}{5}$ seconds.

If initial conditions are given, we can eliminate the arbitrary constants. For example if the pendulum is pulled back so that $y = \frac{\pi}{4}$ and just released (ie. v(0) = 0) then we have:

$$y(0) = \frac{\pi}{4} = k_1 \Rightarrow k_1 = \frac{\pi}{4} \quad v(0) = 0 = k_2 \Rightarrow k_2 = 0$$

The particular solution is then:

$$Y(t) = \begin{bmatrix} y \\ v \end{bmatrix} = \begin{bmatrix} \frac{\pi}{4}\cos(10t) \\ \frac{-5\pi}{2}\sin(10t) \end{bmatrix}$$

phase plane picture

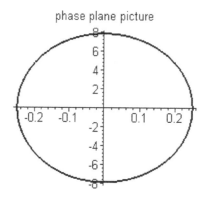

The phase plane picture shows (x(t),y(t)) plotted parametrically. Clearly the motion is periodic, since the trajectory (ie. graph of a particular solution (x(t),y(t))) is a closed curve. The solutions x(t) and y(t) are bounded and remain finite as t increases.

Stability of solutions

Often it is important to know the behavior of the solutions of a system of differential equations dX/dt = AX for t ε [0, oo). Specifically we wish to know if $\|(x(t), y(t))\| = \sqrt{(x(t))^2 + (y(t))^2}$ is bounded or unbounded as $t \to \infty$. Let's suppose that A is a 2 x 2 *invertible* matrix and has NO REPEATED eigenvalues; if we solve $\dfrac{dX}{dt} = AX = \begin{bmatrix} 0 \\ 0 \end{bmatrix}$ then the only solution is x = 0 and y = 0.

We call (0,0) an **equilibrium solution** of dX/dt = AX since if we start with these initial conditions, then the solution of the system is (x(t),y(t)) = (0,0) and we stay where we started!
If we look at the solution of the system in the form

$$X(t) = \begin{bmatrix} x(t) \\ y(t) \end{bmatrix} = c_1 e^{\lambda_1 t} X_1 + c_2 e^{\lambda_2 t} X_2$$

then it is clear that if even one eigenvalue is positive, in general $\|(x(t), y(t))\| \to \infty$ as $t \to \infty$.
Now let's suppose $\lambda = a + bi$; according to Euler we can write:

$$e^{(a+bi)t} = e^{at}[\cos(bt) + i\sin(bt)]$$

and we observe that

$$\left|e^{(a+bi)t}\right| = e^{at}\left|\cos(t) + i\sin(t)\right| \le e^{at}$$

hence it is the **real part** of the eigenvalue λ = **a** + b i which determines the *stability* of the solutions of dX/dt = AX. Simply plot the eigenvalues in the complex plane:

1) If at least one of them lies to the right of the imaginary axis, the solutions become unbounded as $t \to \infty$ and are called UNSTABLE. They head away from the equilibrium solution at (0,0).
2) If they lie to the left of the imaginary axis, then $(x(t), y(t)) \to (0,0)$. The solutions are called **asymptotically stable**. They are bounded and get closer and closer to the equilibrium solution as t increases.
3) If they lie ON the imaginary axis, then the solutions are periodic; they are bounded as t increases and are called *stable*. The solutions orbit the equilibrium solution but don't "fall in" to it.

Glance back at EX1; the matrix A had pure imaginary eigenvalues so the solutions are *stable*. Look at the graph above – as t increases, the trajectory simply orbits the equilibrium solution at the origin. x(t) and y(t) remain bounded but they do NOT approach zero.

If A is any invertible matrix with NO REPEATED eigenvalues - plot the **eigenvalues** in the
complex plane:
1) if even one lies to the right of the imaginary axis, the solutions are UNSTABLE
2) if all lie to the left of the imaginary axis, the solutions are **asymptotically stable**
3) if none lies to the right of the imaginary axis, the solutions are *stable* (at least one eigenvalue
lies ON the imaginary axis)

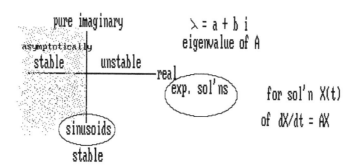

NB. If A is *singular*, then one of the eigenvalues is $\lambda = 0$. The criteria above are *still valid*- as long
as there are NO REPEATED eigenvalues.

EXAMPLE 2 Classify the solutions of the system

$$\frac{d x}{d t} = y$$

$$\frac{d y}{d t} = -101 x - 2 y$$

In matrix form this is

$$\frac{d X}{d t} = \begin{bmatrix} 0 & 1 \\ -101 & -2 \end{bmatrix} \begin{bmatrix} x \\ y \end{bmatrix} = A X$$

Finding the eigenvalues of A:

$$A - \lambda I = \begin{bmatrix} 0 & 1 \\ -101 & -2 \end{bmatrix} - \lambda \begin{bmatrix} 1 & 0 \\ 0 & 1 \end{bmatrix} = \begin{bmatrix} -\lambda & 1 \\ -101 & -2-\lambda \end{bmatrix}$$

Thus the characteristic equation is $-\lambda(-2 - \lambda)+101 = \lambda^2 + 2\lambda + 101 = (\lambda+1)^2 + 10^2 = 0$.
The eigenvalues of A are $\lambda = -1 \pm 10 i$ which lie to the left of the imaginary axis so the solutions of
dX/dt = AX are *asymptotically stable*. As t gets large, $(x(t), y(t)) \rightarrow (0,0)$. The nice thing about
this analysis is that *we don't need to actually solve the system* in order to accurately predict the
behavior of its solutions x(t) and y(t). Below is a typical trajectory; you can see how it spirals in to
the origin in the phase plane.

334

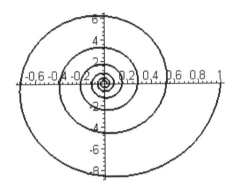

EXAMPLE 3 Determine the stability of the solutions to dX/dt = AX if $A = \begin{bmatrix} 0 & 1 \\ 8 & -2 \end{bmatrix}$.

The characteristic equation polynomial is given by

$$\begin{vmatrix} -\lambda & 1 \\ 8 & -2-\lambda \end{vmatrix} = \lambda^2 + 2\lambda - 8 = (\lambda + 4)(\lambda - 2) = 0$$

Solving this quadratic gives the eigenvalues or spectrum of A: $\sigma(A) = \{-4, 2\}$.

Since one of the eigenvalues lies to the right of the imaginary axis, the solutions are *unstable*. This is qualitative analysis.

If we do actually solve this system, we'd get $X(t) = \begin{bmatrix} e^{-4t} & e^{2t} \\ -4e^{-4t} & 2e^{2t} \end{bmatrix} \begin{bmatrix} c_1 \\ c_2 \end{bmatrix}$. Note that

if $c_2 = 0$ the solution X(t) contains only exp(-4t) and is bounded for all t; however in general X(t) also contains exp(2t), which of course becomes unbounded as $t \to \infty$.

EXERCISES 6.4

Part A. Find the eigenvalues and corresponding eigenspaces for each matrix A.

1. $A = \begin{bmatrix} 0 & 2 \\ -2 & 0 \end{bmatrix}$ ans. $\lambda = 2i, E_{\lambda=2i} = \{ s \begin{bmatrix} 1 \\ i \end{bmatrix} \} \quad \lambda = -2i, E_{\lambda=-2i} = \{ t \begin{bmatrix} 1 \\ -i \end{bmatrix} \}$

2. $A = \begin{bmatrix} 0 & 3 \\ -3 & 0 \end{bmatrix}$

3. $A = \begin{bmatrix} 0 & -5 \\ 1 & -2 \end{bmatrix}$ ans. $\lambda = -1 + 2i, E_{\lambda=-1+2i} = \{ s \begin{bmatrix} 5 \\ 1-2i \end{bmatrix} \} \quad \lambda = -1 - 2i, E_{\lambda=-1-2i} = \{ t \begin{bmatrix} 5 \\ 1+2i \end{bmatrix} \}$

4. $A = \begin{bmatrix} 0 & -25 \\ 1 & -4 \end{bmatrix}$

5. $A = \begin{bmatrix} 0 & -10 \\ 1 & -2 \end{bmatrix}$ ans. $\lambda = -1 + 3i$, $E_{\lambda = -1+3i} = \{ s \begin{bmatrix} 10 \\ -1-3i \end{bmatrix} \}$ $\lambda = -1 - 3i$, $E_{\lambda = -1-3i} = \{ t \begin{bmatrix} 10 \\ -1+3i \end{bmatrix} \}$

6. $A = \begin{bmatrix} -8 & 5 \\ -13 & 8 \end{bmatrix}$

7. $A = \begin{bmatrix} -4 & 2 \\ -10 & 4 \end{bmatrix}$ ans. $\lambda = 2i$, $E_{\lambda = 2i} = \{ s \begin{bmatrix} 1 \\ 1+2i \end{bmatrix} \}$ $\lambda = -2i$, $E_{\lambda = -2i} = \{ t \begin{bmatrix} 1 \\ 1-2i \end{bmatrix} \}$

8. $A = \begin{bmatrix} 3 & -2 \\ 2 & 3 \end{bmatrix}$

9. $A = \begin{bmatrix} 0 & 1 \\ -5 & -2 \end{bmatrix}$ ans. $\lambda = -1 + 2i$, $E_{\lambda = -1+2i} = \{ s \begin{bmatrix} 1 \\ -1+2i \end{bmatrix} \}$ $\lambda = -1 - 2i$, $E_{\lambda = -1-2i} = \{ t \begin{bmatrix} 1 \\ -1-2i \end{bmatrix} \}$

Part B. Given $A = \begin{bmatrix} 0 & 0 & 0 \\ 1 & 0 & -1 \\ 0 & 1 & 0 \end{bmatrix}$;

10. Solve the characteristic equation; show that one eigenvalue is real, the other two are imaginary.
11. Find the eigenspace for each eigenvalue of A.
ans.

$$\lambda = 0, E_{\lambda = 0} = \{ r \begin{bmatrix} 1 \\ 0 \\ 1 \end{bmatrix} \}, \quad \lambda = i, E_{\lambda = i} = \{ s \begin{bmatrix} 0 \\ i \\ 1 \end{bmatrix} \}, \quad \lambda = -i, E_{\lambda = -i} = \{ t \begin{bmatrix} 0 \\ -i \\ 1 \end{bmatrix} \}$$

12. Find the matrices P and D which satisfy AP = PD.

13. Find tr(A); show it satisfies tr(A) = sum of eigenvalues. ans. tr(A) = 0

14. Find det(A); show it satisfies det(A) = product of eigenvalues.
15. Write out the system of 3 linear differential equations in x,y and z arising from dX/dt = AX if

$X = \begin{bmatrix} x \\ y \\ z \end{bmatrix}$.

ans. $\dfrac{dx}{dt} = 0$

$\dfrac{dy}{dt} = x - z$

$\dfrac{dz}{dt} = y$

16. Classify the solutions of dX/dt = AX simply by plotting the $\sigma(A)$ in the complex plane.

17. Solve this system by finding the fundamental matrix solution $\Phi(t)$.

ans. $X(t) = \Phi(t)K = \begin{bmatrix} 1 & 0 & 0 \\ \sin(t) & \cos(t) & -\sin(t) \\ 1-\cos(t) & \sin(t) & \cos(t) \end{bmatrix} \begin{bmatrix} k_1 \\ k_2 \\ k_3 \end{bmatrix}$.

18. Write down the solution of the initial value problem dX/dt = AX if $X(0) = \begin{bmatrix} x(0) \\ y(0) \\ z(0) \end{bmatrix} = \begin{bmatrix} 10 \\ 20 \\ 30 \end{bmatrix}$.

19. Demonstrate that $\dfrac{d}{dt}[\Phi(t)] = A\Phi(t)$.

Part C. The displacement y(t) in a certain spring/mass system is described by the differential equation $\dfrac{d^2y}{dt^2} + 4y = 0$. Initially y(0) = 1 and v(0) = 0.

20. Letting v = dy/dt write this one order two equation as a system of differential equations of the form dY/dt = AY where $Y = \begin{bmatrix} y \\ v \end{bmatrix}$. Hint: y" = dv/dt.

21. Find the eigenvalues and corresponding eigenspaces for A.

ans. $\lambda = 2i, E_{\lambda=2i} = \{ s\begin{bmatrix} 1 \\ 2i \end{bmatrix} \}$ $\lambda = -2i, E_{\lambda=-2i} = \{ t\begin{bmatrix} 1 \\ -2i \end{bmatrix} \}$

22. Find the two matrices P and D which satisfy AP = PD.

23. Write the solution in the form $Y(t) = \Phi(t)K$ where $\Phi(t) = \exp(At)$.

ans. $Y(t) = \begin{bmatrix} \cos(2t) & \frac{1}{2}\sin(2t) \\ -2\sin(2t) & \cos(2t) \end{bmatrix} \begin{bmatrix} k_1 \\ k_2 \end{bmatrix}$

24. Find the particular solution which satisfies the initial conditions x(0) = 5, y(0) = 6.

Part D. Given the system of differential equations $\begin{bmatrix} \dfrac{dx}{dt} \\ \dfrac{dy}{dt} \end{bmatrix} = \begin{bmatrix} 3 & 4 \\ -4 & 3 \end{bmatrix} \begin{bmatrix} x \\ y \end{bmatrix}$;

25. Find the matrices P and D which diagonalize A.   ans. $P = \begin{bmatrix} 1 & 1 \\ i & -i \end{bmatrix}$, $D = \begin{bmatrix} 3+4i & 0 \\ 0 & 3-4i \end{bmatrix}$

26. Are the solutions stable/unstable?

27. Solve the system using the fundamental matrix solution $X(t) = \Phi(t)K$

ans. $X(t) = \begin{bmatrix} e^{3t}\cos(4t) & e^{3t}\sin(4t) \\ -e^{3t}\sin(4t) & e^{3t}\cos(4t) \end{bmatrix} \begin{bmatrix} k_1 \\ k_2 \end{bmatrix}$

28. Find the particular solution which satisfies the initial conditions $x(0) = 7$, $y(0) = 8$.

Part E. The charge $q(t)$ in an RLC series circuit, where $dq/dt = i(t)$ = current is described by the differential equation $\dfrac{d^2 q}{d t^2} + 2\dfrac{dq}{dt} + 50\, q = 0$. Initially $q(0) = 1$ and $i(0) = 0$.

29. Let $i(t) = dq/dt$; then write this order two differential equation as a *system* of differential equations of the form $dY/dt = AY$ where $Y = \begin{bmatrix} q \\ i(t) \end{bmatrix}$. Hint: $q'' = di/dt$.

ans. $\begin{bmatrix} \dfrac{dq}{dt} \\ \dfrac{di}{dt} \end{bmatrix} = \begin{bmatrix} 0 & 1 \\ -50 & -2 \end{bmatrix} \begin{bmatrix} q \\ i(t) \end{bmatrix}$

30. Find the eigenvalues/eigenvectors for A.

31. Find the two matrices P and D which satisfy $AP = PD$.

ans. $P = \begin{bmatrix} 1 & 1 \\ -1+7i & -1-7i \end{bmatrix}$, $D = \begin{bmatrix} -1+7i & 0 \\ 0 & -1-7i \end{bmatrix}$

32. Write the solution in the form $Y = \Phi(t)K$ where $\Phi(t) = \exp(At)$.

33. Find the particular solution, which satisfies the initial conditions. Are the solutions $q(t)$ and $i(t)$ stable?

ans. $Y(t) = e^{-t} \begin{bmatrix} \cos(7t) + \dfrac{1}{7}\sin(7t) & \dfrac{1}{7}\sin(7t) \\ -\dfrac{50}{7}\sin(7t) & \cos(7t) - \dfrac{1}{7}\sin(7t) \end{bmatrix} \begin{bmatrix} 1 \\ 0 \end{bmatrix}$, asymptotically stable

Part F. Classify the solutions of the system of differential equations $dX/dt = AX$ as 1) asymptotically stable, 2) stable or 3) unstable if

34. $A = \begin{bmatrix} 0 & -5 \\ 1 & -2 \end{bmatrix}$

35. $A = \begin{bmatrix} -1 & -5 \\ 0 & -2 \end{bmatrix}$ ans. asymptotically stable

36. $A = \begin{bmatrix} -1 & 3 \\ 0 & -2 \end{bmatrix}$

37. $A = \begin{bmatrix} 0 & 1 \\ -10 & 2 \end{bmatrix}$ ans. unstable

38. $A = \begin{bmatrix} 0 & -1 \\ 9 & 0 \end{bmatrix}$

39. $A = \begin{bmatrix} 0 & 0 \\ 0 & 1 \end{bmatrix}$ ans. unstable

40. $A = \begin{bmatrix} 1 & 1 & 1 \\ 1 & 1 & 1 \\ 2 & 3 & 4 \end{bmatrix}$ Hint: $A > 0$

Part G: given the spring/mass differential equation m y" + c y' + k y = 0, where m, c and k are positive parameters;

41. Let v = dy/dt and write this order two differential equation as a system of the form

dY/dt = AY where $Y = \begin{bmatrix} y \\ v \end{bmatrix}$. ans. $\begin{bmatrix} \dfrac{dy}{dt} \\ \dfrac{dv}{dt} \end{bmatrix} = \begin{bmatrix} 0 & 1 \\ \dfrac{-k}{m} & \dfrac{-c}{m} \end{bmatrix} \begin{bmatrix} y \\ v \end{bmatrix}$

42. Find a formula for the eigenvalues in terms of m, c and k. Show that the solutions y(t) and v(t) are asymptotically stable as long as c is NOT zero.

43. Find the particular solution satisfying y(0) = 1, v(0) = 0 if m = 1, c = 5 and k = 6.

ans. $Y(t) = \begin{bmatrix} 3e^{-2t} - 2e^{-3t} & e^{-2t} - e^{-3t} \\ 6e^{-3t} - 6e^{-2t} & 3e^{-3t} - 2e^{-3t} \end{bmatrix} \begin{bmatrix} 1 \\ 0 \end{bmatrix}$

44. Verify that the solutions y(t) and v(t) from #43 are asymptotically stable for any initial conditions. This spring/mass system is *overdamped*; the mass does NOT oscillate but heads towards equilibrium.

Part H. Given the system $\begin{bmatrix} \dfrac{d\,x}{d\,t} \\[2mm] \dfrac{d\,y}{d\,t} \end{bmatrix} = \begin{bmatrix} -10 & 4 \\ -20 & 8 \end{bmatrix}\begin{bmatrix} x \\ y \end{bmatrix}$;

45. Find the solution of this system of differential equations; then describe the stability of the solutions x(t) and y(t). ans. $X(t) = \begin{bmatrix} 1 & 2 \\ 2 & 5 \end{bmatrix}\begin{bmatrix} e^{-2t} & 0 \\ 0 & 1 \end{bmatrix}\begin{bmatrix} c_1 \\ c_2 \end{bmatrix}$, stable

46. For any non-zero initial conditions, show X(t) approaches an eigenvector of $\lambda = 0$ as $t \to \infty$. Since A here is *singular*, the system dX/dt = AX has *infinitely-many* equilibrium solutions, all lying on the line {(x,y): (x,y) = t(2,5)}, which is the eigenspace of $\lambda = 0$.
NB. Below is the phase plane picture if (x(0),y(0)) = (3,7)

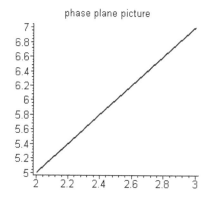

phase plane picture

Part I. $\dfrac{d^2 y}{d\,t^2} + c(y^2 - 1)v + y = 0$ is the non-linear *van der Pol* differential equation (c > 0);

47. Letting v = dv/dt, write the order two differential equation as a system in y and v.

ans. equations

$$\frac{d\,y}{d\,t} = v$$

$$\frac{d\,v}{d\,t} = -y + c(1 - y^2)v$$

48. Show that (y,v) = (0,0) is the only equilibrium solution of this non-linear system.

49. If this non-linear system is linearized around the equilibrium solution, it becomes a LINEAR system of the form dY/dt = JY where $J = \begin{bmatrix} 0 & 1 \\ -1 & c \end{bmatrix}$ and $Y = \begin{bmatrix} y \\ v \end{bmatrix}$. J is called the *Jacobian* matrix; show that the eigenvalues of J are given by $\lambda = \dfrac{c \pm \sqrt{4 - c^2}\,i}{2}$ if c < 2. Determine if the solutions y(t) and v(t) are stable/unstable.   ans. unstable

NB. Below is a graph of a numerical solution to the non-linear system; the equation has an isolated periodic solution called a limit cycle for any c > 0.

vanderPolDEwithc=1.5

50. Show that if c > 2, the solutions y(t) and v(t) are unstable.

6.5 Chapter Six Outline and Review

6.1 Eigenvalues and Eigenvectors

Given a square matrix A; the eigenvalue problem is to find solutions of

$$A X = \lambda X, \quad X \neq \theta, \ \lambda \in \Re$$

where X is the *eigenvector* corresponding to the **eigenvalue** λ.
This linear system is equivalent to

$$(A - \lambda I) X = \theta$$

Characteristic polynomial- $p(\lambda) = \det (A - \lambda I)$
Characteristic equation- $\det (A - \lambda I) = 0$
Eigenvalue- any solution of $p(\lambda) = 0$
Eigenvector- any *non-zero* X satisfying $AX = \lambda X$ for a particular value of λ
To find eigenvectors- for each λ solve the corresponding homogeneous system

$$(A - \lambda I) X = \theta$$

(which MUST be consistent and dependent!)
Calculation checks: det(A) = PRODUCT of eigenvalues, tr(A) = *sum* of eigenvalues
spectrum of A - set of eigenvalues of A (denoted $\sigma(A)$)
spectral radius $\rho(A) = \max \{ | \lambda | : \lambda \in \sigma(A) \}$; it is a non-negative real number
Geometric multiplicity of λ - is the size of the corresponding eigenspace
ie. $dim (null (A - \lambda I))$; it is $\leq$ the algebraic multiplicity of λ (number of times λ is a solution
of $p(\lambda) = 0$). SIMPLE roots always mean geometric multiplicity = algebraic multiplicity = 1
Checks on calculations: det(A) = product of eigenvalues; tr(A) = sum of eigenvalues
If $\lambda = 0$ then A is **singular**
If A > 0 it has a positive eigenvalue r = ρ(A) with a positive eigenvector X

6.2 Diagonalization

Goal- find P and D such that AP = PD
1) Possible if n x n matrix A has n (independent) eigenvectors
2) **Columns** of P are formed from *eigenvectors* of A
3) Main diagonal of D has EIGENVALUES of A and zeros elsewhere (ie. D is a *diagonal* matrix)
Similar matrices- if $B = P^{-1} A P$ then B is **similar** to A (and shares A's eigenvalues)
Thus D is similar to A.
THEOREM Any two similar matrices represent the same linear operator T (with respect to different bases).
Any n x n matrix having n independent eigenvectors is **non-defective**; a defective matrix does NOT have n independent eigenvectors (can happen only when the characteristic equation has a *repeated* root)

6.3 Systems of Linear Differential Equations

Any system of linear differential equations

$$\frac{d\,x}{d\,t} = a\,x + b\,y$$

$$\frac{d\,y}{d\,t} = c\,x + d\,y$$

can be written in matrix form as

$$\begin{bmatrix} \dfrac{d\,x}{d\,t} \\ \dfrac{d\,y}{d\,t} \end{bmatrix} = \begin{bmatrix} a & b \\ c & d \end{bmatrix} \begin{bmatrix} x \\ y \end{bmatrix}$$

or dX/dt = AX where $X = \begin{bmatrix} x \\ y \end{bmatrix} \Rightarrow \dfrac{d}{d\,t}\begin{bmatrix} x \\ y \end{bmatrix} = \begin{bmatrix} \dfrac{d\,x}{d\,t} \\ \dfrac{d\,y}{d\,t} \end{bmatrix} = \dfrac{d\,X}{d\,t}$.

This system is actually *homogeneous* since it can be written as $\dfrac{d\,X}{d\,t} - A\,X = \theta$.

To uncouple the system:
1) let $X = PY$ where P is the *matrix of eigenvectors* of A.
2) then the system is of the form dY/dt = DY which implies

$$Y = e^{D\,t}\,C \quad \text{where } C = \begin{bmatrix} c_1 \\ c_2 \end{bmatrix} \quad \exp(D\,t) = e^{D\,t} = \begin{bmatrix} e^{\lambda_1 t} & 0 \\ 0 & e^{\lambda_2 t} \end{bmatrix}$$

Here $\exp(Dt) = e^{D\,t} = \begin{bmatrix} e^{\lambda_1 t} & 0 \\ 0 & e^{\lambda_2 t} \end{bmatrix}$ is the **matrix exponential** of D; this means

$$X(t) = P\,e^{D\,t}\,C$$

if initial conditions are given, then one finds the constants using

$$X(0) = \begin{bmatrix} x(0) \\ y(0) \end{bmatrix} = P\begin{bmatrix} c_1 \\ c_2 \end{bmatrix} = P\,C \text{ and solving for C}$$

Fundamental matrix solution- denoted by $\Phi(t) = e^{A\,t} = \exp(At)$; it satisfies

1) $\dfrac{d}{d\,t}[\Phi(t)] = A\,\Phi(t)$ and 2) $\Phi(0) = I$ where exp(At) is the matrix exponential

Solution of dX/dt = AX can be written as $X(t) = \Phi(t)\,K$ where K = X(0).

6.4 Complex Eigenvalues
No difference here except λ may be *imaginary*;
1) imaginary eigenvalues will come in complex CONJUGATE pairs $\lambda = a \pm bi$
2) *eigenvectors* corresponding to complex conjugate eigenvalues come in complex conjugate pairs

Euler's formula: $\exp(ati) = \cos(at) + i\sin(at)$

Solving dX/dt = AX if A has **imaginary eigenvalues**
use the FUNDAMENTAL MATRIX SOLUTION:

$$X = P\exp(Dt)P^{-1}K = \Phi(t)K = \exp(At)\begin{bmatrix} k_1 \\ k_2 \end{bmatrix}$$

Stability of solutions of dX/dt = AX; for any matrix A with no REPEATED eigenvalues;
plot eigenvalues in the complex plane

1) if *even one* eigenvalue lies to the right of the imaginary axis, the solutions are UNSTABLE
2) if *all eigenvalues* lie to the left of the imaginary axis, the solutions are asymptotically stable
3) if no eigenvalue lies to the right of the imaginary axis, the solutions are *stable*

Chapter 6 Review

Part A. Given the matrix $A = \begin{bmatrix} 1 & 5 \\ 0 & 4 \end{bmatrix}$

1. Find the characteristic polynomial and the eigenvalues of A. ans. $p(\lambda) = \lambda^2 - 5\lambda + 4$, $\lambda = 1, 4$

2. Find the corresponding eigenvectors for each eigenvalue.
ans. $X_{\lambda=1} = s\begin{bmatrix} 1 \\ 0 \end{bmatrix}$ $X_{\lambda=4} = t\begin{bmatrix} 5 \\ 3 \end{bmatrix}$ $s, t \neq 0$

3. Let [T(X)] = A[X] and pick a particular eigenvector (as a 2 x 1 column matrix) and show that $[T(X)] = \lambda[X]$ for each eigenvalue/eigenvector.

4. Find T(x,y) using A. ans. T(x,y) = (x + 5y, 4y)

5. Find S(x,y) if its standard matrix is $A - \lambda \mathbf{I}$ where $\lambda = 4$. Show that S is singular.
ans. S(x,y) = (-3x+5y,0).

6. Find the eigenvalues for A^{-1}. ans. 1,1/4

7. Find a basis for each eigenspace for #6. ans. $\lambda = \frac{1}{4}$, $X = \begin{bmatrix} 5 \\ 3 \end{bmatrix}$, $\lambda = 1$, $X = \begin{bmatrix} 1 \\ 0 \end{bmatrix}$

8. Find the eigenvalues for A^t. ans. 1, 4

9. Find the eigenvectors for #8. ans. $\lambda = 1$, $X_{\lambda=1} = s\begin{bmatrix} -3 \\ 5 \end{bmatrix}$, $\lambda = 4$, $X_{\lambda=4} = t\begin{bmatrix} 0 \\ 1 \end{bmatrix}$ $s, t \neq 0$

10. Find the eigenvalues and corresponding eigenvectors for A^2. ans. 1,16, same eigenvectors as A

Part B. Given T(x,y,z) = (1/2 x + 1/2 y, 1/2 x + 1/2 y, z);

11. Find A where A is the standard matrix for T.   ans.
$$A = \begin{bmatrix} \frac{1}{2} & \frac{1}{2} & 0 \\ \frac{1}{2} & \frac{1}{2} & 0 \\ 0 & 0 & 1 \end{bmatrix}$$

12. Find the characteristic polynomial of A and its eigenvalues.
ans. $p(\lambda) = -\lambda^3 + 2\lambda^2 - \lambda$, {0, 1}

13. Find the eigenspace for each eigenvalue. ans. $\lambda = 0, E_{\lambda=0} = r\begin{bmatrix} -1 \\ 1 \\ 0 \end{bmatrix}$ $\lambda = 1, E_{\lambda=1} = s\begin{bmatrix} 1 \\ 1 \\ 0 \end{bmatrix} + t\begin{bmatrix} 0 \\ 0 \\ 1 \end{bmatrix}$

14. If W is the eigenspace corresponding to $\lambda = 1$, show that dim(W) = 2 (so W is a plane through the origin). Find the equation of this plane. Note that any vector lying in this plane W satisfies T(X) = X (ie. X is a *fixed point* of T).   ans. basis for W is {(1,1,0),(0,0,1)}, x - y = 0
15. Show that $W^\perp$ is the eigenspace corresponding to $\lambda = 0$.
16. Find the two matrices P and D which satisfy AP = PD.
ans. $P = \begin{bmatrix} -1 & 1 & 0 \\ 1 & 1 & 0 \\ 0 & 0 & 1 \end{bmatrix}, D = \begin{bmatrix} 0 & 0 & 0 \\ 0 & 1 & 0 \\ 0 & 0 & 1 \end{bmatrix}$

17. Show det(A) = product of eigenvalues and tr(A) = sum of eigenvalues.
18. Write A in "factored" form, using P and D. Show that successive powers of A do not change; since A is idempotent (ie. $A^2 = A$), D is also so nothing changes as A is raised to higher and higher powers! ans. $A^n = PDP^{-1}$
19. Find the matrix A' of T with respect to the basis B' = {(-1,1,0),(1,1,0),(0,0,1)}.

Hint: Recall A' is *similar* to A. ans. $A' = \begin{bmatrix} 0 & 0 & 0 \\ 0 & 1 & 0 \\ 0 & 0 & 1 \end{bmatrix}$

Part C. Use the matrix $A = \begin{bmatrix} 1 & 5 \\ 0 & 4 \end{bmatrix}$;

20. Find the diagonal matrix D and the matrix of eigenvectors P.
ans. $D = \begin{bmatrix} 1 & 0 \\ 0 & 4 \end{bmatrix}$ $P = \begin{bmatrix} 1 & 5 \\ 0 & 3 \end{bmatrix}$

21. Check your work by finding AP and PD.
22. Show that $P^{-1} A P = D$.
23. Show det(A) = product of eigenvalues and tr(A) = sum of eigenvalues

24. Let B' = {(1,0),(5,3)} and find the matrix A' which satisfies $[\ T(\ X\ )\ ]_{B'} = A'\ [\ X\ ]_{B'}$.

ans. A' = D

25. Find T(X) if X = (23,12) using the answer to #24. ans. Y = (83,48)

Part D. Given the matrix matrix $A = \begin{bmatrix} 0 & 0 & 1 \\ 0 & 1 & 0 \\ 1 & 0 & 0 \end{bmatrix}$;

26. Find the characteristic polynomial of A. ans. $p(\lambda) = -\lambda^3 + \lambda^2 + \lambda - 1$

27. Find the spectrum of A. ans. $\sigma(A) = \{-1,1\}$

28. Find the eigenspaces corresponding to each eigenvalue. Notice that A has orthogonal eigenvectors. Why?

ans. $\lambda = -1, E_{\lambda=-1} = r \begin{bmatrix} 1 \\ 0 \\ -1 \end{bmatrix}$, $\lambda = 1, E_{\lambda=1} = s \begin{bmatrix} 0 \\ 1 \\ 0 \end{bmatrix} + t \begin{bmatrix} 1 \\ 0 \\ 1 \end{bmatrix}$, A is symmetric

29. If A is not defective, find matrices P and D satisfying AP = PD.

ans. $P = \begin{bmatrix} 1 & 0 & 1 \\ 0 & 1 & 0 \\ -1 & 0 & 1 \end{bmatrix}$ $D = \begin{bmatrix} -1 & 0 & 0 \\ 0 & 1 & 0 \\ 0 & 0 & 1 \end{bmatrix}$

30. What is the geometric multiplicity of $\lambda = -1$? What is the geometric multiplicity of $\lambda = 1$?

ans. 1, 2

31. Find the solution of dX/dt = AX in terms of P and D.

ans. $X(t) = P \exp(Dt)C = \begin{bmatrix} 1 & 0 & 1 \\ 0 & 1 & 0 \\ -1 & 0 & 1 \end{bmatrix} \begin{bmatrix} e^{-t} & 0 & 0 \\ 0 & e^{t} & 0 \\ 0 & 0 & e^{t} \end{bmatrix} \begin{bmatrix} c_1 \\ c_2 \\ c_3 \end{bmatrix}$

Part E. Solve each system of differential equations to get the general solution; if initial conditions are given, find the particular solution which satisfies the system and the given conditions.

32. $\dfrac{dx}{dt} = 3x$

$\dfrac{dy}{dt} = -2y$

ans. $X(t) = P \exp(Dt)C = \begin{bmatrix} 1 & 0 \\ 0 & 1 \end{bmatrix} \begin{bmatrix} e^{3t} & 0 \\ 0 & e^{-2t} \end{bmatrix} \begin{bmatrix} c_1 \\ c_2 \end{bmatrix}$

33. $\dfrac{dx}{dt} = -x + 3y$

$\dfrac{dy}{dt} = -2y$

ans. $X(t) = P \exp(Dt)C = \begin{bmatrix} -3 & 1 \\ 1 & 0 \end{bmatrix} \begin{bmatrix} e^{-2t} & 0 \\ 0 & e^{-t} \end{bmatrix} \begin{bmatrix} c_1 \\ c_2 \end{bmatrix}$

34. $\dfrac{d\,x}{d\,t} = x + 3\,y$

$x(0) = 3,\ y(0) = 4$

$\dfrac{d\,y}{d\,t} = 3\,x + y$

ans. $X(t) = \begin{bmatrix} -1 & 1 \\ 1 & 1 \end{bmatrix} \begin{bmatrix} e^{-2t} & 0 \\ 0 & e^{4t} \end{bmatrix} \begin{bmatrix} \dfrac{1}{2} \\ \dfrac{7}{2} \end{bmatrix}$

Part F. Using the matrix $A = \begin{bmatrix} \dfrac{1}{2} & \dfrac{1}{2} & 0 \\ \dfrac{1}{2} & \dfrac{1}{2} & 0 \\ 0 & 0 & 1 \end{bmatrix}$;

35. Write out the system of differential equations dX/dt = AX.
ans. dx/dt = 1/2x + 1/2y, dy/dt = 1/2x + 1/2y, dz/dt = z

36. Write out the uncoupled system dY/dt = DY by letting X = PY; use $Y = \begin{bmatrix} u \\ v \\ w \end{bmatrix}$.

ans. du/dt = 0, dv/dt = v, dw/dt = w

37. Find the solution of dY/dt = DY.

ans. $Y(t) = \begin{bmatrix} 1 & 0 & 0 \\ 0 & e^t & 0 \\ 0 & 0 & e^t \end{bmatrix} \begin{bmatrix} c_1 \\ c_2 \\ c_3 \end{bmatrix}$

38. Get the solution of dX/dt = AX.   ans. $X(t) = \begin{bmatrix} -1 & e^t & 0 \\ 1 & e^t & 0 \\ 0 & 0 & e^t \end{bmatrix} \begin{bmatrix} c_1 \\ c_2 \\ c_3 \end{bmatrix}$

39. Find the particular solution satisfying x(0) = 1, y(0) = 2, z(0) = 3.

ans.
$X(t) = \begin{bmatrix} -1 & e^t & 0 \\ 1 & e^t & 0 \\ 0 & 0 & e^t \end{bmatrix} \begin{bmatrix} \dfrac{1}{2} \\ \dfrac{3}{2} \\ 3 \end{bmatrix}$

40. Find the fundamental matrix solution $\Phi(t)$ to $dX/dt = AX$. ans. $\dfrac{1}{2}\begin{bmatrix} 1+e^t & e^t-1 & 0 \\ e^t-1 & 1+e^t & 0 \\ 0 & 0 & 2e^t \end{bmatrix}$

41. Show that $\dfrac{d}{dt}[\Phi(t)] = A\,\Phi(t)$; this means that $\Phi(t) = \exp(At)$. If you're paranoid, find $\Phi(0)$.

Part G. Given the order two, linear, homogeneous differential equation $\dfrac{d^2 y}{dt^2} + 16\,y = 0$;

42. Let $v = dy/dt$ and change the one scalar equation into a system of differential equations in y and

$$\dfrac{dy}{dt} = v$$

v. Hint: solve the differential equation for dv/dt. ans.

$$\dfrac{dv}{dt} = -16\,y$$

43. Write the system in matrix multiplication form $dY/dt = AY$ with $Y = \begin{bmatrix} y \\ v \end{bmatrix}$.

ans. $\begin{bmatrix} \dfrac{dy}{dt} \\ \dfrac{dv}{dt} \end{bmatrix} = \begin{bmatrix} 0 & 1 \\ -16 & 0 \end{bmatrix} \begin{bmatrix} y \\ v \end{bmatrix}$

44. Find the P and D matrices for A. ans. $P = \begin{bmatrix} 1 & 1 \\ 4i & -4i \end{bmatrix}$ $D = \begin{bmatrix} 4i & 0 \\ 0 & -4i \end{bmatrix}$

45. Find the solution using the fundamental matrix solution $\Phi(t)$.

ans. $Y(t) = \Phi(t)K = \begin{bmatrix} \cos(4t) & \dfrac{1}{4}\sin(4t) \\ -4\sin(4t) & \cos(4t) \end{bmatrix} \begin{bmatrix} k_1 \\ k_2 \end{bmatrix}$

46. Solve $dY/dt = AY$ subject to initial conditions $y(0) = 1$ and $v(0) = -1$. Check that your solutions satisfy the original system (see # 43). ans. $Y(t) = \begin{bmatrix} \cos(4t) & \dfrac{1}{4}\sin(4t) \\ -4\sin(4t) & \cos(4t) \end{bmatrix} \begin{bmatrix} 1 \\ -1 \end{bmatrix}$

Part H. Given the matrix $A = \begin{bmatrix} 1 & 2 & 3 \\ 1 & 2 & 3 \\ 2 & 4 & 6 \end{bmatrix}$;

47. Find the eigenvalues and corresponding eigenspaces for A.

ans. $\lambda = 0 \Rightarrow E_{\lambda=0} = span(\{ \begin{bmatrix} -3 \\ 0 \\ 2 \end{bmatrix}, \begin{bmatrix} -2 \\ 1 \\ 0 \end{bmatrix} \})$, $\lambda = 9 \Rightarrow E_{\lambda=9} = span(\{ \begin{bmatrix} 1 \\ 1 \\ 2 \end{bmatrix} \})$

48. A is non-defective so get P and D such that AP = PD. ans. $P = \begin{bmatrix} -3 & -2 & 1 \\ 0 & 1 & 1 \\ 1 & 0 & 2 \end{bmatrix}, D = \begin{bmatrix} 0 & 0 & 0 \\ 0 & 0 & 0 \\ 0 & 0 & 9 \end{bmatrix}$

49. Find A^n in terms of P and D. ans. $P D^n P^{-1} = P \begin{bmatrix} 0 & 0 & 0 \\ 0 & 0 & 0 \\ 0 & 0 & 9^n \end{bmatrix} P^{-1}$

50. Perron's Theorem says that if A > 0, then it has a positive eigenvalue λ = r such that r = ρ(A) and λ = r has an eigenvector X > 0. Verify that this is true. ans. r = 9, X = $\begin{bmatrix} 1 \\ 1 \\ 2 \end{bmatrix}$

Part I. Given the matrix equation $X_1 = \begin{bmatrix} j_1 \\ a_1 \end{bmatrix} = \begin{bmatrix} 0 & 1 \\ \frac{5}{4} & \frac{1}{4} \end{bmatrix} \begin{bmatrix} j_0 \\ a_0 \end{bmatrix} = L X_0$ which represents the population

of juvenile and adult seagulls, if $X_2 = L X_1, X_3 = L X_2$ etc.

51. Show $X_n = L^n X_0$; thus the population in the future depends on powers of L.

52. Find the matrices P and D such that LP = PD.   ans. $P = \begin{bmatrix} -1 & 4 \\ 1 & 5 \end{bmatrix}, D = \begin{bmatrix} -1 & 0 \\ 0 & \frac{5}{4} \end{bmatrix}$

53. Let $X_0 = \begin{bmatrix} -10 \\ 10 \end{bmatrix} + \begin{bmatrix} 16 \\ 20 \end{bmatrix}$; don't add these! Apply L to get X_1, X_2, X_3 using the *distributive*

property of matrix multiplication. Observe what is happening to the total seagull population (ie. j + a) by looking at the entries of X as L is applied. Can you predict the future? What eigenvalue dominates the population behavior?

ans. $\lambda = \frac{5}{4}$, $X_3 = \begin{bmatrix} \frac{165}{4} \\ \frac{465}{16} \end{bmatrix}$, total population = j + a increases by 25% at each stage

54. What happens to the seagull population if $X_0 = \begin{bmatrix} 4k \\ 5k \end{bmatrix}$?

ans. Both j and a increase by 25% at each stage.

55. Let $X = PY$ and convert $X_1 = L X_0$ to the Y coordinate system and show $Y_1 = D Y_0$. Find Y_1, Y_2, Y_3 if $X_0 = \begin{bmatrix} 6 \\ 30 \end{bmatrix}$ and observe that the entries of Y depend *directly* on D (thus on the eigenvalues

of L).   ans. $Y_3 = \begin{bmatrix} -10 \\ \dfrac{125}{16} \end{bmatrix}$

Part J. Given the symmetric matrix $A = \begin{bmatrix} 5 & 4 \\ 4 & 11 \end{bmatrix}$; if $X^t A X > 0$ for all $X = \begin{bmatrix} x \\ y \end{bmatrix} \neq \begin{bmatrix} 0 \\ 0 \end{bmatrix}$, then A is

called *positive definite*. A SYMMETRIC matrix is positive definite iff all its eigenvalues are **positive.**

56. Show A is *positive definite*. ans. spectrum of A = {3,13}

57. Test this out by picking any 2-vector (like $X = \begin{bmatrix} -8 \\ 5 \end{bmatrix}$) and show that $X^t A X > 0$.

58. Suppose X is any **eigenvector** of A; show that $X^t A X = \lambda \| X \|^2 > 0$ (which is always *positive* if $\lambda > 0$). Verify this for eigenvector $X = \begin{bmatrix} -20 \\ 10 \end{bmatrix}$.

59. Since $A > 0$ (ie. A is a *positive* matrix), its largest positive eigenvalue r = ρ(A). Also $\lambda = r$ has a positive eigenvector X. Show that this is true.   ans. r = 13 = ρ(A), $X = \begin{bmatrix} 1 \\ 2 \end{bmatrix}$

Part K. The population of barn mice is governed by $X_1 = \begin{bmatrix} j_1 \\ a_1 \end{bmatrix} = \begin{bmatrix} 0 & 4 \\ \frac{1}{4} & 0 \end{bmatrix} \begin{bmatrix} j_0 \\ a_0 \end{bmatrix} = L X_0$;

60. Find the eigenvalues/eigenvectors of L.   ans. $P = \begin{bmatrix} -4 & 4 \\ 1 & 1 \end{bmatrix}, D = \begin{bmatrix} -1 & 0 \\ 0 & 1 \end{bmatrix}$

61. Let $X_0 = \begin{bmatrix} 60 \\ 20 \end{bmatrix}$; then find X1, X2, X3 and X4. Thus show that the states repeat after a period of

two; why is this so? ans. X1 = X3 = $\begin{bmatrix} 80 \\ 15 \end{bmatrix}$, X2 = X4 = $\begin{bmatrix} 60 \\ 20 \end{bmatrix}$ since $L^2 = I$.

62. What initial state would guarantee no change in the mouse population? ans. X0 = $k \begin{bmatrix} 4 \\ 1 \end{bmatrix}$

Part L. Given the matrix $A = \begin{bmatrix} 0 & 0 & 0 \\ 1 & 0 & -13 \\ 0 & 1 & -4 \end{bmatrix}$;

63. Find the eigenvalues of A. ans. $\lambda = 0, -2 \pm 3i$

64. Without solving dX/dt = AX, determine the stability of the solutions. ans. stable

Part M.   Given   the   non-linear   predator-prey   system   of   differential   equations

$\dfrac{dy}{dt} = 6x - xy = f(x, y)$

where x = rabbits, y = foxes.

$\dfrac{dy}{dt} = 2xy - 20\, y = g(x, y)$

This system is non-linear; analyze the solutions by *linearizing* the system:

65. Solve f(x,y) = 0 and g(x,y) = 0 simultaneously to get the two equilibrium solutions.
    ans. (0,0), (10,6)

66. Analyze the non-trivial equilibrium solution by finding the Jacobian matrix J, where

$J = \begin{bmatrix} \dfrac{\partial f}{\partial x} & \dfrac{\partial f}{\partial y} \\ \dfrac{\partial g}{\partial x} & \dfrac{\partial g}{\partial y} \end{bmatrix}$; replace the x-value and y-value in J by the equilibrium solution.

Then write out the linearized version dX/dt = J X.   ans. $\begin{bmatrix} \dfrac{dx}{dt} \\ \dfrac{dy}{dt} \end{bmatrix} = \begin{bmatrix} 0 & -10 \\ 12 & 0 \end{bmatrix} \begin{bmatrix} x \\ y \end{bmatrix}$

67. Find the eigenvalues and describe the stability of the solutions to dX/dt = J X .
    ans. $\lambda = \pm 2\sqrt{30}\,i$, stable NB. the predator-prey system is known to have periodic solutions so
    this is good.

Part N. Given $U = \begin{bmatrix} \dfrac{-3}{5} & 0 & \dfrac{8}{5} \\ 0 & 1 & 0 \\ \dfrac{2}{5} & 0 & \dfrac{3}{5} \end{bmatrix}$;

68. Find the eigenvalues of U. ans. $\sigma(U) = \{-1,1\}$.
69. Show that $U^2 = I$ so U is its own inverse.
70. Describe the stability of the solutions to dX/dt = UX without solving for X(t). ans. unstable
71. Prove that if $U^2 = I$ then the eigenvalues of U are $\{-1,1\}$. Hint: if X is an eigenvector corresponding to λ find $U^2 X$.

Glossary

NB. The number next to the term gives the section number in which the term is introduced.

additivity(S5.2) - axiom for an inner product space which says
$<X + Y,Z> = <X,Z> + <Y,Z>$
adjacency matrix(S2.1) - a matrix (of ones and zeros) which describes whether or not one can travel from city i to city j
algebraic multiplicity(S6.1) - for a root r (and polynomial p(x)), the highest power m for which (x - r) appears as a factor of p(x)

angle (between two vectors)(S5.1) - found using $\cos(\theta) = \dfrac{<X,Y>}{\| X \| \| Y \|}$

augmented matrix(S1.1)- the (unique) matrix which represents a linear system
back-substitution(S1.1) - process by which one solves a linear system whose augmented matrix is in echelon form
basis - an ordered set of vectors in a vector space V which is independent and spans V
Cauchy-Schwarz - inequality which says that $|< X,Y >| \le \| X \| \| Y \|$

Cayley-Hamilton theorem(S6.1)- says that a matrix satisfies its own characteristic equation
circulant matrix(S2.5) - a (square) matrix formed by "pushing" the entries of row one to the right to form row two, etc. as one looks from the top to the bottom of the matrix
co-domain(S4.1) - for the transformation $T | V \rightarrow W$, W is the co-domain; it contains the range of T
column matrix(S2.1)- an n by 1 matrix which has only one column, usually thought of as a vector
column space(S3.5) - span of the columns of a matrix A
consistent(S1.1) - a system which has a solution
consistent and dependent(S1.1) - a system which has infinitely many solutions
coordinates(S3.6) - the unique numbers which multiply the basis vectors to produce a given vector
coordinate matrix(S3.6) - the column matrix which contains the coordinates of a vector (with respect to a particular given basis)
defective(S6.1)- a square matrix (of size n by n) which does not have enough independent eigenvectors to span n-space
dependent(S3.3) - any set of vectors for which a linear combination can produce the zero vector w/o all scalar multipliers being zero
determinant(S2.4) - a number obtained from a square matrix whose value determines if the matrix is invertible or not
determined variable(S1.3) - any variable represented by a pivot column of the augmented matrix
dimension(S3.4) - number of vectors in any basis for a vector space V (ie. the size of a vector space)
dimension (of a matrix)(S2.1) - number of rows by number of columns of a matrix A; written dim(A) = m x n
domain(S4.1) - if $T | V \rightarrow W$ then the domain of T is V (ie. set of all inputs of a transformation T)
echelon form- desired form for an augmented matrix which is essentially triangular
eigenvalue(S6.1) - any number λ which satisfies $A X = \lambda X$ (for nonzero vectors X and square matrix A)
eigenvector(S6.1) - any (nonzero) vector X which satisfies $A X = \lambda X$
finite-dimensional(S3.1) - a vector space with a finite basis
fixed point(S4.3) - any vector X for which T(X) = X
free variable(S1.3) - any variable represented by a non-pivot column of the augmented matrix
geometric multiplicity(S6.1) - the dimension of the eigenspace corresponding to a particular eigenvalue λ

Gram-Schmidt(S5.4) - a process in which an arbitrary basis B is made into an orthonormal one by subtracting away projections

group(S2.5) - a set G together with a binary operation (call it *) such that * is associative, G has an identity element, and every element g in G has an inverse

homogeneity(S5.2)- axiom for an inner product which says that $<kX,Y> = k<X,Y>$

homogeneous(S1.2)- a system of the form $A\,X = \theta$ (ie. all constants are zero)

idempotent(S2.5)- a square matrix A satisfying $A^2 = A$

identity matrix(S2.1) - the matrix I_n satisfying $A\,I_n = I_n\,A = A$ for any square matrix A

image(S4.1)- for $T\,|\,V \rightarrow W$, if X is in V, T(X) is the image of X (and lies in W)

inconsistent(S1.1) - a system with no solution

independent(S3.3) - any set of vectors for which a linear combination can produce the zero vector only if all scalar multipliers are zero

inhomogeneous(S1.2)- a system in which at least one of the constants is not zero

inner product(S5.2) - a function $<>\,|\,V\,x\,V \rightarrow \Re$ which mimics the four properties of the scalar product on n-space(symmetry, additivity, homogeneity and positivity)

inner product space(S5.2) - a vector space which has an inner product defined on it (for short IPS)

invariant subspace- any subspace W of V in which $X \in W \Rightarrow T(\,X\,) \in W$

inverse matrix(S2.2) - a unique matrix (denoted by A^{-1}) such that $A\,A^{-1} = A^{-1}\,A = I_n$

invertible(S2.2) - any square matrix which has an inverse

kernel(S4.3)- for $T\,|\,V \rightarrow W$, the kernel of T is the subspace of V consisting of solutions of $T(\,X\,) = \theta_W$

least squares(S5.5) - the best approximation for the solution of an inconsistent linear system of equations

linear equation(S.1) - any equation of the form $a_1\,x_1 + a_2\,x_2 + ... + a_n\,x_n = b$

linear transformation(S4.1) - any function $T\,|\,V \rightarrow W$ (V and W are vector spaces) for which 1) T(X + Y) = T(X) + T(Y) and 2) T(kX) = kT(X)

linear combination(S3.2) - if $S = \{\,X_1, X_2, ..., X_n\,\}$ then X is a linear combination of the vectors in S if $X = k_1\,X_1 + k_2\,X_2 + ... + k_n\,X_n$

linear system(S1.1) - a set of at least two linear equations to be solved simultaneously

lower triangular(S2.5) - a triangular matrix with all entries above the main diagonal zero

Markov matrix(S2.1)- a square matrix with non-negative entries such that the sum of the entries of each column is one

matrix(S2.1)- any rectangular array of numbers[1]

nilpotent(S2.1)- a square matrix satisfying $A^m = \theta$ (for some positive integer m)

non-pivot column(S1.3)- in an augmented matrix, any column(except the last) which has no leading one in it

norm(S5.1)- length or magnitude of a vector X, denoted by $\parallel X \parallel$

normal(S5.2) - a vector of norm (ie. magnitude or length) one

normed vector space(S5.2)- a vector space which has a norm defined on it, which behaves like length in Euclidean n-space and satisfies 3 axioms(for short NVS)

n-space - set of all n-tuples of the form $(\,x_1, x_2, ..., x_n\,)$ with real components, denoted $\Re^n$

nullity(S3.5) - the dimension of the null space of A

one-to-one(S4.3) - any transformation T for which $T(\,X_1\,) = T(\,X_2\,) \Rightarrow X_1 = X_2$ (or $X_1 \neq X_2 \Rightarrow T(\,X_1\,) \neq T(\,X_2\,)$); no element of rng(T) can have two(or more) pre-images

---

[1] In this book, entries are real except in Section 6.4

onto-(S4.3) a transformation $T \mid V \to W$ for which rng(T) = W (ie. every vector in W has a pre-image in V)

operator(S4.4)- a function T such that $T \mid V \to V$ (ie. domain and co-domain are the same)

orthogonal(S5.1) - two vectors X and Y are orthogonal if <X,Y> = 0

orthogonal complement(S5.1)- given a subspace W, the orthogonal complement $W^{\perp}$ is the vector space of vectors which are orthogonal to those in W

orthogonal matrix(S5.3)- any square matrix A satisfying $A^{-1} = A^{t}$; its columns are unit vectors and mutually orthogonal

overdetermined(S1.1)- any system with more equations than unknowns

permutation matrix(S2.5) - a matrix formed from the identity I by a sequence of row exchanges

perp(S5.1)- indicates the orthogonal complement of a vector space W(ie. $W^{\perp}$)

pivot column(S1.3)- any column of an augmented matrix (except the last) with a leading one in it

positivity(S5.2)- axiom for an inner product which says that $< X, X > \geq 0$ and $< X, X > = 0$ iff $X = \theta$

pre-image(S4.1)- if T(X) = Y then X is the pre-image of Y

projection(S5.1)- a vector formed from two vectors Y and X, indicated by $proj_{X} Y$; it is the scalar multiple of X which is closest to Y

range(S4.3)- for $T \mid V \to W$, the range is a subspace of W which contains all images of vectors from V, denoted rng(T)

rank(S4.3)- the number of nonzero rows(or columns) in the echelon form of A

row-echelon form(S1.1)- a generalized triangular form of a matrix with ones being the first nonzero element in any row, zeros only below this leading one, and leading ones moving right as one looks at the matrix from top to bottom, and rows of zeros at the bottom

row space(S3.5)- the span of the rows of A

scalar multiplication(S3.1)- multiplying a vector by a scalar(ie. real number)

scalar product(S5.1)- a function $\cdot \mid \Re^{n} \, x \, \Re^{n} \to \Re$ which satisfies four axioms (sometimes called the dot product); it takes two vectors and produces a real number

singular(S2.2)- a square matrix A with no inverse

skew-symmetric(S2.5)- a square matrix satisfying $A^{t} = - A$

span(S3.3)- the subspace of all linear combinations of vectors from $S = \{ X_{1}, X_{2}, ..., X_{n} \}$ denoted span(S); this means $span(S)= \{ X := k_{1} X_{1} + k_{2} X_{2} + ... + k_{n} X_{n}, k_{i} \in \Re, X_{i} \in S \}$

spanning set(S3.3)- if all vectors in W (a vector space) are linear combinations of vectors from S, then S is spanning set for W; we write span(S) = W

spectrum(S6.1)- set of eigenvalues of a matrix (or its corresponding transformation T)

square matrix(S2.1) - a matrix with n rows and n columns

square system(S1.1)- has the same number of equations as unknowns

standard basis(S3.4)- the usual basis one uses for a particular vector space V

standard matrix(S4.2)- the matrix A which represents a linear transformation T, if the standard basis is used to produce the columns of A; it satisfies [T(X)] = A[X]

symmetric(S2.1)- a matrix A satisfying $A^{t} = A$

subspace(S3.2)- a subset W of a vector space V which itself is a vector space

trace(S2.1)- the sum of the entries along the main diagonal of a square matrix

transformation(S4.1)- a function from one vector space into another

transpose(S2.1)- interchanging the rows and columns of a matrix, denoted by A^{t}

triangle inequality(S5.2)- for any two vectors in an NVS, $\| X + Y \| \leq \| X \| + \| Y \|$

triangular matrix(S2.5)- a square matrix which has all the entries below (or above) the main diagonal all zero

trivial solution- when all values of the unknowns are zero

trivial space(S3.1)- the vector space $\{\ \theta\ \}$

underdetermined system(S1.1)- a system with fewer equations than unknowns

unimodular(S2.5)- a square matrix satisfying $\det(A) = 1$

upper triangular(S2.5)- a triangular matrix which has all entries below the main diagonal zero

vector space(S3.1)- a set of elements(vectors) in which vector addition and scalar multiplication are defined and behave like the corresponding operations in $\Re^n$

Wrońskian(S2.4)- the determinant of a matrix W formed by a set of functions and their derivative(s)

zero vector(S3.1)- denoted by θ; it is the additive identity in any vector space

Bibliography

1. Fundamental Structures of Algebra by Mostow, Sampson & Meyer
2. Introduction to Topology and Modern Analysis by George Simmons
3. Matrix Analysis by Roger Horn & Charles Johnson